MINITAB COMPUTER SUPPLEMENT

To accompany James T. McClave
and Frank H. Dietrich, II's

STATISTICS

Sixth Edition

Ruth K. Meyer and David D. Krueger
St. Cloud State University

DELLEN
an imprint of
MACMILLAN COLLEGE PUBLISHING COMPANY
NEW YORK

MAXWELL MACMILLAN CANADA
TORONTO

MAXWELL MACMILLAN INTERNATIONAL
NEW YORK OXFORD SINGAPORE SYDNEY

Copyright @ 1994 by Macmillan College Publishing Company, Inc.
Dellen is an imprint of Macmillan College Publishing Company.

Printed in the United States of America

All rights reserved. No part of this book may be reproduced or transmitted in any form or by any means, electronic or mechanical, including photocopying, recording, or any information storage and retrieval system, without permission in writing from the Publisher.

Macmillan College Publishing Company
866 Third Avenue
New York, New York 10022

Macmillan College Publishing Company is
part of the Maxwell Communication
Group of Companies

Maxwell Macmillan Canada, Inc.
1200 Eglinton Avenue East
Suite 200
Don Mills, Ontario M3C 3N1

ISBN 0-02-380833-0

PRINTING: 1 2 3 4 5 6 YEAR: 4 5 6

CONTENTS

Preface		v
1	Introduction to Minitab	1
2	Methods for Describing Sets of Data	35
3	Probability	69
4	Discrete Random Variables	81
5	Continuous Random Variables	97
6	Sampling Distributions	115
7	Inferences Based on a Single Sample: Estimation	137
8	Inferences Based on a Single Sample: Tests of Hypotheses	151
9	Inferences Based on Two Samples: Estimation and Tests of Hypotheses	165
10	Analysis of Variance: Comparing More Than Two Means	181
11	Nonparametric Statistics	207
12	The Chi-Square Test and the Analysis of Contingency Tables	227
13	Simple Linear Regression	235
14	Multiple Regression	249
15	Model Building	277
16	Survey Analysis	305
Appendix		311
Index		333

PREFACE

The *Minitab Computer Supplement* is designed to accompany the Sixth Edition of *Statistics* by James T. McClave and Frank H. Dietrich, II. The supplement is written to introduce students to Minitab, a general purpose statistical computer system.

Minitab was developed at Penn State in 1972 for students in introductory statistics courses. Since then, Minitab has evolved into a comprehensive statistical system that is also used by students in advanced data analysis courses, and by scientists, engineers, and managers in business, government, and industry.

The chapters in the supplement are designed to correspond to the chapters in the text. The final chapter of the supplement describes a survey sampling project. The objectives of the project are to illustrate the use of Minitab in questionnaire evaluation and to provide a review of statistical techniques. We have chosen many examples and exercises from the text to illustrate the use of Minitab in organizing and analyzing data. As new statistical concepts are discussed in each chapter of the text, we introduce appropriate Minitab commands and provide step-by-step descriptions of how to effectively use Minitab. Exercises, many of which are drawn from the text, conclude each chapter. A computer diskette containing the data sets used in the examples and the Appendix is available with this supplement.

This supplement was developed for use with Minitab Release 8 for DOS or Release 9 for VAX/VMS. If you have a different version or the Student Edition of Minitab, most of the commands are compatible.

There are additional texts available for information on Minitab. We recommend the *Minitab Mini-Manual, A Beginner's Guide to Minitab*, 1992, Minitab, Inc., *Minitab Handbook*, Third Edition, Wadsworth Publishing Company, and for more advanced work, *Minitab Reference Manual*, 1992, Minitab, Inc. Information about Minitab software and texts may be obtained from:

Minitab, Inc.
3081 Enterprise Drive
State College, PA 16801

Telephone:	814/238-3280
Fax:	814/238-4383
Telex:	881612

CHAPTER 1
INTRODUCTION TO MINITAB

Minitab[1] is a computer software package that can increase your understanding of statistics and decrease your calculation time. It was originally designed to be an easy-to-use statistical system to help in the teaching of statistics. It has evolved into an excellent package for data analysis. In this chapter we give a general view of how Minitab works, and introduce you to some basic commands to help you get started.

MINITAB COMMANDS INTRODUCED IN THIS CHAPTER

CODE	COPY	DELETE	END	EXECUTE
HELP	INFORMATION	INSERT	JOURNAL	LET
NAME	NEWPAGE	NOJOURNAL	NOOUTFILE	NOTE
OH	OUTFILE	OW	PARSUMS	PRINT
RANK	READ	RETRIEVE	SAVE	SET
SIGN	SORT	STACK	STOP	UNSTACK
WRITE	#			

ACCESSING MINITAB

Minitab is available on a microcomputer, minicomputer, or mainframe. You need to learn several things about your computer system before you can use Minitab. For example, if you use a microcomputer, you need to learn about the operating system, the directory, disk drives, and diskettes. If you use a larger computer system, you need to learn how to 'log on' to the system. Generally the log on procedure requires you to enter an account number, a user name or user identification, and a password. Appendix A of this supplement contains information on the use of Minitab on a DOS microcomputer.

If you are successful with the log on procedure, you are in the computer's operating system. The computer usually gives you information about the system, and then a system prompt. It expects you to respond with a system command. The procedure for accessing Minitab varies, but generally, you need to type the word **Minitab** following the system prompt. The system responds with the version of Minitab that it has in its library, some information on Minitab, and the MTB > prompt.

---

[1]*Minitab is a trademark name of Minitab, Inc.*

For example, if you access Minitab Release 9 on a VAX, the response is similar to the following:

```
MINITAB Statistical Software, Standard Version
Release 9.1 for VAX/VMS
(C) Copyright 1992 Minitab Inc. - All Rights Reserved

Dec. 30, 1993 - Minnesota State University Academic Computer System

Worksheet size: 100013 cells


For information on:            Type:
--------------------------     ----------------
How to use Minitab             HELP
Customer service               HELP OVERVIEW 14
Documentation                  HELP OVERVIEW 15
What's new in this release     NEWS
```

The Data Editor is available with the DOS microcomputer version of Release 8. The Data Editor gives a full screen display of the worksheet for easy data entry or editing. To access the Data Editor, use Alt and D simultaneously; to exit, use Alt and M simultaneously. With Release 8 for DOS microcomputers, you can save and retrieve Lotus 1-2-3 and Symphony worksheets directly within Minitab. Appendix A contains more information on the Data Editor and Lotus 1-2-3.

Minitab can be used in batch mode or interactive mode. If you write a Minitab program to solve a problem and then run the program to obtain the results, you are using batch mode. In interactive mode, you communicate directly with the computer. That is, whenever you type a Minitab command and press the enter or return key, you receive a response. Almost all Minitab commands work the same way in batch mode and interactive mode. We use the interactive mode for all examples in this supplement.

> **Comment** We use Release 8 to illustrate all examples in this supplement. *If you are using Minitab Release 9, most of the commands have the same arguments and subcommands, but some of the output may differ slightly. Refer to the Release 9 Reference Manual or the HELP facility for specific information on commands that differ from the examples in this supplement.*

MINITAB WORKSHEET

Minitab works with data in a worksheet of columns and rows. Usually, a column contains the data for one variable, with an observation in each row. Columns are denoted C1, C2, C3, ..., and rows within columns are numbered 1, 2, 3, and so on. The worksheet is maintained in the computer, but you can imagine it looks like the following.

```
                       COLUMNS
     ROWS       C1       C2       C3     .  .  .
       1
       2
       3              data entry area
       .
       .
       .
```

The maximum size of your worksheet is displayed on the screen when you start Minitab. Most systems allow a worksheet containing approximately 100 columns.

Minitab can store individual numbers, such as an average or standard deviation, in constant locations denoted K1, K2, K3, ... Constants are created by the LET command or any command that produces an answer with one number. For more advanced work, Minitab has matrices denoted M1, M2, M3, ... Each matrix can store one table of numbers. The number of available matrices depends on your computer.

When you begin a Minitab session, the worksheet is blank. During the session, data are put in the worksheet, and then can be saved for future use. If you are using Release 9, Minitab asks if you want to save the file before the session ends.

MINITAB COMMANDS

A Minitab session begins with the prompt MTB >. Minitab expects you to enter a command on the same line as the prompt. We call this the command line. All command lines are free format; that is, all text may be entered in upper or lower case letters anywhere on the line.

A command line always starts with a command, such as READ, followed by one or more arguments. An argument may be a column number (C1) or column name ('SALES'), constant (89.78), stored constant (K5), or a computer file ('PROBLEM2.MTW').

The arguments must be in the proper order on the command line. The letter C represents a column and the letter K represents a constant. For example, C1 represents the first column of the worksheet, and K1, the first stored constant. The letter E represents a column, constant, or matrix. For example, the command

```
MTB > PRINT E,...,E
```

says that the PRINT command prints columns, constants, or matrices.

Minitab only uses the first four letters of the command and the arguments. Extra text may be added on some command lines for annotation. Symbols, such as #, +, - or $, or numbers, should not be used. For example, the command

```
MTB > READ THE FOLLOWING DATA IN COLUMNS C1, C2, AND C3
```

is used by Minitab as

```
MTB > READ C1 C2 C3
```

A set of consecutive columns or stored constants may be abbreviated with a dash. For example, if you want to read data in columns C1 to C3, the simplest form of the command is

```
MTB > READ C1-C3
```

You can access individual rows by specifying a row number as a subscript in parentheses following the column number or name. For example, C1(9) denotes the ninth row of C1. The subscript can be any expression which produces a positive integer. C1(COUNT(C2)) denotes the row of C1 corresponding to the number of observations in C2.

Minitab has about 200 different commands, some of which have options and subcommands. Whenever we introduce a command or subcommand, we separate it from the rest of the text. The optional arguments are placed in parentheses. Refer to Appendix C for a summary list of the commands used in this supplement.

Subcommands

Many Minitab commands have subcommands to specify options or provide additional output. To use one or more subcommands, place a semicolon at the end of the command line. Minitab responds with a SUBC> prompt on the next line. You can enter a subcommand on the line following the prompt, and end the line with a semicolon if you want another subcommand. The last subcommand line ends with a period.

For example, to get a count and percent of some qualitative data in column C1, the Minitab lines would be as follows:

```
MTB > TALLY C1;
SUBC> COUNTS;
SUBC> PERCENTS.
```

If you forget to place a period on the last subcommand line, you are requested to enter another subcommand. You can just enter a period following the SUBC> prompt. If you make a mistake, enter ABORT on a new subcommand line. This cancels the command and all subcommands, and you can then enter the correct command. For example, if you enter the incorrect column number, the command can be canceled as follows:

```
MTB > TALLY C3;
SUBC> ABORT.
MTB >
```

OBTAINING INFORMATION ABOUT MINITAB

The HELP facility contains the most recent information about Minitab. INFORMATION gives the columns and constants that are defined in the current worksheet.

HELP (COMMAND (SUBCOMMAND))

A general overview as well as specific information about commands and subcommands can be obtained with the HELP facility.

To access HELP with Release 8 on a microcomputer, you can enter HELP on the command line or press F1 or ALT+H to access a menu of help overview topics. The following section gives some examples of the HELP facility.

1. An overview of Minitab

    ```
    MTB > HELP OVERVIEW
    General Help on Minitab topics
    Choose the overview topic of interest by pressing F1 or
    by typing HELP OVERVIEW n from the Session window.  For
    example, HELP OVERVIEW 3 provides information on Session
    commands.

    0  How to use Help              7  Alpha Data
    1  Introduction to Minitab      8  Missing Value Code
    2  Menus, dialogs and screens   9  Files
    3  Session Commands            10  Versions of Minitab
    4  Subcommands                 11  PC Details
    5  Stored Constants            12  Documentation and Support
    6  Matrices                    13  Quality Control Macros
    ```

2. Information on commands

    ```
    MTB > HELP COMMANDS

    To get a list of the Minitab commands in one of the categories
    below, type HELP COMMANDS followed by the appropriate number,
    for example, HELP COMMANDS 1 for General Information.

     1 General Information          12 Time Series
     2 Input and Output of Data     13 Statistical Process Control
     3 Editing and Manipulating     14 Exploratory Data Analysis Data
     4 Arithmetic                   15 Distributions & Random Data
     5 Plotting Data                16 Sorting
     6 Basic Statistics             17 Matrices
     7 Regression                   18 Miscellaneous
     8 Analysis of Variance         19 Stored Commands and Loops
     9 Multivariate Analysis        20 Design of Experiments
    10 Nonparametrics               21 QC Macros
    11 Tables                       22 How Commands are Explained
    ```

3. For information on a specific command, type HELP followed by the command. For example,

    ```
    MTB > HELP TALLY

    TALLY the data in columns C,...,C
    (Stat > Tables > Tally)
    Subcommands:

    COUNTS         CUMCOUNTS         ALL
    PRECENTS       CUMPERCENTS
    STORE (experimental)

    Displays one-way tables for each variable listed.  The
    variables must contain integers from -9999 to +9999 or missing
    value (*).

    Tally's output consists of multiple columns displayed across
    the screen. The default output contain frequency counts for
    ```

6   Chapter 1

every distinct value of the variable. Percents, cumulative counts and cumulative percents can also be added.

4. To obtain information on a subcommand, type HELP followed by the command and subcommand. For example,

```
MTB > HELP TALLY PERCENT
PERCENT   distribution of distinct values in the column(s)

Gives percentages for each distinct value in the input
column(s), starting at the smallest distinct value.
```

The INFORMATION command can be used to obtain information about your current worksheet. This command is particularly useful if you have retrieved a saved file and want to verify that the desired data are in the current worksheet.

INFORMATION (C,...,C)

This command prints a list of all columns and stored constants in the current worksheet. The output includes column names and counts of values and missing values. Columns containing alpha data are labeled A. Use the option to obtain information on specific columns.

MINITAB SESSION

Perhaps the best way to introduce Minitab is to show you an example of a Minitab session. The following example summarizes the performance of some financial assets over the years 1985 to 1993.

■ **Example 1** The table gives rates of return on stocks and bonds, short term investments, and inflation rates for the years 1985 through 1993. The rate of return is a measure of the amount of money an investor makes on an investment. The inflation rate is the change in the Consumer Price Index (CPI).

	Year ending June 30								
	1985	1986	1987	1988	1989	1990	1991	1992	1993
Wilshire 5000 Stock Index	31.2	35.3	20.2	-5.9	19.5	12.8	7.0	13.9	16.3
Salomon Bros Bond Index	29.9	19.9	5.6	8.2	12.2	7.7	10.8	14.2	12.0
91-Day U.S. Treasury Bills	9.3	7.3	5.7	6.1	8.2	8.2	6.9	4.6	3.1
Inflation Rate	3.7	1.7	3.7	3.9	5.2	4.7	4.7	3.1	3.0

Use Minitab to find the mean rate of return of each financial asset and the mean inflation rate over the nine year period. Graphically compare the performance of stocks and bonds.

Introduction to Minitab

Solution A Minitab session begins with the MTB > prompt. The commands are described fully later in this guide. Note that Minitab commands are entered following the MTB > prompt, and that data are entered following the DATA > prompt.

```
MTB > NAME C1 'YEAR' C2 'STOCKS' C3 'BONDS'
MTB > NAME C4 'T-BILLS' C5 'INF-RATE'
MTB > SET 'YEAR'
DATA> 1985:1993
DATA> END
MTB > SET 'STOCKS'
DATA> 31.2 35.3 20.2 -5.9 19.5 12.8 7.0 13.9 16.3
DATA> END
MTB > SET 'BONDS'
DATA> 29.9 19.9 5.6 8.2 12.2 7.7 10.8 14.2 12.0
DATA> END
MTB > SET 'T-BILLS'
DATA> 9.3 7.3 5.7 6.1 8.2 8.2 6.9 4.6 3.1
DATA> END
MTB > SET 'INF-RATE'
DATA> 3.7 1.7 3.7 3.9 5.2 4.7 4.7 3.1 3.0
DATA> END
MTB > INFORMATION

     COLUMN    NAME       COUNT
     C1        YEAR         9
     C2        STOCKS       9
     C3        BONDS        9
     C4        T-BILLS      9
     C5        INF-RATE     9

CONSTANTS USED: NONE

MTB > PRINT C1-C5

  ROW   YEAR   STOCKS   BONDS   T-BILLS   INF-RATE

   1    1985    31.2    29.9      9.3       3.7
   2    1986    35.3    19.9      7.3       1.7
   3    1987    20.2     5.6      5.7       3.7
   4    1988    -5.9     8.2      6.1       3.9
   5    1989    19.5    12.2      8.2       5.2
   6    1990    12.8     7.7      8.2       4.7
   7    1991     7.0    10.8      6.9       4.7
   8    1992    13.9    14.2      4.6       3.1
   9    1993    16.3    12.0      3.1       3.0

MTB > SAVE 'FINANCE'

Worksheet saved into file: FINANCE.MTW
MTB > MEAN 'STOCKS'
   MEAN    =       16.700
MTB > MEAN 'BONDS'
   MEAN    =       13.389
MTB > MEAN 'T-BILLS'
   MEAN    =       6.6000
MTB > MEAN 'INF-RATE'
   MEAN    =       3.7444
```

8   Chapter 1

```
MTB > MPLOT 'STOCKS' VS 'YEAR' AND 'BONDS' VS 'YEAR';
SUBC> TITLE 'STOCKS AND BONDS';
SUBC> TITLE 'Mean Rate of Return';
SUBC> YLABEL 'RATE';
SUBC> XLABEL 'YEAR';
SUBC> XSTART = 1984;
SUBC> XINCREMENT = 2.
```

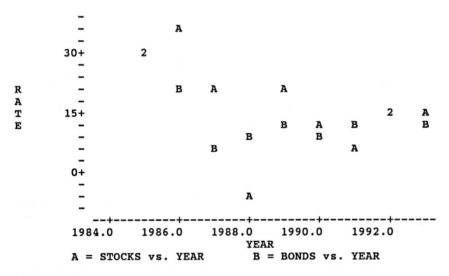

```
MTB > STOP
```

The NAME commands give descriptive names to the columns. After the columns are named, any command accepts either the column names or column numbers as arguments. For example, we use the column name for the argument with SET,

```
MTB > SET 'YEAR'
```

and the column numbers with PRINT,

```
MTB > PRINT C1-C5
```

The SET command tells the computer to enter the data following the DATA > prompt in the specified column. The END statement marks the end of the data. INFORMATION lists the columns, including names and counts, of the current worksheet. We use PRINT C1-C5 to check that the data have been correctly entered in C1 through C5. Minitab uses the names of the columns on the output.

MEAN is an example of a Minitab command that gives one number for an answer. The result is automatically printed. The mean investment returns for stocks and bonds were higher over the nine year period than the mean return for treasury bills.

Next we did a plot to compare the rates of return for stocks and bonds. MPLOT produces a multiple plot of the rates for the nine years. We can control the output with the subcommands. Titles and axes labels are added with subcommands. XSTART begins the scale on the horizontal axis with 1984, and XINCREMENT allows two years between tick (+) marks. If we connect the stock and bond data points, labeled A and B, we can compare the trends in the rates of return over time. The '2s' on the plot indicate that the 1985 and 1992 rates for stocks and bonds were about the same. The Minitab session is terminated with the STOP command.

∎

MINITAB FILES

Minitab uses several types of files to store data, commands, and output. You can save a copy of the program, outfile, and worksheet. Each type of file has its own file extension. This extension is a three letter ending that describes the type of file. You have the option of adding your own file extension, or Minitab will automatically add an extension when a file is created. For example, Minitab adds the extension .MTW to a file containing the data in the Minitab worksheet.

Program File

The JOURNAL command can be used to document a Minitab session or to save a copy of a program. On most computers, the input commands and data are saved in a file. JOURNAL is not available with the Student Version of Minitab.

JOURNAL ('FILENAME')

This command copies all subsequent input lines to a computer file, or adds input lines to the end of an existing file. The default filename extension is MTJ, and the default filename is MINITAB.MTJ.

NOJOURNAL

This command is used to cancel JOURNAL. The commands and data following NOJOURNAL are not saved.

For example, to save a program under the filename EXAMPLE1, the line in the program would be

    MTB > JOURNAL 'EXAMPLE1'

All lines in the program following JOURNAL are saved under the filename EXAMPLE1.MTJ. The file can be edited with your computer system editor or a word processor. The EXECUTE command runs a stored program.

> EXECUTE 'FILENAME'
>
> Minitab executes all the commands in the file. The default extension is .MTB. Use the full filename to execute a program having a different extension. More than one EXECUTE statement can be used in a program.

For example, the program file EXAMPLE1.MTJ can be executed by entering the command,

```
MTB > EXECUTE 'EXAMPLE1.MTJ'
```

Comment We suggest that you always use the filename extension to run a JOURNAL program file. The default extension used by EXECUTE is .MTB, the same extension that Minitab uses for macros that are described in Chapter 6.

Output File

The OUTFILE command saves the output in a computer file. You can then print the file, or edit the file with the system editor or a word processor.

> OUTFILE 'FILENAME'
>
> This command saves the output from all subsequent commands in a computer file. The default filename extension is LIS. If a filename is reused, the new output is appended to the end of the existing file.
>
> NOOUTFILE
>
> Minitab stops saving output in a file.

For example, to save the output in a file named EXAMPLE1.LIS, enter

```
MTB > OUTFILE 'EXAMPLE1'
```

prior to the program commands. Output from commands entered before OUTFILE or after NOOUTFILE is not saved.

Worksheet File

It is always an excellent idea to save the worksheet in a computer file. Quite often you need to run more than one program using the same set of data, or you may need to correct errors at a later time. You can save the worksheet file with SAVE or WRITE. SAVE stores the entire worksheet, including column names and constants. The file created by the SAVE command cannot be edited, printed, or used with other software; Minitab must be used to edit or print the data. The RETRIEVE command is used to enter the saved file into the worksheet. You need the PORTABLE subcommand if the worksheet is to be transferred to a different type of computer.

SAVE THE WORKSHEET (IN 'FILENAME')
   PORTABLE
   LOTUS (PC only)

All columns, column names, stored constants, and matrices are saved in a file called FILENAME. The default filename extension is MTW, or MTP if the PORTABLE subcommand is used. The default filename is MINITAB.MTW or MINITAB.MTP.

The PORTABLE subcommand permits a worksheet to be transferred to a different type of computer. For example, a worksheet can be created and saved on a microcomputer and transferred to a minicomputer. On a microcomputer, use LOTUS (not available with the student version) to save a Minitab worksheet as a Lotus or Symphony spreadsheet file.

RETRIEVE THE WORKSHEET STORED IN 'FILENAME'
   PORTABLE
   LOTUS (PC only)

The RETRIEVE command enters data saved by the SAVE command. Any data in the current Minitab worksheet is erased when you use RETRIEVE.

Use PORTABLE if the worksheet was saved on a different type of computer, and LOTUS (not available with the student version) if the microcomputer file is a Lotus or Symphony worksheet.

For example, the command

```
MTB > SAVE 'EXAMPLE1'
```

saves the worksheet as EXAMPLE1.MTW. You can retrieve the worksheet on the same type of computer with the command,

```
MTB > RETRIEVE 'EXAMPLE1'
```

The default filename extension MTW is optional.

The data file created by WRITE can be used with other software programs and with Minitab. For example, you may use the computer system editor or a word processor to edit the data. The file can be entered in the current worksheet with READ, SET, or INSERT. Use READ if there are several columns in the file and SET if there is only one column in the file. The INSERT command is used to append files. The descriptive details of READ, SET, and INSERT are given in the next section.

WRITE THE DATA IN 'FILENAME' FROM C,...,C
FORMAT (SPECIFICATION)

The file contains column data. Column names and stored constants are not saved. Minitab adds missing values if column lengths are not equal. The default file extension when using the WRITE command is DAT. Use FORMAT to control the form of the output. The SPECIFICATION is described in the next section with the SET command.

READ DATA FROM 'FILENAME' IN C,...,C

SET DATA FROM 'FILENAME' IN C

INSERT DATA 'FILENAME' BETWEEN ROWS K AND K+1 OF C,...,C

The READ, SET, and INSERT commands input data saved by WRITE, or input a file created by the system editor or some other program.

For example, the command

    MTB > WRITE 'EXAMPLE1' C1 C2 C3

saves the data from columns C1, C2, and C3 in a file called EXAMPLE1.DAT. The data in this file can be entered in a Minitab worksheet using the command

    MTB > READ 'EXAMPLE1' IN C4 C5 C6

Notice the data can be read in different columns than those used in the WRITE command. The filename extension DAT is optional.

A standard ASCII file created by other software can be entered in a Minitab worksheet. The READ, SET, and INSERT commands enter rows of data from files. For example,

    MTB > READ 'FILENAME' C,...,C

inputs the file called FILENAME in the specified columns of the worksheet. When used in a Minitab command, filenames must be enclosed in single quotation marks.

DATA

Most Minitab commands work only with numerical data. Some commands, for example, the input, output, and editing commands, also work with alpha data. Alpha data, such as the names of companies, are usually used as row labels. Alpha data may contain up to 80 characters, including letters, numbers, or punctuation symbols. Numbers used for alpha characters cannot be used in computations. Alpha and numeric data cannot be contained in the same column.

Entering Data

All types of data are entered in columns using SET or READ, or added to columns using INSERT. SET enters one column of data at a time. The FORMAT subcommand specifies the form of alpha and numeric data. The SPECIFICATION is similar to a Fortran format.

SET THE FOLLOWING DATA ('FILENAME') IN C
   NOBS = K
   FORMAT (SPECIFICATION)

This command puts data in one column of the worksheet. If a filename is given, data are read from the file. Otherwise the command is followed by a DATA> prompt. Numbers are usually entered free format, separated by commas or blanks, on one or more lines. The NOBS subcommand specifies the number of observations. It is ignored if an END statement is encountered before the specified number of observations is read.

The FORMAT specification, enclosed in parentheses, is required for alpha data and is an option for numeric data. The allowable formats include:

Fw	Field width for entering numbers
Fw.d	Field width with specified decimal place
Aw	Alpha width for entering character data
X	Skip a space
Tn	Move to position n
n	Repeat factor
/	Move to next data line

The formats can be combined in several ways. The basic format is Fw or field width, which says to enter a number from a field of width w.

END

This statement is used on the DATA> line after the data have been entered.

For example, the Minitab commands

```
MTB > SET C1;
SUBC> FORMAT(4(A2,1X)).
DATA> CA TX FL NC
DATA> GA VA WA AZ
DATA> END
```

enter the alpha data in column C1 of the Minitab worksheet. Four state abbreviations are read per line; each abbreviation of two alpha characters is entered in a row of C1.

■ **Example 2** The number of seats that a state may have in the House of Representatives is based on the state's census. According to the 1990 census, eight states have gained one or more seats in the House over the previous ten years. The following table gives the 1990 census figures, the ten-year percentage change in population, and the increase in the number of seats in the House of Representatives for each of the eight states. Use SET to enter the data in a Minitab worksheet.

State	Population	Change	Increase
California	29,839,250	26.1	+7
Texas	17,059,805	19.9	+3
Florida	13,003,362	33.4	+4
North Carolina	6,657,630	13.2	+1
Georgia	6,508,419	19.1	+1
Virginia	6,216,568	16.3	+1
Washington	4,887,941	18.3	+1
Arizona	3,677,985	35.3	+1

Solution We use the FORMAT subcommand with SET to input the alpha data. The SPECIFICATION option defines the spacing for the states' names on the DATA> line. The 4(A10,1X) says to enter each state name using 10 spaces followed by a blank space. This is repeated four times on each line. The numerical data are entered free format.

```
MTB > NAME C1 'STATES' C2 '1990 POP' C3 'CHANGE' C4 'US REPS'
MTB > SET 'STATES';
SUBC> FORMAT (4(A10,1X)).
DATA> CALIFORNIA TEXAS      FLORIDA    N.CAROLINA
DATA> GEORGIA    VIRGINIA   WASHINGTON ARIZONA
DATA> END
MTB > SET '1990 POP'
DATA> 29839250 17059805 13003362 6657630 6508419
DATA> 6216568 4887941 3677985
DATA> END
MTB > SET 'CHANGE'
DATA> 26.1 19.9 33.4 13.2 19.1 16.3 18.3 35.3
DATA> END
MTB > SET 'US REPS'
DATA> 7 3 4 1 1 1 1 1
DATA> END
MTB > PRINT C1-C4

   ROW      STATES     1990 POP   CHANGE   US REPS
     1   CALIFORNIA    29839250     26.1         7
     2   TEXAS         17059805     19.9         3
     3   FLORIDA       13003362     33.4         4
```

```
    4    N.CAROLINA    6657630    13.2    1
    5    GEORGIA       6508419    19.1    1
    6    VIRGINIA      6216568    16.3    1
    7    WASHINGTON    4887941    18.3    1
    8    ARIZONA       3677985    35.3    1

MTB > SAVE 'CENSUS'

Worksheet saved into file: CENSUS.MTW
MTB > STOP
```

The NAME command assigns names to the columns. We use PRINT C1-C4 to check that the data have been correctly entered. Note that a FORMAT subcommand is not needed to print the alpha data. The NAME and PRINT commands are described later in this chapter. ∎

Comment Several lines can be used to enter numerical data in a column. Just press the enter key when you get near the edge of the screen. Minitab responds with DATA>, and you can continue typing numbers following the prompt.

Abbreviations can be used with SET to enter a sequence of numbers, repeated numbers, or repeated sequences of numbers.

1. Sequence of numbers, where a is the starting number and b is the ending number.

    DATA> a:b

    For example, to put the integers (1 2 3 4 5) in C1, the program lines would be

    ```
    MTB > SET C1
    DATA> 1:5
    DATA> END
    ```

2. Sequence of numbers, where the step between the numbers is c.

    DATA> a:b/c

    The following program lines put the integers (2 4 6 8 10) in C1.

    ```
    MTB > SET C1
    DATA> 2:10/2
    DATA> END
    ```

3. Repeated sequence of numbers, where r is the number of times the sequence a:b is repeated.

    DATA> r(a:b)

    If you want to repeat the sequence 1, 2, 3, 4 three times in column C1 (1 2 3 4 1 2 3 4 1 2 3 4), the program lines would be

    ```
    MTB > SET C1
    DATA> 3(1:4)
    DATA> END
    ```

16   Chapter 1

4. Repeated numbers, where r is the number of times each number is repeated.

   DATA> (a:b)r

   To repeat each year between 1970 and 1972 three times in column C1 (1970 1970 1970 1971 1971 1971 1972 1972 1972), enter

   MTB > SET C1
   DATA> (1970:1972)3
   DATA> END

5. Combinations of the above procedures can be used. For example,

   MTB > SET C1
   DATA> 3(2:4)2
   DATA> END

   will enter the sequence (2 2 3 3 4 4 2 2 3 3 4 4 2 2 3 3 4 4) in C1.

READ enters data in rows across several columns. It is a useful command for entering a table of numbers.

READ THE FOLLOWING DATA ('FILENAME') IN C,...,C
  NOBS = K
  FORMAT (SPECIFICATION)

This command enters data row by row in the specified columns. If a filename is given, data are read from the file. Otherwise, Minitab responds to READ with a DATA> prompt. Each line of data you enter on the DATA> line is put across one row of the worksheet. Use END after all data have been entered.

Enter alpha data with the FORMAT subcommand. The SPECIFICATION is described with SET.

Comment   A missing data value is denoted by an * in a numerical column and by a blank in an alpha column. Most commands exclude missing values in calculations. A diagnostic is given if a command does not accept missing data.

■ **Example 3**   Consider the following financial data concerning New York Stock Exchange trading reported on January 2, 1993. The SALES are the numbers of shares, in hundreds, traded during the week. The HIGH and LOW are the highest and lowest selling prices, and CLOSE is the week's final selling price. CHANGE is the change between the final price and the previous week's final price. Enter the alpha and numerical data in a Minitab worksheet.

Stocks	Sales	High	Low	Close	Change
AT&T	29,219	53 1/8	51	51	-1 3/4
Sears	10,077	46 5/8	44 1/4	45 1/2	+1 1/8
Fed. Express	2,277	55 7/8	52 3/4	54 1/2	+ 5/8
General Mills	2,134	71 5/8	68 1/2	68 1/2	-3 1/8

Solution The first two program lines name the columns with the descriptive names given in the table. We use the SET command to enter the names of the stocks in the first column. It's often easier to enter a single column of alpha data with SET. The SPECIFICATION for the alpha data (A5) says to enter each name on a separate line using five characters. The READ command enters the table of financial data. The data are entered, row by row, following the DATA> prompt. Fractions are entered as decimals.

```
MTB > NAME C1 'STOCKS' C2 'SALES' C3 'HIGH'
MTB > NAME C4 'LOW' C5 'CLOSE' C6 'CHANGE'
MTB > SET 'STOCKS';
SUBC> FORMAT (A5).
DATA> AT&T
DATA> SEARS
DATA> FEDEX
DATA> GMILL
DATA> END
MTB > READ C2-C6
DATA> 29219   53.125   51     51    -1.75
DATA> 10077   46.625   44.25  45.5   1.125
DATA>  2277   55.875   52.75  54.5    .625
DATA>  2134   71.625   68.5   68.5  -3.125
DATA> END
      4 ROWS READ
MTB > PRINT C1-C6

ROW   STOCKS   SALES     HIGH      LOW    CLOSE    CHANGE

  1   AT&T     29219    53.125    51.00   51.00    -1.750
  2   SEARS    10077    46.625    44.25   45.50     1.125
  3   FEDEX     2277    55.875    52.75   54.50     0.625
  4   GMILL     2134    71.625    68.50   68.50    -3.125

MTB > SAVE 'FINDATA'

Worksheet saved into file:   FINDATA.MTW
MTB > STOP
```

The PRINT output allows us to check the accuracy of the data. The SAVE command stores the column names and data in a file named FINDATA.MTW. ■

INSERT adds rows of data at the top, between two rows, or at the bottom of one or more columns. The FORMAT subcommand specifies the form of alpha and numeric data. The SPECIFICATION is similar to a Fortran format.

18 *Chapter 1*

> INSERT DATA ('FILENAME') (BETWEEN K AND K+1), COLUMNS C,...,C
> NOBS = K
> FORMAT (SPECIFICATION)
>
> This command enters data in the specified rows. If a filename is given, data are read from the file. Otherwise, Minitab responds with a DATA> prompt. Enter the data on the DATA> line and END after the data have been entered. Use K = 0 to insert data at the top of a column, and omit K and K+1 to insert data at the end of a column. The NOBS subcommand specifies the number of observations. The SPECIFICATION for FORMAT is described with the SET command.

■ **Example 4** The SAVE command in Example 3 stored the financial data in the file FINDATA. Retrieve the Minitab worksheet and print the data. Add the following financial data to the table.

Stocks	Sales	High	Low	Close	Change
Jostens	4,907	27 1/8	24 7/8	26 7/8	+1 1/4
Perkins	192	18 5/8	17 5/8	17 5/8	-3/4

Solution The RETRIEVE command enters the financial data in the worksheet. The INFORMATION command gives the columns, names of columns, and numbers of observations. The letter A before C1 indicates that alpha data are saved in that column. INSERT adds the names and the numerical data for Jostens and Perkins. We need a FORMAT statement for the alpha data.

```
MTB > RETRIEVE 'FINDATA'
   WORKSHEET SAVED   1/10/1994

Worksheet retrieved from file: FINDATA.MTW
MTB > INFORMATION

    COLUMN    NAME      COUNT
A   C1        STOCKS    4
    C2        SALES     4
    C3        HIGH      4
    C4        LOW       4
    C5        CLOSE     4
    C6        CHANGE    4

CONSTANTS USED: NONE
MTB > PRINT C1-C6
    ROW    STOCKS    SALES    HIGH      LOW      CLOSE    CHANGE

      1    AT&T      29219    53.125    51.00    51.00    -1.750
      2    SEARS     10077    46.625    44.25    45.50     1.125
      3    FEDEX      2277    55.875    52.75    54.50     0.625
      4    GMILL      2134    71.625    68.50    68.50    -3.125

MTB > INSERT STOCK NAMES TO 'STOCKS';
SUBC> FORMAT (A5).
DATA> JOSTN
DATA> PERKN
DATA> END
```

```
MTB > INSERT DATA 'SALES'-'CHANGE'
DATA> 4907   27.125   24.875   26.875   1.25
DATA> 192    18.625   17.625   17.625  -.75
DATA> END
      2 ROWS READ
MTB > PRINT 'STOCKS'-'CHANGE'

 ROW   STOCKS   SALES      HIGH       LOW    CLOSE   CHANGE
   1   AT&T     29219    53.125    51.000    51.00   -1.750
   2   SEARS    10077    46.625    44.250    45.50    1.125
   3   FEDEX     2277    55.875    52.750    54.50    0.625
   4   GMILL     2134    71.625    68.500    68.50   -3.125
   5   JOSTN     4907    27.125    24.875    26.875   1.250
   6   PERKN      192    18.625    17.625    17.625  -0.750

MTB > SAVE 'FINDATA2'

Worksheet saved into file: FINDATA2.MTW
MTB > STOP
```

The additional stock data are added to the ends of the columns. The PRINT commands give the financial data before and after inserting the information on Jostens and Perkins.

■

Manipulating Data

Several Minitab commands for managing data are COPY, STACK, UNSTACK, CODE, and SIGN. You can copy data to new columns and constants with COPY, combine blocks of columns and constants with STACK, separate with UNSTACK, and code data with CODE and SIGN.

> COPY C,...,C TO C,...,C
> COPY K,...,K TO C
> COPY K,...,K TO K,...,K
> COPY C TO K,...,K
>   USE ROWS K,...,K
>   USE ROWS WHERE C = K,...,K
>   OMIT ROWS K,...,K
>   OMIT ROWS WHERE C = K,...,K
>
> The COPY command copies data from columns and constants to new columns and constants. The USE subcommand selects rows to copy, and OMIT indicates which rows not to copy.

For example, suppose we have a column C1 containing missing data: -48 51 0 * 23 * 91 27. We can copy the nonmissing data to another column.

20   Chapter 1

```
MTB > COPY C1 TO C2;
SUBC> OMIT C1='*'.
```

Column C2 would contain -48 51 0 23 91 27.

STACK (E,...,E),...,(E,...,E) PUT IN (C,...,C)
    SUBSCRIPTS IN C

This command combines corresponding columns and constants in blocks enclosed in parenthesis. If each block contains only one constant or column, the parentheses can be omitted.

The SUBSCRIPT subcommand creates a column of subscripts. The first block is given the subscript 1, the second block 2, and so on.

UNSTACK (C,...,C) IN (E,...,E),...,(E,...,E)
    SUBSCRIPTS IN C

This command separates one or more columns in blocks of columns or stored constants. The subcommand SUBSCRIPTS specifies the blocks. The rows with the smallest subscript are stored in the first block, the second smallest in the second block, and so on.

Consider the following program.

```
MTB > READ C1-C4
DATA> -2 23 -6 45
DATA> -6 37 -8 62
DATA> -3 41 -9 85
DATA> END
      3 ROWS READ
MTB > STACK C1-C4 IN C5
MTB > PRINT C5

C5
-2   -6   -3   23   37   41   -6   -8   -9   45   62   85

MTB > STACK (C1 C2) AND (C3 C4) IN (C6 C7);
SUBC> SUBSCRIPTS IN C8.
MTB > PRINT C1-C4 C6-C8

ROW   C1    C2    C3    C4    C6    C7   C8

  1   -2    23   -6    45    -2    23    1
  2   -6    37   -8    62    -6    37    1
  3   -3    41   -9    85    -3    41    1
  4                                -6    45    2
  5                                -8    62    2
  6                                -9    85    2
```

Introduction to Minitab   21

The first STACK command combines C1 through C4 in C5. The next STACK combines the data in the block (C1 C2) with data in block (C3 C4). The results are put in C6 and C7, and the subscripts denoting the blocks in C8.

CODE (K,...,K) TO K,...,(K,...K,) TO K FOR C,...,C PUT IN C,...,C

This command codes the values specified on the command line; all other values remain unchanged. The results are stored in new columns. Only numeric columns can be coded.

For example, the command

```
MTB > CODE (1:25) TO 1, (26:50) TO 2, (51:75) to 3 FOR C1 PUT IN C2
```

codes the values in C1 to the values in C2 as shown below.

C1	C2
23	1
49	2
67	3
54	3
10	1
0	0
19	1

SIGN OF E (PUT IN E)

The SIGN command gives a summary table of negative, zero, and positive values. The option converts negative, zero, and positive values to -1, 0, and +1, respectively, and stores the new values.

For example, suppose C1 contains the numbers: 3, 9, -5, 0, and 7. The command

```
MTB > SIGN C1, PUT IN C2
```

converts the data in C1 to the following data in column C2. There are three positive values, one zero value, and one negative value in the data set.

C1	C2
3	+1
9	+1
-5	-1
0	0
7	+1

Sorting and Ranking Columns of Data

The SORT command alphabetizes data and orders data values. The command allows you to carry along additional columns.

SORT THE DATA IN C,...,C PUT IN C,...,C
  BY C,...,C
  DESCENDING C,...,C

This command orders alpha or numeric data in one or more columns, and carries along corresponding data in additional columns. The reordered data are stored in the last group of specified columns. Missing values are sorted last in numeric columns, and first in alpha columns.

The columns listed with the BY subcommand determine the order of sorting by multiple columns. If BY is not used, the data are sorted by the first column. The default is ascending order; use the DESCENDING subcommand to sort data in descending order.

■ **Example 5** Example 2 of this supplement gives data on the United States 1990 census, the ten-year percentage change in population, and the increase in the number of seats in the House of Representatives for each of eight states. Alphabetize the states, retaining the 1990 census for each state.

Solution The census data are stored in a file named CENSUS. INFORMATION gives the column information on the current worksheet.

```
MTB > RETRIEVE 'CENSUS'
WORKSHEET SAVED   1/10/1994

Worksheet retrieved from file: CENSUS.MTW

MTB > INFORMATION

    COLUMN    NAME       COUNT
A   C1        STATES       8
    C2        1990 POP     8
    C3        CHANGE       8
    C4        US REPS      8
```

CONSTANTS USED: NONE

```
MTB > PRINT C1-C4

ROW        STATES     1990 POP   CHANGE   US REPS

  1     CALIFORNIA    29839250    26.1       7
  2     TEXAS         17059805    19.9       3
  3     FLORIDA       13003362    33.4       4
  4     N.CAROLINA     6657630    13.2       1
  5     GEORGIA        6508419    19.1       1
  6     VIRGINIA       6216568    16.3       1
  7     WASHINGTON     4887941    18.3       1
  8     ARIZONA        3677985    35.3       1

MTB > SORT 'STATES' CARRY '1990 POP' PUT IN C5 C6
MTB > PRINT C5 C6

ROW           C5              C6

  1     ARIZONA          3677985
  2     CALIFORNIA      29839250
  3     FLORIDA         13003362
  4     GEORGIA          6508419
  5     N.CAROLINA       6657630
  6     TEXAS           17059805
  7     VIRGINIA         6216568
  8     WASHINGTON       4887941

MTB > STOP
```

■

The RANK command ranks the data values in order of magnitude, from least to greatest. Many statistical procedures, such as nonparametric tests, use the relative ranks of the data rather than the actual numerical values.

RANK C, PUT IN C

This command uses a 1 for the smallest value, 2 for the second smallest, and so on. If there are ties, the average rank is assigned.

If C1 contains the values shown below, the command

```
MTB > RANK C1 PUT VALUES IN C2
```

ranks the values in C1 and stores the ranks in C2.

```
        C1              C2
        78               2
        97               5
        84               3.5
        84               3.5
        69               1
```

Oftentimes we want a cumulative count of a column of data values. The PARSUMS command calculates partial sums.

> PARSUMS OF C, PUT IN C
>
> This command calculates and stores partial sums. The partial sum for row i is equal to the sum of the rows up to and including row i.

For example, if C1 contains the values shown below, the command

```
MTB > PARSUMS C1 PUT VALUES IN C2
```

calculates and stores the partial sums in C2.

C1	C2
6	6
9	15
8	23
3	26
6	32

Printing Data

The Minitab PRINT command is used to print data. Use this command to view the worksheet on the screen and to verify that the data have been correctly entered in a column or constant location.

> PRINT E,...,E
>   FORMAT (SPECIFICATION)
>
> Use this command to print columns, constants, matrices, or combinations, that are stored in the worksheet. The SPECIFICATION is described with the SET command. Minitab chooses the form of the output if a FORMAT subcommand is not given. If more than one column is listed on the command line, the columns are printed vertically across the screen, and the row numbers are printed down the left side of the screen. If only one column is listed, it is printed horizontally across the screen.

If you want the output of a command on a new page of paper, the NEWPAGE command can be used on most computers. For example, use NEWPAGE to print tables and/or graphs on one page.

> **NEWPAGE**
>
> This command starts the output of the next command on a new page.

For example, if you enter,

```
MTB > NEWPAGE
MTB > PRINT C1-C3
```

Minitab would print columns C1, C2, and C3 beginning at the top of a page.

OH and OW control the height and width of Minitab output. OH determines the number of lines printed to the screen between pausing. The entire output, however, is saved in an outfile if output is being stored. OW determines the width of the output. This command is useful if you want narrow graphs for reports.

> **OH = K LINES (PC ONLY)**
>
> This command controls the amount of output on the screen. After every K number of lines of output, Minitab pauses and prints "Continue?". Type Yes or press the enter key to continue printing output, or type No to stop the printing for that command. The default is K = 0, which means that all the output is printed on the screen without pauses.
>
> **OW = K SPACES**
>
> This command controls the width of the Minitab output. Most commands will allow values of K from 30 to 132.

ARITHMETIC

You can use LET to compute algebraic expressions, do comparison or logical operations, and correct errors. The LET arguments can be columns, stored constants, or numbers. LET does not automatically print a result; you need a PRINT statement. The Minitab commands ADD, SUBTRACT, MULTIPLY, DIVIDE, and RAISE can also be used for calculations. A list of arithmetic functions is given in Appendix C of this supplement.

> LET E = ARITHMETIC EXPRESSION
>
> The arithmetic expression may use columns, constants, and the arithmetic symbols:
>
> | | |
> |---|---|
> | + | add |
> | - | subtract |
> | * | multiply |
> | / | divide |
> | ** | raise to a power |
>
> The expression may also use column commands, such as MEAN and SUM. The column name or number following the column command must be enclosed in parentheses. No extra text may be used with LET unless it follows a # sign.
>
> Comparison operations include the following:
>
> | | |
> |---|---|
> | = or EQ | equal to |
> | ~= or NE | not equal to |
> | < or LT | less than |
> | > or GT | greater than |
> | <= or LE | less than or equal to |
> | >= or GE | greater than or equal to |
>
> Logical operations include the following:
>
> | | |
> |---|---|
> | & or AND | |
> | \| or OR | |
> | ~ or NOT | |
>
> Use the HELP facility for detailed information on comparison and logical expressions.

■ **Example 6** Suppose the variable X can have the values 5, 3, 1, 3 and the variable Y can have the values 4, 7, 5, 8. Use the LET command to do the following arithmetic for corresponding values of X and Y: X + Y, (2X + 3Y)/5, add the Y values, and square the X values.

Solution The sum of the Y values is stored as a constant K1. The constant can be printed with the columns containing the data and other results.

```
MTB > NAME C1 'X' C2 'Y'
MTB > READ DATA IN C1 AND C2
DATA> 5 4
DATA> 3 7
DATA> 1 5
DATA> 3 8
DATA> END
     4 ROWS READ
MTB > LET C3 = 'X' + 'Y'
MTB > LET C4 = (2*'X' + 3*'Y')/5
MTB > LET K1 = SUM ('Y')    # STORES THE SUM AS A CONSTANT
```

```
MTB > LET C5 = 'X'**2      # SQUARES THE X VALUES
MTB > PRINT K1 C1-C5
K1         24.0000

  ROW    X    Y    C3    C4    C5

   1     5    4     9   4.4    25
   2     3    7    10   5.4     9
   3     1    5     6   3.4     1
   4     3    8    11   6.0     9

MTB > STOP
```

■

MISCELLANEOUS TOPICS

This section gives information on some additional capabilities of Minitab. Topics include annotating programs, naming columns, manipulating data, ending a session, and rules on syntax.

Documentation

You can include notes or comments in a Minitab program with the NOTE command or # symbol. Minitab ignores everything on the command line after NOTE or #.

NOTE
#

These commands are used to annotate a program. Everything you type on a line after NOTE or # is ignored by Minitab. The # symbol may be used anywhere on a command, subcommand, or data line.

For example, # may be used with READ to identify the problem.

```
MTB > READ THE DATA IN C1-C4    # EXERCISE 26, PAGE 44
```

Naming Columns

Columns can be assigned names with the NAME command. After a column is named, Minitab uses the name of the column on all output.

> NAME C 'NAME1' C 'NAME2' ...
>
> This command assigns a name, in upper or lower case letters, to each specified column. Anytime a column name is used, it must be enclosed in single quotes (apostrophes). After a column has been named, Minitab commands accept either the column number or name.
>
> Some restrictions on the names are:
> 1. The name may be 1 to 8 characters long.
> 2. The name may not begin or end with a space.
> 3. The single quote (') and # symbols may not be used in a name.
> 4. Column names must be different, but a column name may be changed with a new NAME line.

Comment With Release 9 on the VAX, you can also name constants and matrices.

Consider the following examples.

1. If C1 contains the survey number, C2 the age of the respondent, and C3 the salary of the respondent, the following command names the three columns:

   MTB > NAME C1 'SURVEY' C2 'AGE' C3 'SALARY'

2. After the columns have been named, the READ command can be used to enter data:

   MTB > READ 'SURVEY' 'AGE' 'SALARY'

3. After the data have been entered, the following command can be used to calculate the average salary:

   MTB > MEAN 'SALARY'

   Comment We recommend that you always use the NAME command. It makes the program and output easier to read and understand.

Ending a Session

The STOP command terminates a Minitab session. The worksheet is erased with this command.

> **STOP**
>
> The STOP command terminates the Minitab program and puts you in the computer's operating system. If you use Minitab in the batch mode, STOP is the last Minitab command of a program.

Following the STOP command on a microcomputer, Minitab gives the version of Minitab and the available storage.

```
MTB > STOP
*** Minitab Release 9.1 *** Minitab, Inc. ***
Worksheet size: 100013 cells
```

ERROR CORRECTION

This section covers some common errors and the corrections. If you are using Minitab in batch mode, all errors can be corrected with the computer system editor or a word processor.

1. Typing error on the line you are entering.

   Use the backspace key to erase the error. The name of this key depends on your keyboard. You may need to experiment to determine which key works for your system.

2. An error message appears on the screen.

   Often this is the result of entering a command line incorrectly. The error message is usually self-explanatory. Once you find the mistake, retype the correct line. For example, if SET is misspelled the following is printed:

   ```
   MTB > SEET C1
   *ERROR* NAME NOT FOUND IN DICTIONARY
   ```

3. A data value was entered incorrectly, and the line has already been entered.

   The LET command may be used to change a single number. The form is

   ```
   MTB > LET Cj(i) = correct number
   ```

   where j is the column number and i is the row number containing the error. For example,

   ```
   MTB > LET C1(5) = 87.3
   ```

   replaces the incorrect value in the fifth row of column C1 with 87.3.

4. One or more rows of data were accidentally omitted from one or more columns.

   INSERT, described on page 18, adds data at the top, between two rows, or at the bottom of specified columns.

   For example, the command

   ```
   MTB > INSERT DATA BETWEEN ROWS 2 AND 3 OF 'DAY' AND 'COST'
   DATA> 3 46.25
   DATA> END
   ```

   adds the cost $46.25 for the third day.

DAY	COST	DAY	COST
1	12.50	1	12.50
2	36.25	2	36.25
4	18.75	3	46.25
		4	18.75

5. Too many data values have been entered.

   Use the DELETE command to remove extra data.

DELETE ROWS K,...,K FROM C,...,C

This command deletes the rows of data indicated on the command line, and adjusts the remaining rows to fill in the deleted rows. Use a colon to delete a sequence of rows.

For example, use the command

```
MTB > DELETE ROWS 2 AND 6 FROM 'DAY' AND 'COST'
```

to omit $16.50 and $23.80 for days 1 and 5 from the following data:

DAY	COST	DAY	COST
1	12.50	1	12.50
1	16.50	2	36.25
2	36.25	3	46.25
3	46.25	4	18.75
4	18.75		
5	23.80		

More Syntax Rules

1. Each command must begin on a new line. If the command does not fit on one line, use the ampersand symbol (&) or two plus (+ +) symbols at the end of the line. Minitab responds with CONT>, and you can continue the command following the prompt.

```
MTB > READ THE SURVEY NUMBER IN C1, AGE OF THE      &
CONT> RESPONDENT IN C2, AND SALARY IN C3
```

2. Each command expects certain arguments. For example, SET looks for one column or column name. The command line

   ```
   MTB > SET THE 1984 SALARIES IN C1
   ```

   causes an error because of the number 1984.

3. Care is advised when reusing columns or constants for storing data. Any previous contents are erased.

4. Numbers on a data line can be separated by a comma or one or more spaces. For example,

   ```
   MTB > SET C1
   DATA> 57      7,150    2     10
   ```

   C1 contains the five numbers 57, 7, 150, 2, and 10.

EXERCISES

1. Consider the following exam scores for 14 students.

Student	Exam 1	Exam 2
1	75	88
2	66	73
3	92	98
4	86	79
5	95	92
6	79	85
7	81	74
8	63	80
9	76	64
10	45	69
11	85	78
12	77	81
13	57	70
14	69	72

   a. Use the SET command to input the data.
   b. Use the READ command to input the data.
   c. What is the advantage of using SET instead of READ?

2. The columns on the left contain data entry errors. Write the commands to correct the errors to obtain the columns on the right.

32   Chapter 1

a.
C1	C2		C1	C2
75	88		75	88
66	73		66	73
92	98		92	89
86	79		68	79

b.
C1	C2		C1	C2
67	92		67	92
88	85		88	85
88	85		58	61
58	61			

c.
C1	C2		C1	C2
67	92		67	92
88	85		66	73
58	61		88	85
			58	61

3. Find the error or errors on each one of the following command lines.

   a. MTB > SET SALES IN COLUMN 1
   b. MTB > READ 1989 SALES IN C4 AND NUMBER OF EMPLOYEES IN C5
   c. MTB > NAME C4 '1989 SALES' C5 'EMPLOYEES'
   d. MTB > PRINT SALES

4. The following command lines have added text. Give the shortest version of the command lines.

   a. MTB > SET THE COMPANY SALES IN COLUMN C3
   b. MTB > DESCRIBE THE DATA IN C1, C2, C3, AND C4
   c. MTB > PLOT THE MONTHLY SALES IN C2 VERSUS TIME IN C1

5. The following table gives the 1993 quarterly sales for five companies.

Company	Q1	Q2	Q3	Q4
Peters	45	78	65	81
Sinco	68	91	76	95
Mairo	33	45	43	52
Beck	59	76	88	74
Relle	49	73	67	80

   a. Write a Minitab program to enter the table in a worksheet, calculate the total 1990 sales for each company, and print a table showing the company name and total sales. Use the NAME command in the program.
   b. Add the Minitab commands to save the program, the output, and the data.
   c. Run the program.

CLASS DATA SET

This data set can be set up as a class project, and used for statistical analysis throughout the course. We refer to this data base in some chapter exercises.

After an introductory session on Minitab, have each student respond to the following suggested variables:

Gender
Number of credits earned prior to this quarter/semester
Number of credits this quarter/semester
Marital status
Age
Distance student lives from class
Number of hours student works
Grade point average
Type of car
Major program

The students can create the data base individually or as a class project. Some of the items produce qualitative data, which have to be numerically coded. Each student can save the worksheet for later use.

Appendix B contains a data set obtained from a sample of 200 students. The data set is included on the data disk available with this guide. The SAVE command was used to save the class data set on the disk. To access the data, type:

```
MTB > RETRIEVE 'CLASSDAT.MTP';
SUBC> PORTABLE.
```

CHAPTER 2
METHODS FOR DESCRIBING SETS OF DATA

A role of statistics is to provide techniques to collect, describe, and analyze data for the decision process. This chapter demonstrates some Minitab commands for organizing and summarizing qualitative and quantitative data sets in tables and graphs, and some numerical measures for describing characteristics of quantitative data. We include measures that are typically used in statistics, such as the measures of central tendency and measures of variability.

NEW COMMANDS

BOXPLOT	COUNT	DESCRIBE	DOTPLOT	HISTOGRAM
MAXIMUM	MEAN	MEDIAN	MINIMUM	N
NMISS	PLOT	RANGE	RCOUNT	RMAX
RMEAN	RMEDIAN	RMIN	RN	RNMISS
RRANGE	RSSQ	RSTDEV	RSUM	SSQ
STEM	STDEV	SUM	TALLY	

QUALITATIVE DATA

Qualitative data include nominal and ordinal data. Nominal or categorical data are measurements that identify or classify the category of each unit in the sample or population. Some examples of nominal data are the brands of fax machines purchased by local businesses, the airlines used for business travel, the types of cars available for lease, and the gender of each executive board member. Nominal data are usually reported as labels or names, and converted to numerical codes for use in computer analyses. For example, gender may be coded male = 1 and female = 0.

Ordinal data are measurements which represent some order or ranking with respect to the variable of interest. Thus, ordinal data contain all the information of nominal data plus an ordering of the data. For example, the brands of fax machines rated on a scale of 1 (of least value) to 5 (of most value), airlines ranked according to service, and types of cars from compact to full-size available for lease. Ordinal data are simply a ranking or ordering of the data; arithmetic calculations are meaningless.

The TALLY command summarizes nominal and ordinal data. The output includes a frequency and a percent frequency distribution. The frequency is the number of observations in each category, and the percent is the percentage of observations in each category.

> TALLY THE DATA IN C,...,C
>   COUNTS
>   CUMCOUNTS
>   PERCENTS
>   CUMPERCENTS
>   ALL
>
> The TALLY command produces a one way distribution for each column listed on the command line. If no subcommands are used, counts or frequencies of the distinct values in the input column or columns are printed.
>
> The COUNTS subcommand to produce counts or frequencies is required if other subcommands are used. CUMCOUNTS prints the cumulative frequencies, starting at the smallest value. PERCENTS prints the percentage or percent frequency of each value. CUMPERCENTS prints the cumulative percentage, starting at the smallest value. The ALL subcommand prints all of the above.

■ **Example 1** Consider the class data set provided in Appendix B of this supplement. The students were asked to state whether they own a United States model, a foreign car, or whether they do not own a car. The following table gives a sample of 30 students. Use Minitab to construct a frequency and a percent frequency distribution.

Student	Type of Car	Student	Type of Car
1	US	16	US
2	US	17	US
3	Foreign	18	Foreign
4	US	19	Foreign
5	US	20	US
6	US	21	US
7	None	22	None
8	US	23	US
9	US	24	Foreign
10	US	25	US
11	None	26	US
12	US	27	Foreign
13	US	28	US
14	US	29	US
15	None	30	None

Solution The TALLY command requires a distinct value representing the type of car for each student. We use the numerical code, 0 for U.S. car, 1 for a foreign car, and 2 if no car.

```
MTB > NAME C1 'CARTYPE'
MTB > # CODE IS U.S. 0, FOREIGN 1, NO CAR 2
MTB > SET 'CARTYPE'
DATA> 0 0 1 0 0 0 2 0 0 2 0 0 0 2 0 0 1 1 0
```

```
DATA> 0 2 0 1 0 0 1 0 0 2
DATA> END
MTB > TALLY 'CARTYPE';
SUBC> COUNT;
SUBC> PERCENT.

   CARTYPE  COUNT  PERCENT
         0     20    66.67
         1      5    16.67
         2      5    16.67
        N=     30

MTB > SAVE 'CAR'

Worksheet saved into file: CAR.MTW
MTB > STOP
```

The output includes the frequency (COUNT) and percent frequency (PERCENT) distributions for this set of nominal data. The percent frequency = 100(frequency/n), where n is the number of observations. From the output, we see that about 67% of the students have a United States model, 17% have a foreign model, and 17% do not own a car.

■

■ **Example 2** Appendix B of this supplement contains a data set on residential homes sold in a Minnesota community in 1988. Construct frequency and percent frequency distributions for the location of the homes within the community and the type of financing used to purchase the homes.

Solution The data set is stored in a file named HOMES on the data disk available with this supplement. Codes have been defined in the data file for the nominal data. For location within the community, area 1 is within the city, area 2 is the suburbs, and area 3 is the country surrounding the city. For the type of financing, Assumed Seller's Financing 1, Cash 2, Contract for Deed 3, Conventional Loan 4, FHA Loan 5, VA Loan 6, and Other type of Financing 7.

```
MTB > RETRIEVE 'HOMES';
  WORKSHEET SAVED   1/10/1994

Worksheet retrieved from file: HOMES.MTW
MTB > INFORMATION

COLUMN    NAME        COUNT    MISSING
C1        AREA          197
C2        BEDROOMS      197
C3        LIST PR       197
C4        SOLD PR       197
C5        FINANCE       197
C6        DAYS          197        2
C7        MTH SOLD      197
C8        DAY SOLD      197

CONSTANTS USED: NONE
```

```
MTB > TALLY 'AREA';
SUBC> COUNT;
SUBC> PERCENT.

    AREA   COUNT  PERCENT
       1     104    52.79
       2      45    22.84
       3      48    24.37
      N=     197

MTB > TALLY 'FINANCE';
SUBC> COUNT;
SUBC> PERCENT.

 FINANCE   COUNT  PERCENT
       1      10     5.08
       2      14     7.11
       3      19     9.64
       4      58    29.44
       5      79    40.10
       6      16     8.12
       7       1     0.51
      N=     197

MTB > STOP
```

The output gives the frequencies and percent frequencies for location and type of financing. The largest percent of homes sold were within the city (about 53%), and the largest percent of the homes had FHA loans (about 40%).

■

QUANTITATIVE DATA

Quantitative data include interval and ratio data. Interval data are numerical data in which the difference between the data with respect to the variable of interest can be determined, but there is no zero point or origin. Examples of interval data are the time of arrival of airlines and scores on the Scholastic Aptitude Test. The difference between the times of arrival of airlines can be determined, but there is no zero time of arrival.

Ratio data, the highest level of data, are measurements in which the ratio is meaningful. For example, the number of absentees per day in a major corporation, the daily stock market returns, the change in the annual rate of inflation, and the length of time between computer sales are ratio measurements. For ratio data, the zero point or origin has meaning. For example, zero absentees per day, zero stock market returns, and zero change in the inflation rate have meaning.

Minitab organizes quantitative data in columns either as unstacked or stacked data. If each set of data is put in a separate column, the data are unstacked. If different sets of data are put in the same column, and the code identifying each set is in another column, the data are stacked. Minitab uses the subcommand BY with several commands if the data are stacked. The column specified with BY is the column containing the code. Example 7 of this chapter illustrates stacked data.

Stem and Leaf Display

Graphical methods for quantitative data summarize the relevant information contained in a set of data. One method that results in little loss of original data is the stem and leaf display. STEM AND LEAF partitions each measurement into two parts, a stem and a leaf.

STEM AND LEAF OF C,...,C
INCREMENT = K
TRIM OUTLIERS
BY C

A stem and leaf display is constructed for each column of data specified on the command line. The subcommand INCREMENT controls the scaling of the display, where K is the distance between stems. The TRIM subcommand removes outliers from the display, and labels outliers as low or high values. The BY subcommand is used if data for several groups are stacked in a column, and a separate display is needed for each group. The column containing the code is given on the BY subcommand line. TRIM and BY cannot be used together.

The first column in the stem and leaf display gives a cumulative count of the observations beginning with the lowest stem up to the stem containing the median, and a cumulative count of the observations beginning with the highest stem down to the stem containing the median. The count for the stem containing the median is in parentheses. The second column contains the stems and the third column contains the leaves. Minitab uses 1, 2, or 5 lines per stem depending on the number of observations and range of the data.

■ **Example 3** Refer to Table 2.1 on page 24 of the text. The Environmental Protection Agency EPA determined the following mileage ratings of 100 cars of a certain new model. Write a Minitab program to construct a stem and leaf display. Interpret.

EPA Mileage Ratings

36.3	41.0	36.9	37.1	44.9	36.8	30.0	37.2	42.1	36.7
32.7	37.3	41.2	36.6	32.9	36.5	33.2	37.4	37.5	33.6
40.5	36.5	37.6	33.9	40.2	36.4	37.7	37.7	40.0	34.2
36.2	37.9	36.0	37.9	35.9	38.2	38.3	35.7	35.6	35.1
38.5	39.0	35.5	34.8	38.6	39.4	35.3	34.4	38.8	39.7
36.3	36.8	32.5	36.4	40.5	36.6	36.1	38.2	38.4	39.3
41.0	31.8	37.3	33.1	37.0	37.6	37.0	38.7	39.0	35.8
37.0	37.2	40.7	37.4	37.1	37.8	35.9	35.6	36.7	34.5
37.1	40.3	36.7	37.0	33.9	40.1	38.0	35.2	34.8	39.5
39.9	36.9	32.9	33.8	39.8	34.0	36.8	35.0	38.1	36.9

Chapter 2

Solution We save the mileage ratings in a file named EPA.

```
MTB > NAME C1 'MPG'
MTB > SET 'MPG'
DATA> 36.3 32.7 40.5 36.2 38.5 36.3 41.0 37.0 37.1 39.9
DATA> 41.0 37.3 36.6 37.9 39.0 36.8 31.8 37.2 40.3 36.9
DATA> 36.9 41.2 37.6 36.0 35.5 32.5 37.3 40.7 36.7 32.9
DATA> 37.1 36.6 33.9 37.9 34.8 36.4 33.1 37.4 37.0 33.8
DATA> 44.9 32.9 40.2 35.9 38.6 40.5 37.0 37.1 33.9 39.8
DATA> 36.8 36.5 36.4 38.2 39.4 36.6 37.6 37.8 40.1 34.0
DATA> 30.0 33.2 37.7 38.3 35.3 36.1 37.0 35.9 38.0 36.8
DATA> 37.2 37.4 37.7 35.7 34.4 38.2 38.7 35.6 35.2 35.0
DATA> 42.1 37.5 40.0 35.6 38.8 38.4 39.0 36.7 34.8 38.1
DATA> 36.7 33.6 34.2 35.1 39.7 39.3 35.8 34.5 39.5 36.9
DATA> END
MTB > SAVE 'EPA'

Worksheet saved into file: EPA.MTW
MTB > STEM AND LEAF 'MPG'

Stem-and-leaf of MPG      N = 100
Leaf Unit = 0.10

     1    30 0
     2    31 8
     6    32 5799
    12    33 126899
    18    34 024588
    29    35 01235667899
    49    36 01233445566777888999
   (21)   37 000011122334456677899
    30    38 0122345678
    20    39 00345789
    12    40 0123557
     5    41 002
     2    42 1
     1    43
     1    44 9

MTB > STOP
```

The Minitab output is similar to the stem and leaf display given in Figure 2.2 on page 25 of the text. The leftmost column of numbers gives a cumulative count of the data. There is one observation on the first stem, two on the first two stems, six on the first three stems, and so on. The count in parentheses indicates that the median is contained on that stem. The second column is the stem and the third column is the leaf. In this example, the leaf unit is the tenths digit. The display shows the symmetrical shape of the data. ∎

Histograms

A histogram is a graph of a frequency or relative frequency distribution. The frequency distribution for quantitative data consists of classes or intervals, such as "3 to under 5", and the frequencies or counts of the values that fall within each class. The relative frequency is the proportion of data values that fall in each class. The number of classes is determined by the number of data values and the range of the data.

In this section, we describe two ways to construct a histogram. The HISTOGRAM command produces a frequency distribution and a graph similar to the stem and leaf display. The PLOT command can be used to produce either a frequency or relative frequency graph.

HISTOGRAM OF C,...,C
    INCREMENT = K
    START AT K (END AT K)
    SAME
    BY C

This command gives a frequency distribution and graphical representation for each column listed on the command line. It groups the data in intervals, and displays the frequency of each interval both numerically and as a horizontal line of asterisks (*). Observations falling on a boundary are put in the interval with the larger midpoint.

The classes for the distribution can be specified with the subcommands INCREMENT and START. INCREMENT specifies the distance between midpoints. START specifies the midpoint of the starting interval, and END, the midpoint of the last interval. The SAME subcommand uses the same midpoints for all histograms. The subcommand BY provides a separate histogram for each code in C.

■ **Example 4** Refer to the sample of 100 EPA mileage ratings given in Example 3 on page 39 of this chapter. Construct a histogram of the data.

Solution The mileage ratings were saved in a file named EPA in Example 3. We use the INFORMATION command to verify that the column of data has been retrieved in the current worksheet.

```
MTB > RETRIEVE 'EPA'
 WORKSHEET SAVED   1/10/1994

Worksheet retrieved from file: EPA.MTW
MTB > INFORMATION

COLUMN      NAME        COUNT
C1          MPG          100

CONSTANTS USED: NONE

MTB > PRINT 'MPG'

MPG
   36.3    32.7    40.5    36.2    38.5    36.3    41.0    37.0    37.1    39.9    41.0
   37.3    36.5    37.9    39.0    36.8    31.8    37.2    40.3    36.9    36.9    41.2
   37.6    36.0    35.5    32.5    37.3    40.7    36.7    32.9    37.1    36.6    33.9
   37.9    34.8    36.4    33.1    37.4    37.0    33.8    44.9    32.9    40.2    35.9
```

```
38.6  40.5  37.0  37.1  33.9  39.8  36.8  36.5  36.4  38.2  39.4
36.6  37.6  37.8  40.1  34.0  30.0  33.2  37.7  38.3  35.3  36.1
37.0  35.9  38.0  36.8  37.2  37.4  37.7  35.7  34.4  38.2  38.7
35.6  35.2  35.0  42.1  37.5  40.0  35.6  38.8  38.4  39.0  36.7
34.8  38.1  36.7  33.6  34.2  35.1  39.7  39.3  35.8  34.5  39.5
36.9
```

```
MTB > HISTOGRAM 'MPG'

Histogram of MPG    N = 100

Midpoint   Count
      30       1   *
      32       5   *****
      34      12   ************
      36      31   *******************************
      38      31   *******************************
      40      15   ***************
      42       4   ****
      44       1   *
```

The HISTOGRAM command gives a different frequency distribution than that shown in Table 2.2 on page 27 of the text. To obtain a distribution similar to that given in the text, we use subcommands, INCREMENT and START, with the HISTOGRAM command. START is the midpoint of the first class: ($29.95 + $31.45)/2 = $30.70. INCREMENT is the difference between two adjacent lower class boundaries; for example, $32.95 - $31.45 = $1.50.

```
MTB > HISTOGRAM 'MPG';
SUBC> START 30.70;
SUBC> INCREMENT 1.50.

Histogram of MPG    N = 100

Midpoint   Count
   30.70       1   *
   32.20       5   *****
   33.70       9   *********
   35.20      14   **************
   36.70      33   *********************************
   38.20      18   ******************
   39.70      12   ************
   41.20       6   ******
   42.70       1   *
   44.20       1   *

MTB > STOP
```

The HISTOGRAM command gives a frequency distribution of the data and a pseudo-histogram with the asterisks (*) to the right of the count. In many cases this is sufficient graphical output for summarizing the data. Notice that Minitab groups the data in this histogram more than in the stem and leaf of the previous example. You can control the grouping with the INCREMENT subcommand. ■

The PLOT command produces a histogram with the data classes defined on the horizontal axis and frequencies or relative frequencies on the vertical axis. Rectangles can be constructed by hand on the Minitab output. The heights of the rectangles represent the frequencies or relative frequencies.

```
PLOT C VERSUS C
  XINCREMENT = K
  XSTART AT K (END AT K)
  YINCREMENT = K
  YSTART AT K (END AT K)
  SYMBOL = 'SYMBOL'
  TITLE = 'TEXT'
  FOOTNOTE = 'TEXT'
  XLABEL = 'TEXT'
  YLABEL = 'TEXT'
```

The first column specified on the command line is plotted on the vertical axis and the second column on the horizontal axis. Each point is plotted with an '*' or the count if more than one point falls on the same position. A '+' is used if the count is over nine.

Subcommands are used to specify the scales. XINCREMENT and YINCREMENT set the distances between the tick (+) marks on corresponding axes. XSTART and YSTART specify the first points, and the END options specify the last points, plotted on the axes. SYMBOL changes the symbol used to plot each point.

A maximum of three titles and two footnotes may be added to the plot with TITLE and FOOTNOTE. XLABEL and YLABEL name the two axes. Any characters may be used as text. The maximum number of characters printed for TITLE, FOOTNOTE, and XLABEL equals the width of the plot (57 by default). The maximum number of characters for YLABEL equals the height of the plot (17 by default).

To use the PLOT command for histograms, the column containing the frequencies or relative frequencies is listed first on the command line and then the column of midpoints.

■ **Example 5** Consider the EPA mileage ratings given in the previous example. Use PLOT to construct a relative frequency histogram.

Solution To construct the relative frequency histogram, we use the midpoints and frequencies given on the latest HISTOGRAM output of Example 4 on page 42 of this chapter. The LET command calculates the relative frequency (frequency/n) for each class.

```
MTB > NAME C1 'MIDPOINT' C2 'FREQ' C3 'REL FREQ'
MTB > READ 'MIDPOINT' 'FREQ'
DATA> 30.70  1
DATA> 32.20  5
DATA> 33.70  9
DATA> 35.20  14
DATA> 36.70  33
DATA> 38.20  18
DATA> 39.70  12
```

```
DATA> 41.20 6
DATA> 42.70 1
DATA> 44.20 1
DATA> END
      10 ROWS READ
MTB > LET 'REL FREQ' = 'FREQ'/SUM('FREQ')
MTB > PRINT C1-C3

 ROW  MIDPOINT   FREQ   REL FREQ

  1     30.7      1      0.01
  2     32.2      5      0.05
  3     33.7      9      0.09
  4     35.2     14      0.14
  5     36.7     33      0.33
  6     38.2     18      0.18
  7     39.7     12      0.12
  8     41.2      6      0.06
  9     42.7      1      0.01
 10     44.2      1      0.01

MTB > PLOT 'REL FREQ' VS 'MIDPOINT';
SUBC> TITLE 'EPA MILEAGE RATINGS';
SUBC> YLABEL 'RELATIVE FREQ';
SUBC> XLABEL 'MILEAGE'.
                             EPA MILEAGE RATINGS
R       -
E   0.36+
L       -
A       -                           *
T       -
I       -
V   0.24+
E       -
        -                                   *
F       -
R       -                  *
E   0.12+                                        *
Q       -
        -        *
        -
        -  *                                          *    *
    0.00+
        --------+----------+----------+----------+----------+--------
              32.5       35.0       37.5       40.0       42.5
                                   MILEAGE

MTB > STOP
```

Rectangles can be added by hand to complete the relative frequency histogram similar to that shown in Figure 2.1 on page 25 of the text. The asterisks locate the centers of the tops of the rectangles. ∎

Dot Plot

A Minitab dot plot uses a horizontal axis and groups the data as little as possible. Each observation is represented as a dot on the horizontal axis.

Methods for Describing Sets of Data   45

> DOTPLOT OF DATA IN C,...,C
>   INCREMENT = K
>   START AT K (END AT K)
>   SAME
>   BY C
>
> A dot plot is made for each column listed on the command line. The subcommands INCREMENT and START allow you to specify your own scales. SAME plots each column with the same scale. Use BY if data for several groups are stacked in one column, and you want a separate display for each group. The codes for the groups must be in the column specified by the BY subcommand.

■ **Example 6** Construct a dot plot of the EPA mileage ratings given in Example 3 on page 39 of this chapter. Compare the dot plot with the stem and leaf display and histogram constructed in previous examples.

Solution In Example 3, the mileage ratings were saved in a file named EPA.

```
MTB > RETRIEVE 'EPA'
 WORKSHEET SAVED   1/10/1994

Worksheet retrieved from file: EPA.MTW
MTB > INFORMATION

COLUMN     NAME        COUNT
C1         MPG          100

CONSTANTS USED: NONE

MTB > DOTPLOT 'MPG'
```

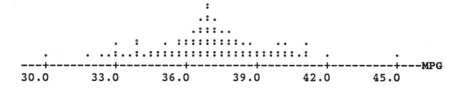

```
MTB > STOP
```

The DOTPLOT command uses a horizontal axis for the processing times, whereas both HISTOGRAM and STEM AND LEAF use a vertical axis. For this data set, DOTPLOT does not group the data. Each observation is plotted as a dot corresponding to its location on the horizontal axis. This plot shows the symmetrical shape of the data and the extreme mileage ratings of 30 and 45 mpg.

■

■ **Example 7** Each year the Bureau of Justice Statistics publishes a *Sourcebook of Criminal Justice Statistics*. The following table gives the total expenditures (in thousands of dollars) for the local justice system and the population (in thousands) on April 1, 1990 for each state and the District of Columbia.

State	Expenditure	Population	State	Expenditure	Population
Alabama	$ 643,016	4,041	Montana	$ 117,287	799
Alaska	334,148	550	Nebraska	247,984	1,578
Arizona	1,169,547	3,665	Nevada	431,946	1,202
Arkansas	271,045	2,351	New Hampshire	227,431	1,109
California	11,191,558	29,760	New Jersey	2,562,284	7,730
Colorado	800,904	3,294	New Mexico	355,156	1,515
Connecticut	973,884	3,287	New York	8,641,418	17,990
Delaware	201,463	666	North Carolina	1,234,074	6,629
Dist Columbia	726,156	607	North Dakota	78,505	639
Florida	3,727,778	12,938	Ohio	2,016,729	10,847
Georgia	1,547,021	6,478	Oklahoma	544,326	3,146
Hawaii	322,614	1,108	Oregon	691,039	2,842
Idaho	164,310	1,007	Pennsylvania	2,230,924	11,882
Illinois	2,658,624	11,431	Rhode Island	249,997	1,003
Indiana	791,236	5,544	South Carolina	657,726	3,487
Iowa	444,440	2,777	South Dakota	93,420	696
Kansas	550,052	2,478	Tennessee	944,351	4,877
Kentucky	560,597	3,685	Texas	3,565,587	16,987
Louisiana	828,321	4,220	Utah	283,897	1,723
Maine	207,449	1,228	Vermont	101,144	563
Maryland	1,439,882	4,781	Virginia	1,453,777	6,187
Massachusetts	1,822,763	6,016	Washington	1,031,411	4,867
Michigan	2,416,554	9,295	West Virginia	174,458	1,793
Minnesota	821,228	4,375	Wisconsin	1,021,085	4,892
Mississippi	315,355	2,573	Wyoming	111,986	454
Missouri	920,313	5,117			

a. Enter the alpha and numeric data in a worksheet. Calculate the per capita justice system expenditures for each state and the District of Columbia.
b. Construct dot plots of the justice system expenditures, population, and per capita expenditures. Describe the distributions.
c. Construct and discuss separate dot plots for the justice system expenditures, population, and per capita expenditures for the states and the District of Columbia east and west of the Mississippi River.
d. Construct dot plots of the justice system expenditures, population, and per capita expenditures for the 50 states. Compare these dot plots with those constructed in part b.

Solution

a. The FORMAT subcommand is used to enter the two letter postal abbreviation for the states and the District of Columbia. The specification (17(A2,1X)) says to enter 17 names per line following DATA>. Each name has two characters followed by a blank space. The LET command calculates per capita expenditures.

Methods for Describing Sets of Data 47

```
MTB > NAME C1 'STATE+DC' C2 'EXPEND' C3 'POP' C4 'PERCAP'
MTB > SET 'STATE+DC';
SUBC> FORMAT (17(A2,1X)).
DATA> AL AK AZ AR CA CO CT DE DC FL GA HI ID IL IN IA KS
DATA> KY LA ME MD MA MI MN MS MO MT NE NV NH NJ NM NY NC
DATA> ND OH OK OR PA RI SC SD TN TX UT VT VA WA WV WI WY
DATA> END
MTB > SET 'EXPEND'
DATA>  643016    334148   1169547    271045  11191558    800904    973884
DATA>  201463    726156   3727778   1547021    22614    164310   2658624
DATA>  791236    444440    550052    560597    828321    207449   1439882
DATA> 1822763   2416554    821228    315355    920313    117287    247984
DATA>  431946    227431   2562284    355156   8641418   1234074     78505
DATA> 2016729    544326    691039   2230924    249997    657726     93420
DATA>  944351   3565587    283897    101144   1453777   1031411    174458
DATA> 1021085    111986
DATA> END
MTB > SET 'POP'
DATA>  4041    550   3665   2351  29760   3294   3287    666    607
DATA> 12938   6478   1108   1007  11431   5544   2777   2478   3685
DATA>  4220   1228   4781   6016   9295   4375   2573   5117    799
DATA>  1578   1202   1109   7730   1515  17990   6629    639  10847
DATA>  3146   2842  11882   1003   3487    696   4877  16987   1723
DATA>   563   6187   4867   1793   4892    454
MTB > END
MTB > LET 'PERCAP' = 'EXPEND'/'POP'
MTB > PRINT C1-C4
```

ROW	STATE+DC	EXPEND	POP	PERCAP
1	AL	643016	4041	159.12
2	AK	334148	550	607.54
3	AZ	1169547	3665	319.11
4	AR	271045	2351	115.29
5	CA	11191558	29760	376.06
6	CO	800904	3294	243.14
7	CT	973884	3287	296.28
8	DE	201463	666	302.50
9	DC	726156	607	1196.30
10	FL	3727778	12938	288.13
11	GA	1547021	6478	238.81
12	HI	322614	1108	291.17
13	ID	164310	1007	163.17
14	IL	2658624	11431	232.58
15	IN	791236	5544	142.72
16	IA	444440	2777	160.04
17	KS	550052	2478	221.97
18	KY	560597	3685	152.13
19	LA	828321	4220	196.28
20	ME	207449	1228	168.93
21	MD	1439882	4781	301.17
22	MA	1822763	6016	302.99
23	MI	2416554	9295	259.98
24	MN	821228	4375	187.71
25	MS	315355	2573	122.56
26	MO	920313	5117	179.85
27	MT	117287	799	146.79
28	NE	247984	1578	157.15
29	NV	431946	1202	359.36
30	NH	227431	1109	205.08
31	NJ	2562284	7730	331.47
32	NM	355156	1515	234.43

48   Chapter 2

```
33        NY      8641418    17990    480.35
34        NC      1234074     6629    186.16
35        ND        78505      639    122.86
36        OH      2016729    10847    185.93
37        OK       544326     3146    173.02
38        OR       691039     2842    243.15
39        PA      2230924    11882    187.76
40        RI       249997     1003    249.25
41        SC       657726     3487    188.62
42        SD        93420      696    134.22
43        TN       944351     4877    193.63
44        TX      3565587    16987    209.90
45        UT       283897     1723    164.77
46        VT       101144      563    179.65
47        VA      1453777     6187    234.97
48        WA      1031411     4867    211.92
49        WV       174458     1793     97.30
50        WI      1021085     4892    208.73
51        WY       111986      454    246.67
```

The District of Columbia has the largest and West Virginia the smallest per capita justice system expenditures.

b. Dot plots of several columns can be constructed with one command line. Minitab uses the maximum and minimum data values to determine the scale.

```
MTB > DOTPLOT 'EXPEND' 'POP' 'PERCAP'
```

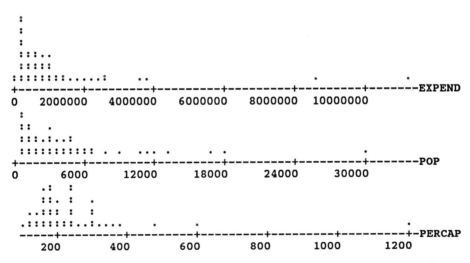

The distributions of justice system expenditures, population, and per capita expenditures are skewed to the right. There are some unusually high data values.

c. To separate the states and the District of Columbia east and west of the Mississippi River, we need a code to identify location. The BY subcommand with DOTPLOT constructs separate dot plots. This is an example of stacked data with respect to the location of states. The justice system expenditures are stacked in one column and a code identifying the location in another column.

```
MTB > # CODE FOR WEST 0, EAST 1
MTB > NAME C5 'CODE'
MTB > SET 'CODE'
DATA> 1 0 0 0 0 0 1 1 1 1
DATA> 1 0 0 1 1 0 0 1 0 1
DATA> 1 1 1 0 1 0 0 0 0 1
DATA> 1 0 1 1 0 1 0 0 1 1
DATA> 1 0 1 0 0 1 1 0 1 1 0
DATA> END
MTB > DOTPLOT 'EXPEND';
SUBC>    BY 'CODE'.

              :.
CODE        ::..
  0         ::::::.                .                                                  .
         +---------+---------+---------+---------+---------+--------EXPEND
              :
CODE        :  . .
  1         :.::: .:.....:         .                   .
         +---------+---------+---------+---------+---------+--------EXPEND
         0      2000000    4000000    6000000    8000000   10000000

MTB > DOTPLOT 'POP';
SUBC>    BY 'CODE'.

CODE           .
  0         :.. :
            ::::: .:..                .                       .
         +---------+---------+---------+---------+---------+--------POP
CODE        . .      .
  1         ::...:.:.::  . .   ... .              .
         +---------+---------+---------+---------+---------+--------POP
         0       6000      12000     18000     24000     30000

MTB > DOTPLOT 'PERCAP';
SUBC>    BY 'CODE'.

CODE             :.  :
  0          :::::::  .. ..          .                                                 .
         -----+---------+---------+---------+---------+---------+--PERCAP
CODE              .
  1             .:.  :  :
              ...::: :..:  .         .                                   .
         -----+---------+---------+---------+---------+---------+--PERCAP
             200       400       600       800      1000      1200
```

Minitab requires a DOTPLOT command line with BY for each variable if the scales differ. There are some significant differences in the distributions of states east and west of the Mississippi River.

d. We copy the data omitting the District of Columbia, the ninth observation, to construct dot plots for the 50 states.

```
MTB > NAME C11 '50STATE' C12 '50EXPEND' C13 '50POP'
MTB > NAME C14 '50PERCAP' C15 '50CODE'
MTB > COPY C1-C5 C11-C15;
SUBC>   OMIT DC IN ROW 9.
```

50   *Chapter 2*

```
MTB > SAVE 'CRIME'

Worksheet saved into file: CRIME.MTW
MTB > DOTPLOT C12-C14
```

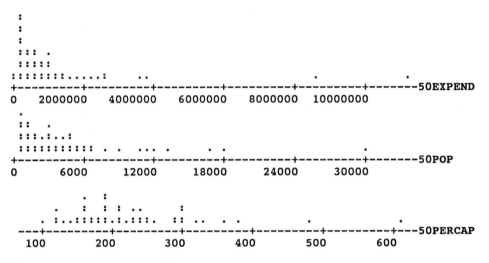

```
MTB > STOP
```

The per capita expenditure dot plot without the District of Columbia is substantially different than the dot plot which includes the District. The scale on the horizontal axis changes. The District of Columbia has an unusually large justice system per capita expenditure when compared with the 50 states. ■

■ **Example 8**   In Example 2 of this chapter, frequency and relative frequency distributions of the location of homes and type of financing were given for the data set in Appendix B. Use HISTOGRAM to construct horizontal bar graphs of the two qualitative variables.

Solution   The data are saved in a file named HOMES. Codes have been defined for the qualitative variable, location within the community. Area 1 is within the city, area 2 is the suburbs, and area 3 is the country surrounding the city. For the type of financing, Assumed Seller's Financing = 1, Cash = 2, Contract for Deed = 3, Conventional Loan = 4, FHA Loan = 5, VA Loan = 6, and Other type of Financing = 7.

```
MTB > RETRIEVE 'HOMES';
  WORKSHEET SAVED   1/10/1994

Worksheet retrieved from file: HOMES.MTW
MTB > INFORMATION

COLUMN      NAME        COUNT      MISSING
C1          AREA         197
C2          BEDROOMS     197
C3          LIST PR      197
C4          SOLD PR      197
C5          FINANCE      197
C6          DAYS         197           2
```

```
C7          MTH SOLD      197
C8          DAY SOLD      197

CONSTANTS USED: NONE
MTB > HISTOGRAM 'AREA'

Histogram of AREA    N = 197
Each * represents 5 obs.

Midpoint    Count
       1      104    ********************
       2       45    *********
       3       48    **********

MTB > HISTOGRAM 'FINANCE'

Histogram of FINANCE   N = 197
Each * represents 2 obs.

Midpoint    Count
       1       10    *****
       2       14    *******
       3       19    **********
       4       58    *****************************
       5       79    ****************************************
       6       16    ********
       7        1    *

MTB > STOP
```

The Minitab bar graph gives the frequencies for each class. Each asterisk in the AREA bar chart represents five observations, and each asterisk in the FINANCE bar chart represents 2 observations. Minitab chooses the scale depending on the number of observations and the number of classes. ∎

MINITAB NUMERICAL DESCRIPTIVE MEASURES

Numerical descriptive measures characterize or give information about a set of data. Individual numerical measures can be calculated for columns or rows in a Minitab worksheet.

Commands for Data Stored in Columns

The following commands calculate numerical measures for data stored in columns. A brief description of each measure is given for each command. There is not a Minitab command to calculate the mode.

> | COMMAND | EXPLANATION |
> |---|---|
> | N OF C (PUT IN K) | Number of nonmissing values |
> | NMISS OF C (PUT IN K) | Number of missing values |
> | MEAN OF C (PUT IN K) | Sample average |
> | MEDIAN OF C (PUT IN K) | Center value in ordered array |
> | STDEV OF C (PUT IN K) | Sample standard deviation |
> | MAXIMUM OF C (PUT IN K) | Largest value |
> | MINIMUM OF C (PUT IN K) | Smallest value |
> | SUM OF C (PUT IN K) | Sum of all the values |
> | SSQ OF C (PUT IN K) | Sum of squares of the values |
> | COUNT OF C (PUT IN K) | Total number of observations |
> | RANGE OF C (PUT IN K) | Largest value minus smallest value |
>
> Each command calculates and prints a single number. The option stores the result as a constant.

The DESCRIBE command provides a table of summary measures. In many situations this is the only command you need to numerically summarize a data set.

> DESCRIBE THE DATA IN C,...,C
>   BY C
>
> This command prints the statistics N, N*, MEAN, MEDIAN, TRMEAN, STDEV, SEMEAN, MIN, MAX, Q1, Q3 for each column given on the command line. N is the number of nonmissing observations and N* is the number of missing observations. If there are no missing observations, N* is omitted on the output.
>
> The 90% trimmed mean TRMEAN removes the smallest 5% and the largest 5% of the observations and averages the rest. The standard error of the mean is SEMEAN = $STDEV/\sqrt{n}$. Q1 is the first quartile and Q3 is the third quartile.
>
> The subcommand BY is used for stacked data. The column containing the code is used with BY.

■ **Example 9** Refer to the EPA mileage ratings given in Example 3 on page 39 of this chapter. Numerically summarize the mileage ratings. Write a brief description of each measure.

Solution The mileage ratings are saved in a file named EPA.

```
MTB > RETRIEVE 'EPA'
WORKSHEET SAVED  1/10/1994
```

```
Worksheet retrieved from file: EPA.MTW
MTB > PRINT 'MPG'

MPG
   36.3   32.7   40.5   36.2   38.5   36.3   41.0   37.0   37.1   39.9   41.0
   37.3   36.5   37.9   39.0   36.8   31.8   37.2   40.3   36.9   36.9   41.2
   37.6   36.0   35.5   32.5   37.3   40.7   36.7   32.9   37.1   36.6   33.9
   37.9   34.8   36.4   33.1   37.4   37.0   33.8   44.9   32.9   40.2   35.9
   38.6   40.5   37.0   37.1   33.9   39.8   36.8   36.5   36.4   38.2   39.4
   36.6   37.6   37.8   40.1   34.0   30.0   33.2   37.7   38.3   35.3   36.1
   37.0   35.9   38.0   36.8   37.2   37.4   37.7   35.7   34.4   38.2   38.7
   35.6   35.2   35.0   42.1   37.5   40.0   35.6   38.8   38.4   39.0   36.7
   34.8   38.1   36.7   33.6   34.2   35.1   39.7   39.3   35.8   34.5   39.5
   36.9

MTB > N 'MPG'
    N       =         100
MTB > NMISS 'MPG'
    NMISSING=           0
```

There are 100 nonmissing and 0 missing mileage ratings.

```
MTB > MEAN 'MPG'
    MEAN    =        36.994
MTB > MEDIAN 'MPG'
    MEDIAN  =        37.000
```

The mean, $\bar{x} = 37.0$ miles per gallon is the same as the median of 37.0; this is a characteristic of symmetrical data sets.

```
MTB > STDEV 'MPG'
    ST.DEV. =         2.4179
```

The sample standard deviation s is 2.42 miles per gallon. The square of the standard deviation is the variance: $s^2 = 5.85$.

```
MTB > MAXIMUM 'MPG'
    MAXIMUM =        44.900
MTB > MINIMUM 'MPG'
    MINIMUM =        30.000
```

The largest mileage rating is 44.9 and the lowest is 30.0 miles per gallon.

```
MTB > SUM 'MPG'
    SUM     =      3699.40
```

The sum of the mileage ratings is 3699.4 (36.3 + 32.7 + ... + 36.9).

```
MTB > SSQ 'MPG'
    SSQ     =       137434
```

The sum of the squares of the mileage ratings is 137,434 ($36.3^2 + 32.7^2 + ... + 36.9^2$).

```
MTB > COUNT 'MPG'
    COUNT   =         100
```

The total number of missing and nonmissing values is 100.

54 *Chapter 2*

```
MTB > RANGE 'MPG'
     RANGE   =      14.900
```

The range of 14.9 miles per gallon measures the numerical distance from the lowest to the largest mileage rating. The standard deviation will usually fall between the range/6 and range/4. In this example, $s = 2.42$ is slightly less than the range/6 = 2.48.

```
MTB > DESCRIBE 'MPG'

              N       MEAN      MEDIAN     TRMEAN     STDEV     SEMEAN
MPG          100      36.994    37.000     36.992     2.418     0.242

              MIN      MAX        Q1         Q3
MPG          30.000   44.900    35.625     38.375

MTB > STOP
```

In addition to some of the column measures given above, DESCRIBE gives the TRMEAN, SEMEAN, Q1, and Q3.

TRMEAN is a 90% trimmed mean. It is the average of the data values after removing the smallest 5% (rounded to the nearest integer) and the largest 5%. Since 5% of 100 = 5, the smallest 5 and the largest 5 mileages are removed, and the mean of the remaining 90 mileages is calculated. The trimmed mean of 36.992 is about the same as the untrimmed mean for this symmetric data set.

The standard error of the mean is SEMEAN = $2.418/\sqrt{100}$ = .242. Q1 and Q3 are the first and third quartiles. About 25 percent of the observations are less than Q1 = 35.6 and about 75 percent are less than Q3 = 38.4. ∎

To calculate the mode, arrange the data in ascending order and search for the value that occurs most often. The SORT command orders the data values from least to greatest.

■ **Example 10** Sort the mileage ratings given in Example 3 on page 39 of this chapter. Find the mode.

Solution The mileages are saved in a file called EPA.

```
MTB > RETRIEVE 'EPA'
   WORKSHEET SAVED   1/10/1994

Worksheet retrieved from file: EPA.MTW
MTB > NAME C2 'ORDER'
MTB > SORT 'MPG' PUT IN 'ORDER'
MTB > PRINT 'ORDER'

ORDER
   30.0    31.8    32.5    32.7    32.9    32.9    33.1    33.2    33.6    33.8    33.9
   33.9    34.0    34.2    34.4    34.5    34.8    34.8    35.0    35.1    35.2    35.3
   35.5    35.6    35.6    35.7    35.8    35.9    35.9    36.0    36.1    36.2    36.3
   36.3    36.4    36.4    36.5    36.5    36.6    36.6    36.7    36.7    36.7    36.8
   36.8    36.8    36.9    36.9    36.9    37.0    37.0    37.0    37.0    37.1    37.1
   37.1    37.2    37.2    37.3    37.3    37.4    37.4    37.5    37.6    37.6    37.7
   37.7    37.8    37.9    37.9    38.0    38.1    38.2    38.2    38.3    38.4    38.5
```

```
38.6   38.7   38.8   39.0   39.0   39.3   39.4   39.5   39.7   39.8   39.9
40.0   40.1   40.2   40.3   40.5   40.5   40.7   41.0   41.0   41.2   42.1
44.9
```

```
MTB > STOP
```

The mode is the measurement that occurs most frequently. We peruse the ordered data and find that the mileage rating of 37.0 occurs four times; it is the mode. In this data set, the mean, median, and mode are about the same.

■

Interpreting the Standard Deviation

The standard deviation is a measure of the average deviation of the data values from the mean. There are two general rules for interpreting the standard deviation. Both rules consider the proportion of data values that fall within a certain number of standard deviations of the mean.

Chebyshev's Rule applies to any sample of measurements regardless of the shape of the distribution. In general, the rule says that at least $(1 - 1/k^2)$ of the measurements fall within k standard deviations of the mean. For example, at least 3/4 or 75% of the measurements fall within 2 standard deviations of the mean, and at least 8/9 or 89% of the measurements fall within 3 standard deviations of the mean.

The following Empirical Rule applies to data that have a symmetrical or mound-shaped distribution:

1. Approximately 68% of the measurements fall within $(\bar{x} - s, \bar{x} + s)$.
2. Approximately 95% of the measurements fall within $(\bar{x} - 2s, \bar{x} + 2s)$.
3. Essentially all of the measurements fall within $(\bar{x} - 3s, \bar{x} + 3s)$.

■ **Example 11** Consider the mileage ratings given in Example 3 on page 39 of this chapter. The graphs in previous examples showed that the data set is symmetric. Find the proportion of measurements that fall within $\bar{x} \pm s$, $\bar{x} \pm 2s$, $\bar{x} \pm 3s$. Compare these with the Empirical Rule.

Solution The mileages are saved in a file called EPA. We use LET to calculate the intervals, and to count the measurements that fall within the intervals. To count, we use the comparison ($<=$, $>=$) and logical feature (AND) of LET. If a measurement falls within the interval, the result is set to 1; if not, the result is set to 0. A TALLY gives the resulting counts.

```
MTB > RETRIEVE 'EPA'
  WORKSHEET SAVED   1/10/1994

Worksheet retrieved from file: EPA.MTW
MTB > LET K1 = MEAN('MPG')-STDEV('MPG')
MTB > LET K2 = MEAN('MPG')+STDEV('MPG')
MTB > PRINT K1 K2
K1        34.5761
K2        39.4119
MTB > # COMPARISON AND LOGICAL FEATURE OF LET
MTB > LET C2 = ('MPG' >= K1) AND ('MPG' <= K2)
```

```
MTB > TALLY C2   # CODED 1 IF WITHIN 1 ST DEV

       C2   COUNT
        0     32
        1     68
       N=    100

MTB > LET K1 = MEAN('MPG')-2STDEV('MPG')
MTB > LET K2 = MEAN('MPG')+2STDEV('MPG')
MTB > PRINT K1 K2
K1      32.1582
K2      41.8298
MTB > LET C2 = ('MPG' >= K1) AND ('MPG' <= K2)
MTB > TALLY C2   # CODED 1 IF WITHIN 2 ST DEV

       C2   COUNT
        0      4
        1     96
       N=    100

MTB > LET K1 = MEAN('MPG')-3*STDEV('MPG')
MTB > LET K2 = MEAN('MPG')+3*STDEV('MPG')
MTB > PRINT K1 K2
K1      29.7403
K2      44.2477
MTB > LET C2 = ('MPG' >= K1) AND ('MPG' <= K2)
MTB > TALLY C2   # CODED 1 IF WITHIN 3 ST DEV

       C2   COUNT
        0      1
        1     99
       N=    100

MTB > STOP
```

(handwritten annotation near the first LET command: "2 * STDEV")

The output of the TALLY commands shows that 68% of the measurements fall within $\bar{x} \pm s$, 96% fall within $\bar{x} \pm 2s$, and 99% fall within $\bar{x} \pm 3s$. Since the distribution of EPA mileage ratings is mound-shaped, these proportions agree with the Empirical Rule. ∎

The z-score represents the distance, in terms of standard deviations that a measurement is from the mean of the data set. The sample z-score for a measurement is $z = (x - \bar{x})/s$. It can be used to identify unusual measurements, called outliers. An outlier is an observation that is unusually large or small when compared with other measurements in a sample. It may be an incorrect measurement or may be from a different population than the rest of the sample. A numerical or graphical analysis can detect outliers. Use the LET command to calculate the z-score.

A box plot, based on quartiles, describes the distribution of data values and locates outliers. The first quartile Q1 is the 25th percentile (the value that exceeds 25% of the data values), the second quartile Q2 or the median is the 50th percentile, and the third quartile Q3 is the 75th percentile.

> BOXPLOT OF DATA IN C
>   INCREMENT = K
>   START AT K (END AT K)
>   BY C
>
> A box plot is constructed for the data stored in the column specified on the command line. The INCREMENT subcommand sets the distance between the tick (+) marks on the plot. The START subcommand specifies the first point and END, the last point plotted on the axes. Use the BY subcommand for stacked data if you want a separate box plot for each group of data. The column containing the code for the groups must be specified on the BY subcommand line.

BOXPLOT displays the main features of a set of data. The box represents the middle half of each data set. The ends of the box are approximately located at the quartiles, Q1 and Q3, and the median is marked with a '+'. Special symbols on either side of the box indicate the extent of the data and the location of extreme values.

The interquartile range IQR is the distance between the upper and lower quartiles. Inner fences are located at a distance 1.5(IQR) below Q1 and above Q3. Outer fences are located at a distance 3(IQR) below Q1 and above Q3. Dashed "whiskers" run from the edge of the box to the two most extreme values that are within the inner fences. A value between the inner and outer fence is plotted with an '*' and is a possible outlier. An extreme value beyond the outer fences is plotted with a '0' and is a probable outlier.

■ **Example 12** Refer to the sample of 100 mileage ratings given in Example 3 on page 39 of this chapter.

a. Construct a box plot of the data. Are there any outliers?
b. Calculate z-scores for the data.

Solution

a. The data are saved in a file named EPA.

```
MTB > RETRIEVE 'EPA'
  WORKSHEET SAVED   1/10/1994

Worksheet retrieved from file: EPA.MTW
MTB > INFORMATION

COLUMN      NAME       COUNT
C1          MPG          100

CONSTANTS USED: NONE
```

MTB > BOXPLOT 'MPG'

```
                    ----------
           *    -------------I  +  I------------           *
                    ----------
    ----+---------+---------+---------+---------+---------+--MPG
      30.0      33.0      36.0      39.0      42.0      45.0
```

The mileage ratings of about 30 and 45 are possible outliers.

b. We use the LET command to calculate z-scores, and SORT to order the z-scores in ascending order.

```
MTB > NAME C3 'Z-SCORE'
MTB > LET 'Z-SCORE'=('MPG'-MEAN('MPG'))/STDEV('MPG')
MTB > SORT 'Z-SCORE' PUT IN 'Z-SCORE'
MTB > PRINT 'Z-SCORE'

Z-SCORE
 -2.89260  -2.14815  -1.85864  -1.77592  -1.69321  -1.69321  -1.61049
 -1.56913  -1.40370  -1.32098  -1.27963  -1.27963  -1.23827  -1.15555
 -1.07283  -1.03148  -0.90740  -0.90740  -0.82468  -0.78333  -0.74197
 -0.70061  -0.61789  -0.57654  -0.57654  -0.53518  -0.49382  -0.45246
 -0.45246  -0.41110  -0.36974  -0.32839  -0.28703  -0.28703  -0.24567
 -0.24567  -0.20431  -0.20431  -0.16295  -0.16295  -0.12159  -0.12159
 -0.12159  -0.08024  -0.08024  -0.08024  -0.03888  -0.03888  -0.03888
  0.00248   0.00248   0.00248   0.00248   0.04384   0.04384   0.04384
  0.08520   0.08520   0.12655   0.12655   0.16791   0.16791   0.20927
  0.25063   0.25063   0.29199   0.29199   0.33335   0.37471   0.37471
  0.41606   0.45742   0.49878   0.49878   0.54014   0.58150   0.62285
  0.66421   0.70557   0.74693   0.82965   0.82965   0.95372   0.99508
  1.03644   1.11915   1.16051   1.20187   1.24323   1.28459   1.32594
  1.36730   1.45002   1.45002   1.53274   1.65681   1.65681   1.73953
  2.11175   3.26978

MTB > STOP
```

The lowest and highest measurements identified in the box plot have z-scores of -2.89 and 3.27. These mileages could be incorrectly recorded measurements, or unusually low and high mileage cars. ■

Commands for Data Stored in Rows

A row command calculates a summary measure for each row of a set of columns and stores the results in a column. To calculate measures for rows, use an R before each column command. You need a PRINT command to view the results.

```
RN  E,...,E   NUMBER OF NONMISSING VALUES IN C
RNMISS  E,...,E   NUMBER OF MISSING VALUES IN C
RMEAN  E,...,E   MEAN IN C
RMEDIAN  E,...,E   MEDIAN IN C
RSTDEV  E,...,E   STANDARD DEVIATION IN C
RMAX  E,...,E   MAXIMUM IN C
RMIN  E,...,E   MINIMUM IN C
RSUM  E,...,E   SUM IN C
RSSQ  E,...,E   SUM OF SQUARES IN C
RCOUNT  E,...,E   NUMBER OF COLUMNS IN C
RRANGE  E,...,E   RANGE IN C
```

■ **Example 13** Results of surveys reporting the use of drugs, alcohol, and cigarettes among college students are given in the *Sourcebook of Criminal Justice Statistics* published by the Bureau of Justice Statistics. The following table gives the percentages of students' usage for the previous 30 days. The number of respondents to the survey ranged from 1,080 to 1,410 for the years 1982 through 1991.

	1982	1983	1984	1985	1986	1987	1988	1989	1990	1991
Marihuana	26.8	26.2	23.0	23.6	22.3	20.3	16.8	16.3	14.0	14.1
Inhalants	0.8	0.7	0.7	1.0	1.1	0.9	1.3	0.8	1.0	0.9
Hallucinogens	4.3	2.7	2.6	2.0	3.6	3.4	2.8	3.7	2.5	2.0
Cocaine	7.9	6.5	7.6	6.9	7.0	5.0	4.7	3.0	2.3	2.3
Heroin	0.0	0.0	0.0	0.0	0.0	0.1	0.1	0.1	0.0	0.1
Other Opiates	0.9	1.1	1.4	0.7	0.6	0.8	0.8	0.7	0.5	0.6
Stimulants	9.9	7.0	5.5	4.2	3.7	2.3	1.8	1.3	1.4	1.0
Sedatives	5.4	2.3	2.2	1.4	1.3	1.3	1.2	0.4	0.2	0.3
Tranquilizers	1.4	1.2	1.1	1.4	1.9	1.0	1.1	0.8	0.5	0.6
Alcohol	82.8	80.3	79.1	80.3	79.7	78.4	77.0	76.2	74.5	74.7
Cigarettes	24.4	24.7	21.5	22.4	22.4	24.0	22.6	21.1	21.5	23.2

Calculate the mean, standard deviation, maximum, minimum, and range of usage for the types of drugs during the years, 1982 to 1991.

Solution   We enter the names of the drugs using the FORMAT subcommand with SET, and use READ to enter the table of drug use.

```
MTB > NAME C1 'DRUG' C2 '1982' C3 '1983' C4 '1984' C5 '1985' C6 '1986'
MTB > NAME C7 '1987' C8 '1988' C9 '1989' C10 '1990' C11 '1991'
MTB > SET 'DRUG';
SUBC> FORMAT (A9,1X).
DATA> MARIHUANA
DATA> INHALANT
DATA> HALLUCINO
DATA> COCAINE
DATA> HEROIN
DATA> OPIATES
DATA> STIMULANT
```

```
DATA> SEDATIVES
DATA> TRANQUILS
DATA> ALCOHOL
DATA> CIGARETTE
DATA> END
MTB > READ '1982'-'1991'
DATA> 26.8  26.2  23.0  23.6  22.3  20.3  16.8  16.3  14.0  14.1
DATA>  0.8   0.7   0.7   1.0   1.1   0.9   1.3   0.8   1.0   0.9
DATA>  4.3   2.7   2.6   2.0   3.6   3.4   2.8   3.7   2.5   2.0
DATA>  7.9   6.5   7.6   6.9   7.0   5.0   4.7   3.0   2.3   2.3
DATA>  0.0   0.0   0.0   0.0   0.0   0.1   0.1   0.1   0.0   0.1
DATA>  0.9   1.1   1.4   0.7   0.6   0.8   0.8   0.7   0.5   0.6
DATA>  9.9   7.0   5.5   4.2   3.7   2.3   1.8   1.3   1.4   1.0
DATA>  5.4   2.3   2.2   1.4   1.3   1.3   1.2   0.4   0.2   0.3
DATA>  1.4   1.2   1.1   1.4   1.9   1.0   1.1   0.8   0.5   0.6
DATA> 82.8  80.3  79.1  80.3  79.7  78.4  77.0  76.2  74.5  74.7
DATA> 24.4  24.7  21.5  22.4  22.4  24.0  22.6  21.1  21.5  23.2
DATA> END
     11 ROWS READ
MTB > SAVE 'DRUGS'

Worksheet saved into file: DRUGS.MTW
MTB > NAME C12 'MEAN' C13 'STDEV' C14 'MAXIMUM' C15 'MINIMUM' C16 'RANGE'
MTB > RMEAN C2-C11 'MEAN'
MTB > RSTDEV C2-C11 'STDEV'
MTB > RMAX C2-C11 'MAXIMUM'
MTB > RMIN C2-C11 'MINIMUM'
MTB > RRANGE C2-C11 'RANGE'
MTB > PRINT C1 C12-C16
```

ROW	DRUG	MEAN	STDEV	MAXIMUM	MINIMUM	RANGE
1	MARIHUANA	20.34	4.77963	26.8	14.0	12.8
2	INHALANT	0.92	0.18738	1.3	0.7	0.6
3	HALLUCINO	2.96	0.76187	4.3	2.0	2.3
4	COCAINE	5.32	2.17756	7.9	2.3	5.6
5	HEROIN	0.04	0.05164	0.1	0.0	0.1
6	OPIATES	0.81	0.26854	1.4	0.5	0.9
7	STIMULANT	3.81	2.91755	9.9	1.0	8.9
8	SEDATIVES	1.60	1.51877	5.4	0.2	5.2
9	TRANQUILS	1.10	0.41366	1.9	0.5	1.4
10	ALCOHOL	78.30	2.67416	82.8	74.5	8.3
11	CIGARETTE	22.78	1.26474	24.7	21.1	3.6

```
MTB > STOP
```

The row commands store the numerical measures in specified columns. For example, the mean alcohol usage ranged from 74.5 to 82.8% with a mean of 78.3% over the time period, 1982 to 1991. ∎

EXERCISES

1. A sociologist conducted a survey of citizens over the age of 60 years who have no private insurance and whose net worth is too high for Medicaid. The ages of 25 uninsured senior citizens are given in Exercise 2.15 on page 35 of the text. Use graphical and numerical methods to describe the ages of uninsured senior citizens. Discuss.

Ages of Uninsured Senior Citizens

68	73	66	76	86
74	61	89	65	90
69	92	76	62	81
63	68	81	70	73
60	87	75	64	82

2. A large international corporation customarily rents several homes for relocating employees in Phoenix. The Personnel Division of the corporation is interested in studying the monthly rents of homes within the city. The following prices of 40 three bedroom homes were randomly selected from Phoenix in August, 1989.

Rental Prices

$625	$795	$365	$ 595	$600
600	600	575	540	850
475	450	625	595	650
550	465	445	735	620
750	465	625	975	425
475	720	500	800	985
535	575	545	1,175	650
525	565	850	625	770

a. Construct a frequency distribution of the rental prices. Describe the distribution.
b. Construct a dot plot of the data. Compare the dot plot with the graph of part a.
c. The corporation was interested in determining the proportion of homes with rents greater than $700. Estimate this proportion based on the sample of 40 homes.

3. The growing number of refugees is a problem which seriously affects national economies. Refugees are defined by Freedom House "as people who are forced to flee or are expelled from their homelands due to war, civil conflict, pestilence, natural disaster or persecution...." The following numbers of refugees are cumulative 1988 totals reported in *The Wall Street Journal*, September 28, 1989.

Country	Refugees	Country	Refugees
Chile	4,800	Soviet Union	73,163
Cuba	5,973	Yugoslavia	21,566
El Salvador	26,215	Afghanistan	1,032,391
Guatemala	40,580	Bangladesh	48,500
Haiti	828,000	Iran	362,834
Nicaragua	54,969	Iraq	508,237
Angola	395,711	Palestinians	1,398,100
Burundi	186,000	Sri Lanka	94,883
Chad	41,700	South Yemen	55,000
Ethiopia	1,042,656	Burma	24,300
Mozambique	1,147,013	Cambodia	357,295
Namibia	81,403	China	112,000
Czechoslovakia	2,948	Indonesia	8,000
Hungary	6,467	Laos	93,446

62   Chapter 2

Poland	191,153	Philippines	90,000
Romania	26,495	Vietnam	85,172

a. Construct a stem and leaf display of the data. Discuss the distribution of refugees from the 32 countries.
b. Enter a code identifying the countries as Asian or non-Asian. Construct separate graphical displays with the same scale of the numbers of refugees from Asian and from non-Asian countries. What does the Minitab output suggest about the numbers of refugees from Asia as compared with other nations?
c. Construct a histogram of the numbers of refugees. Compare the output with the stem and leaf display.

4. Consider the 1988 top twenty-five magazine advertisers reported in the *Leading National Advertisers*. The advertising dollars (in thousands) are listed for 1987 and 1988.

Advertiser	1987	1988
Philip Morris Cos.	$271,178	$270,251
General Motors Corp.	153,926	190,799
RJR Nabisco	105,674	131,463
Ford Motor Co.	125,529	125,532
Chrysler Corp.	100,433	104,527
Proctor & Gamble Co.	79,501	79,279
AT&T	76,270	66,193
Time Warner	53,029	64,989
Nestle SA	56,616	63,477
Grand Metropolitan PLC	54,948	59,545
Unilever NV	58,259	59,449
Franklin Mint	31,733	48,981
Honda Motor Co.	42,701	47,159
Bristol-Myers Squibb	40,154	43,616
Revlon Group	36,562	42,488
Sony Corp.	31,862	41,771
General Electric Co.	37,727	36,747
Schering-Plough Corp.	29,202	36,617
Toyota Motor Corp.	27,105	36,476
American Brands	38,545	35,888
Nissan Motor Co.	19,361	35,826
E.I. du Pont de Nemours & Co.	41,676	34,676
Sara Lee Corp.	24,268	34,107
U.S. Government	44,164	33,436
Sears, Roebuck & Co.	21,608	33,373

a. Graphically and numerically compare the expenditures for the two years.
b. Calculate and construct graphical displays of the changes in advertising expenditures from 1987 to 1988. Discuss the distribution of changes in advertising expenditures. What information is provided by studying expenditure changes that is not provided by studying yearly expenditures?

5. The Minnesota Real Estate Research Center compiles information on homes sold in several areas of Minnesota. The following table gives the 1989 selling prices of 20 homes randomly selected from homes sold in St. Cloud and Rochester.

St. Cloud		Rochester	
$ 85,000	$104,400	$103,925	$129,900
46,000	90,600	66,000	47,800
78,900	53,500	69,900	96,000
123,000	119,500	53,000	92,330
116,000	54,000	125,500	54,900
46,600	64,500	61,500	144,000
99,875	71,900	64,000	89,000
64,000	69,900	80,750	85,900
52,000	111,900	74,500	135,000
52,500	64,500	129,195	58,000

a. Construct dot plots of the selling prices for the homes in St. Cloud and Rochester. Compare the selling prices for the two cities.

b. Stack the data. Use the BY subcommand to obtain dot plots of the selling prices for the two cities. Compare the output with that of part a.

6. Refer to Exercise 2.19 on page 38 of the text. The U.S. Department of Education reported that the national dropout rate for high school students fell more than 1% between 1982 and 1984. The following table gives the dropout rate for each state and the District of Columbia in 1982 and 1984.

State	1982	1984	State	1982	1984	State	1982	1984
Ala.	36.6	37.9	Ky.	34.1	31.6	N.D.	16.1	13.7
Alaska	35.7	25.3	La.	38.5	43.3	Ohio	22.5	20.0
Ariz.	36.6	35.4	Maine	27.9	22.8	Okla.	29.2	26.9
Ark.	26.6	24.8	Md.	25.2	22.2	Ore.	27.6	26.1
Calif.	39.9	36.8	Mass.	23.6	25.7	Pa.	24.0	22.8
Colo.	29.1	24.6	Mich.	28.4	27.8	R.I.	27.3	31.3
Conn.	29.4	20.9	Minn.	11.8	10.7	S.C.	37.4	35.5
Del.	25.3	28.9	Miss.	38.7	37.6	S.D.	17.3	14.5
D.C.	43.1	44.8	Mo.	25.8	23.8	Tenn.	32.2	29.5
Fla.	39.8	37.8	Mont.	21.3	17.9	Texas	36.4	35.4
Ga.	35.0	36.9	Neb.	18.1	13.7	Utah	25.0	21.3
Hawaii	25.1	26.8	Nev.	35.2	33.5	Vt.	20.4	16.9
Idaho	25.6	24.2	N.H.	23.0	24.8	Va.	26.2	25.3
Ill.	23.9	25.5	N.J.	23.5	22.3	Wash.	23.9	24.9
Ind.	28.3	23.0	N.M.	30.6	29.0	W.Va.	33.7	26.9
Iowa	15.9	14.0	N.Y.	36.6	37.8	Wisc.	16.9	15.5
Kansas	19.3	18.3	N.C.	32.9	30.7	Wyo.	27.6	24.0

a. Construct and interpret a stem and leaf display for the 1982 and for the 1984 dropout rates.
b. Suppose the Department of Education is interested in comparing the dropout rates for states east and west of the Mississippi River. Use the BY subcommand to construct two stem and

64   Chapter 2

leaf displays for 1984 dropout rates for the states east and west of the river. Compare the displays.

b. Construct dot plots of the 1982 and 1984 dropout rates. Compare the dot plots with the displays in part a.

c. Calculate the paired differences between the rates in 1984 and 1982. Construct a stem and leaf display of the differences. Interpret.

e. Use DESCRIBE to numerically describe the dropout rates for both years and for the differences. Compare the rates. 2*5 six — e

f. Calculate the intervals $\bar{x} \pm s$, $\bar{x} \pm 2s$, and $\bar{x} \pm 3s$ for the differences. Determine the number of observations that fall within each interval. Compare these results with the Empirical Rule.

g. Obtain box plots of the 1982 and 1984 dropout rates. Are there any extreme rates? If so, do the states having the extreme rates change from 1982 to 1984?

7. Minitab allows us to study the effects of a change in one or more data values on numerical descriptive measures. Consider the following data sets.

C1	C2	C3
50	100	180
190	190	190
200	200	200
176	176	176
162	162	162
230	230	230
274	274	274

The first data value in each column is different. Use DESCRIBE to obtain a table of numerical measures for the three columns. Compare the means, medians, and standard deviations.

8. The yearly housing inventory index measures the number of months it takes to sell all available residential properties at the current yearly rate of sales. A low index indicates a strong market demand, shorter time on the market, and higher prices. A high index indicates a large supply of available homes and a decline in home prices.

The *1988 Minnesota Housing Report* (Minnesota Real Estate Research Center, St. Cloud State University) provides the 1986, 1987, and 1988 unsold housing inventory index for five Minnesota regions. Calculate the average and standard deviation of the yearly indexes for each region.

Region	1986	1987	1988
Range	17.3	9.8	7.7
Lakes	16.7	11.9	13.9
Metro	4.8	6.3	4.9
Southeast	10.3	8.1	6.9
Central	13.3	11.2	9.8

9. The Consumer Price Index (CPI) measures the change over time in the price of foods and services purchased by wage earners and clerical workers. The *Statistical Abstract of the United States 1992* reports the annual percent change in consumer prices for several countries for the years 1981 through 1990.

Country	1881	1982	1983	1984	1985	1986	1987	1988	1989	1990
U.S.	10.3	6.1	3.2	4.3	3.5	1.9	3.7	4.1	4.8	5.4
Canada	12.5	10.8	5.8	4.3	4.0	4.2	4.4	4.0	5.0	4.8
Japan	4.9	2.7	1.9	2.2	2.0	0.6	-0.1	0.7	2.3	3.1
Austria	6.8	5.4	3.3	5.6	3.2	1.7	1.4	2.0	2.5	3.3
Belgium	7.6	8.7	7.7	6.3	4.9	1.3	1.6	1.2	3.1	3.4
Denmark	11.7	10.1	6.9	6.3	4.7	3.6	4.0	4.6	4.8	2.7
France	13.4	11.8	9.6	7.4	5.8	2.7	3.1	2.7	3.6	3.4
Italy	18.7	16.3	15.0	10.6	8.6	6.1	4.6	5.0	6.6	6.1
Spain	14.6	14.4	12.2	11.3	8.8	8.8	5.2	4.8	6.8	6.7
U.K.	11.9	8.6	4.6	5.0	6.1	3.4	4.1	4.9	7.8	9.5
W.Germ.	6.3	5.3	3.3	2.4	2.2	-0.1	0.2	1.3	2.8	2.7

For example, the 1981 number for the United States shows that the price of consumer goods increased 10.3% from 1980 to 1981.

a. Calculate the mean and standard deviation of the changes in consumer prices for each year. Construct dot plots for each year. What is the worldwide trend in consumer prices?
b. Calculate the mean, median, and standard deviation for each country. Generally countries desire a low and constant growth in consumer prices. A low standard deviation indicates a constant growth. Which three countries have the lowest increases in consumer prices? Which three countries have the most constant growth?

10. A large corporation records the daily productivity of each assembly line employee. The following table gives the daily productivity of a random sample of 13 employees for six days.

	Days					
Employee	1	2	3	4	5	6
Hodel, Lois	48	50	0	49	44	44
Eich, Peter	42	46	47	43	30	0
Zirbes, Tom	48	45	43	38	35	33
Notch, Mike	44	47	45	46	40	29
Barker, Hugh	44	32	42	50	38	43
Sakry, Carol	42	48	49	50	48	41
Fisher, Dean	44	44	49	43	44	41
Jurek, Roy	48	46	49	47	40	48
Wenz, Robin	50	38	47	43	35	28
Piehl, Gina	46	43	48	50	36	32
Coborn, Bob	42	46	49	50	45	39
Gohman, Lora	43	41	48	48	44	43
Theis, Mark	42	48	49	45	39	37

a. Which three workers had the highest average productivity? The lowest average productivity?
b. Consistent productivity is an important employee trait. Are there any workers who are more consistent than others? Discuss.
c. Is there a change in the productivity of all employees over time? Discuss. Hint: Use DOTPLOT and DESCRIBE.
d. Stack the productivity measures in a column. Construct a histogram and a box plot of the

stacked data. Describe the distribution of productivity measures.

11. A psychologist developed a technique to improve rote memory. Twenty high school students were randomly selected and taught the new technique. The following data from Exercise 2.30 on page 45 of the text are the number of word phrases out of 100 that were memorized correctly by the students.

Number of Word Phrases

91	64	98	66	83
87	83	86	80	93
83	75	72	79	90
80	90	71	84	68

a. Use graphical and numerical methods to describe the data. Discuss. Are there any outliers?
b. Find and interpret the mode.
c. Add the constant 3 to each number of word phrases. Recalculate the mean and standard deviation. What is the effect on these measures of adding a constant to each data value?
d. Multiply each number of word phrases by the constant 3. Recalculate the mean and standard deviation. What is the effect on these measures of multiplying each data value by a constant?

12. Consider Exercise 2.36 on page 46 of the text. The table contains the price per acre of farmland for a sample of states east and west of the Mississippi.

State	Price per Acre	State	Price per Acre
Ariz.	$ 265	Nebr.	$ 444
Calif.	1,726	Nev.	229
Colo.	435	N.H.	1,419
Conn.	3,208	N.J.	3,525
Del.	1,642	N.Mex.	163
Fla.	1,527	N.Dak.	360
Kansas	466	Pa.	1,510
Mass.	2,372	R.I.	3,335
Md.	2,097	S.Dak.	250
Mont.	222	Wyo.	177

a. Find the mean, median, variance, and standard deviation of price per acre for the 20 states.
b. Define a code to identify states east and west of the Mississippi. Find the mean price per acre for the eastern and western states. Compare.
c. Rank the states in terms of price per acre carrying along the code for the location of the states. Which state had the highest price per acre? The lowest? In which region of the country are these states? Discuss.
d. Construct dot plots of the prices per acre for all the states and for the eastern and western states. Discuss the dot plots.

13. The marketing department of a manufacturer of minicomputer systems surveyed 40 most recent customers to determine the number of hours of down time they had experienced during the previous month. The following data are reported in Exercise 2.86 on page 79 of the text.

Methods for Describing Sets of Data 67

Customer Number	Down Time	Customer Number	Down Time	Customer Number	Down Time
230	12	244	2	257	18
231	16	245	11	258	28
232	5	246	22	259	19
233	16	247	17	260	34
234	21	248	31	261	26
235	29	249	10	262	17
236	38	250	4	263	11
237	14	251	10	264	64
238	47	252	15	265	19
239	0	253	7	266	18
240	24	254	20	267	24
241	15	255	9	268	49
242	13	256	22	269	50
243	8				

a. Construct a box plot of the measurements. What does the box plot reveal about the frequency distribution of the amount of down time? Are there any outliers?
b. Use DESCRIBE to obtain numerical descriptive measures of the 40 measurements. Discuss the measures.
c. Find $\bar{x} \pm s$, $\bar{x} \pm 2s$, and $\bar{x} \pm 3s$. Determine the number of observations that fall within each interval. Compare these results with the Empirical Rule.

14. The Department of Health and Human Services collects data on the energy assistance program for low income people. The following table gives the 1991 fiscal year's budget for low-income energy assistance payments (in thousands of dollars) and the populations (in thousands) on April 1, 1990 for the 50 states and the District of Columbia.

State	Assistance	Population	State	Assistance	Population
Alabama	$12,149	4,041	Montana	$10,397	799
Alaska	7,755	550	Nebraska	13,021	1,578
Arizona	5,876	3,665	Nevada	2,760	1,202
Arkansas	9,270	2,351	New Hampshire	11,225	1,109
California	65,178	29,760	New Jersey	55,053	7,730
Colorado	22,725	3,294	New Mexico	7,356	1,515
Connecticut	29,646	3,287	New York	179,755	17,990
Delaware	3,935	666	North Carolina	26,789	6,629
Dist Columbia	4,604	607	North Dakota	11,295	639
Florida	19,224	12,938	Ohio	72,590	10,847
Georgia	15,199	6,478	Oklahoma	11,168	3,146
Hawaii	1,531	1,108	Oregon	17,613	2,842
Idaho	8,864	1,007	Pennsylvania	96,555	11,882
Illinois	82,055	11,431	Rhode Island	9,761	1,003
Indiana	37,152	5,544	South Carolina	9,649	3,487
Iowa	26,330	2,777	South Dakota	9,173	696
Kansas	12,092	2,478	Tennessee	18,585	4,877
Kentucky	19,334	3,685	Texas	31,982	16,987

68   Chapter 2

Louisiana	12,421	4,220	Utah	10,561	1,723
Maine	19,206	1,228	Vermont	8,413	563
Maryland	22,670	4,781	Virginia	27,651	6,187
Massachusetts	69,302	6,016	Washington	28,971	4,867
Michigan	77,904	9,295	West Virginia	12,795	1,793
Minnesota	56,126	4,375	Wisconsin	50,521	4,892
Mississippi	10,416	2,573	Wyoming	4,228	454
Missouri	32,776	5,117			

a. Enter the alpha and numeric data in a worksheet. Calculate the per capita low income energy assistance payments for the states.

b. Construct dot plots of the energy assistance payments, population, and per capita payments. Describe the distributions.

c. Construct separate dot plots for the states and the District of Columbia located east and west of the Mississippi River. Compare the dot plots.

d. Obtain numerical descriptive measures of the energy assistance payments, population, and per capita payments for the fifty states and the District of Columbia, and for the fifty states without the District of Columbia. Compare the measures.

e. Calculate z-scores for the per capita payments for the states and the District of Columbia. Rank the data from largest to smallest per capita payments. Interpret the smallest and largest z-scores.

f. The states will actually get about 15% more than was budgeted for fiscal 1991 because additional emergency payments were made when heating fuel prices increased 20% over 1990 prices. Calculate the adjusted energy assistance payments and per capita assistance for each state and the District of Columbia. Calculate the mean and standard deviation of the adjusted per capita payments. Compare these measures with the budgeted measures.

15. Consider the class data set given in Appendix B of this supplement and answer the following questions.

   a. Graphically and numerically summarize the distribution of ages of the students.
   b. Compare the ages of male and female students.
   c. Repeat parts a and b for the number of hours students work.

CHAPTER 3

PROBABILITY

A goal of statistics is to use sample information to make an inference about the population. Probability, as a measure of uncertainty associated with chance events, is used to assess the reliability of a statistical inference. In this chapter, we use Minitab to simulate experiments and to construct bivariate probability tables.

NEW COMMANDS

BASE   RANDOM   TABLE

EXPERIMENTS, EVENTS, AND PROBABILITY

An experiment is a process which results in a chance outcome, called an event. For example, investing in IBM stock, starting a credit card business, or running for public office are experiments that result in events which cannot be predicted with certainty.

Probability measures the uncertainty surrounding the occurrence of an event. Generally the probability of an event is approximated by the proportion of times the event occurs in repeated experiments. Minitab can be used to simulate such experiments. The RANDOM command generates simple random samples of n observations from an infinite population with a specified probability distribution. A simple random sample is a set of n observations selected from a population such that every set of n observations has an equal chance of being selected. Each time RANDOM is used, Minitab uses a different random starting point for the sequence of random numbers. If you wish to generate the same set of numbers, or control the starting point of a sequence, use the BASE command.

```
RANDOM K OBSERVATIONS IN EACH OF C,...,C
  BERNOULLI P = K
  BINOMIAL FOR N = K P = K
  CHISQUARE V = K
  DISCRETE X IN C P(X) IN C
  EXPONENTIAL MU = K
  F WITH VONE = K AND VTWO = K
  INTEGER FOR A = K B = K
  LOGNORMAL MU = K SIGMA = K
  NORMAL WITH MU = K AND SIGMA = K
  T WITH V = K
  POISSON MU = K
  UNIFORM A = K B = K
```

This command generates a random sample of K observations in each column. The sampling is from a specified probability distribution defined in the subcommand. Some allowable subcommands are given above. If you do not use a subcommand, Minitab generates data from a standard normal distribution. Use the HELP facility to obtain specific information.

```
BASE = K
```

Use this command to control the starting point for a sequence of random numbers. The generator continues from the specified starting point for the Minitab session.

■ **Example 1** Table 1 in Appendix A of the text is an example of a random number table. A random number table is constructed such that every number occurs with equal probability. Use Minitab to generate a random number table. Construct a histogram of the random numbers. Do the numbers occur with equal probability?

Solution The RANDOM command and INTEGER subcommand provide 1,000 random numbers.

```
MTB > RANDOM 1000 C1;
SUBC> INTEGER 0 TO 9.
MTB > PRINT C1
C1
  2  5  0  9  3  9  4  3  4  5  9  3  2  2  4
  8  9  4  9  9  0  7  4  2  3  1  2  2  4  8
  5  6  5  9  3  2  8  1  6  6  3  9  7  9  2
  8  9  9  4  6  1  8  8  5  1  7  3  3  9  6
  5  5  1  5  0  5  7  8  2  9  0  4  3  0  5
  5  9  8  7  8  5  0  2  9  2  2  5  4  6  1
  6  2  9  1  6  2  3  8  1  6  9  1  3  1  4
  7  7  0  3  0  4  2  2  6  2  7  5  4  0  4
  0  2  7  8  0  9  9  1  9  1  5  8  6  5  8
  4  1  6  1  7  8  7  5  3  1  0  3  3  6  4
  1  7  7  9  7  1  9  8  1  6  4  4  5  3  0
```

4	8	8	6	9	1	4	5	1	3	6	3	3	2	2
0	3	0	4	7	5	4	9	2	7	9	1	5	2	9
1	7	0	5	8	7	5	2	1	0	8	1	5	5	8
0	2	5	6	5	4	7	8	1	9	0	7	0	0	1
1	0	0	6	0	8	9	6	3	2	7	7	5	8	2
3	2	0	9	2	0	7	7	6	2	3	7	3	2	7
5	0	1	2	2	8	2	4	4	3	6	4	3	2	3
4	2	4	9	9	1	9	5	3	1	3	2	0	9	2
9	5	0	3	1	4	3	2	5	0	8	4	6	5	4
2	3	7	6	5	1	8	2	9	1	7	7	1	7	4
3	1	4	4	7	0	6	9	4	0	3	2	5	9	1
2	2	5	0	0	7	7	8	1	9	2	3	4	2	8
5	2	3	2	9	4	8	5	9	1	0	4	0	1	9
5	7	0	3	1	9	9	7	8	6	4	7	1	5	7
4	0	9	1	5	0	6	3	6	8	1	1	2	8	3
1	1	7	8	3	6	8	0	7	8	0	8	9	3	0
6	6	3	9	2	7	3	9	9	9	1	5	8	1	5
5	0	0	5	2	3	3	6	7	4	3	1	3	6	0
8	5	6	1	4	0	1	4	4	7	6	7	6	9	4
6	8	1	4	8	4	6	0	1	4	0	2	8	8	2
7	5	1	0	6	0	6	7	4	3	0	0	6	0	0
7	1	6	6	2	6	2	9	9	3	5	3	0	1	0
9	9	6	4	3	1	3	8	6	9	4	7	8	6	0
0	6	4	9	4	9	5	3	6	3	9	4	4	3	0
8	4	4	0	6	2	0	8	1	0	0	1	0	3	8
4	0	7	3	0	2	5	9	8	3	5	7	8	3	1
6	4	2	6	5	8	6	8	3	7	1	0	5	2	0
0	8	4	4	2	1	1	3	1	5	9	6	2	2	5
4	4	3	9	6	8	1	3	2	3	9	9	2	9	6
1	9	2	7	7	8	1	0	9	4	3	1	9	3	6
9	4	2	3	5	9	2	3	1	9	0	0	7	3	9
7	0	9	9	6	2	4	3	9	9	1	1	9	4	8
8	8	9	6	7	1	0	3	7	0	6	6	9	0	3
1	6	3	2	5	9	3	1	0	1	2	9	1	0	9
6	9	9	4	2	8	7	6	1	8	9	8	3	3	8
3	7	5	1	2	7	4	6	6	1	6	7	2	6	0
1	5	6	3	4	2	9	2	6	7	4	5	0	3	1
0	2	5	4	8	4	1	0	6	2	3	0	8	6	0
0	2	5	5	4	4	4	4	7	7	7	0	3	2	0
1	9	8	4	4	1	4	7	3	5	7	8	4	3	1
7	6	0	0	6	9	8	7	2	5	8	6	9	1	3
1	6	7	5	0	5	2	4	9	4	1	7	0	3	3
8	2	2	7	2	0	0	1	8	7	3	5	0	0	8
2	5	3	4	1	6	8	9	5	2	7	2	4	9	2
0	9	9	7	4	0	5	7	5	8	1	2	5	5	2
5	8	6	5	2	4	5	1	6	3	1	6	7	8	3
6	9	4	2	4	7	8	1	6	9	1	0	7	7	7
2	2	4	0	1	4	2	7	8	5	9	7	4	6	5
4	4	9	0	2	6	7	9	1	1	1	2	1	3	2
1	0	4	8	4	2	9	8	7	6	8	6	4	0	9
4	1	2	3	3	1	2	9	4	4	3	4	7	8	9
2	6	4	0	5	9	9	9	4	5	3	7	9	0	5
8	4	7	4	8	5	0	4	7	4	8	3	6	9	2
3	3	0	2	5	2	9	1	0	3	4	1	6	9	8
4	3	3	3	8	4	8	3	2	7	8	3	6	8	4
5	9	0	0	0	2	6	4	9	2					

72   *Chapter 3*

```
MTB > HISTOGRAM C1

Histogram of C1    N = 1000
Each * represents 5 obs.

Midpoint    Count
       0      110   **********************
       1      105   *********************
       2      107   *********************
       3      104   *********************
       4      111   ***********************
       5       85   *****************
       6       90   ******************
       7       89   ******************
       8       88   ******************
       9      111   ***********************

MTB > STOP
```

The histogram shows that the random numbers occur with about the same probabilities. The probability of each number occurring is .1; thus, each number should appear approximately 100 times.

■

■ **Example 2** Example 3.1 on page 97 of the text describes an experiment in which two fair coins are tossed and the number of heads recorded. Suppose you are asked to observe the proportion of heads as you toss from one to fifty coins. Simulate this experiment using Minitab. Plot the proportion of heads versus the number of tosses; that is, from $n = 1$ to 50.

Solution A Bernoulli process is characterized by a series of independent trials, each of which results in one of two outcomes, success and failure. The probability of success, denoted p, remains constant from trial to trial. The probability of failure is $(1-p)$. We can simulate the coin tossing experiment by a Bernoulli process with $p = .5$. With the RANDOM and the BERNOULLI subcommand, a head, viewed as a success, is assigned a 1 and a failure is assigned a 0.

```
MTB > NAME C1 'TOSS' C2 'HEAD' C3 'SUM HEAD' C4 'P(HEAD)'
MTB > SET 'TOSS'
DATA> 1:50
DATA> END
MTB > RANDOM 50 OBSERVATIONS IN 'HEAD';
SUBC> BERNOULLI P = .5.
MTB > PARSUM 'HEAD', PUT IN 'SUM HEAD'
MTB > LET 'P(HEAD)' = 'SUM HEAD'/'TOSS'
```

```
MTB > PRINT C1-C4

ROW     TOSS    HEAD    SUM HEAD    P(HEAD)

  1       1       1         1      1.00000
  2       2       1         2      1.00000
  3       3       0         2      0.66667
  4       4       1         3      0.75000
  5       5       0         3      0.60000
  6       6       1         4      0.66667
  7       7       0         4      0.57143
  8       8       0         4      0.50000
  9       9       0         4      0.44444
 10      10       1         5      0.50000
 11      11       1         6      0.54545
 12      12       1         7      0.58333
 13      13       1         8      0.61538
 14      14       0         8      0.57143
 15      15       1         9      0.60000
 16      16       1        10      0.62500
 17      17       1        11      0.64706
 18      18       0        11      0.61111
 19      19       0        11      0.57895
 20      20       1        12      0.60000
 21      21       0        12      0.57143
 22      22       1        13      0.59091
 23      23       0        13      0.56522
 24      24       0        13      0.54167
 25      25       0        13      0.52000
 26      26       1        14      0.53846
 27      27       0        14      0.51852
 28      28       1        15      0.53571
 29      29       0        15      0.51724
 30      30       0        15      0.50000
 31      31       0        15      0.48387
 32      32       1        16      0.50000
 33      33       0        16      0.48485
 34      34       0        16      0.47059
 35      35       1        17      0.48571
 36      36       1        18      0.50000
 37      37       0        18      0.48649
 38      38       1        19      0.50000
 39      39       0        19      0.48718
 40      40       1        20      0.50000
 41      41       0        20      0.48780
 42      42       0        20      0.47619
 43      43       0        20      0.46512
 44      44       1        21      0.47727
 45      45       0        21      0.46667
 46      46       1        22      0.47826
 47      47       1        23      0.48936
 48      48       0        23      0.47917
 49      49       0        23      0.46939
 50      50       0        23      0.46000
```

```
MTB > PLOT 'P(HEAD)' VS 'TOSS';
SUBC> TITLE 'COIN TOSSING EXPERIMENT';
SUBC> TITLE 'Proportion of Heads'.
```

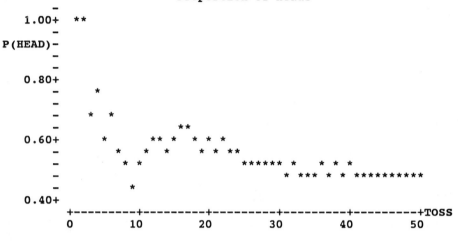

```
MTB > STOP
```

As the number of tosses increases, the proportion of heads approaches .5. We expect the process to stabilize at $p = .5$ as the number of tosses becomes very large.

BIVARIATE TABLE

The TABLE command can be used to construct a bivariate table. The command organizes the data in cells, and prints summary statistics.

TABLE THE DATA CLASSIFIED BY C,...,C
  COUNTS
  ROWPERCENT
  COLPERCENT
  TOTPERCENT
  CHISQUARE (OUTPUT CODE K)
  MEANS FOR C,...,C
  MEDIANS FOR C,...,C
  SUMS FOR C,...,C
  MINIMUMS FOR C,...,C
  MAXIMUMS FOR C,...,C
  STDEV FOR C,...,C
  STATS FOR C,...,C
  DATA FOR C,...,C
  N FOR C,...,C
  NMISS FOR C,...,C
  PROPORTION OF K (THROUGH K) FOR C,...,C
  NOALL IN MARGINS
  ALL FOR C,...,C

The TABLE command prints one-way, two-way, and multi-way tables. If no subcommands are used, a count of the number of observations in each cell is given. The COUNTS subcommand is needed to get a count if other subcommands are used. ROWPERCENT provides the percent of the row total in a cell. COLPERCENT provides the percent of the column total in a cell. TOTPERCENT gives the percent of all observations in a cell. CHISQUARE does a test of independence between the rows and columns of each two-way table.

Most of the subcommands on numerical descriptive measures have been previously described as commands. The STATS subcommand prints the N, MEAN, and STDEV of each column. DATA groups the data in cells. The default output includes marginal statistics for the rows and columns. NOALL cancels this output. ALL prints marginal statistics for those variables listed with this subcommand. These variables must also be listed in the TABLE command line.

■ **Example 3** This example considers part of a survey in which employees of a large research and development corporation are classified according to gender and highest degree obtained. The corporation, wanting to provide an in-house continuing education program for its employees, surveyed the employees to determine the interest in such a program. Two questions on the survey related to gender and current educational level. The following table gives the responses for 200 employees.

Gender: Male = 1   Female = 2
Highest Degree: High School = 1   Bachelor's = 2
                  Master's = 3   Ph.D. = 4

Gender	Degree	Gender	Degree	Gender	Degree	Gender	Degree
2	2	2	2	2	4	2	1
1	4	2	3	1	2	1	2
1	4	1	2	2	2	2	4
1	2	1	3	2	2	2	1
1	4	1	2	2	2	2	2
1	2	2	2	2	1	1	1
1	2	1	3	1	2	2	3
2	4	1	3	2	1	1	2
2	1	2	1	2	1	1	2
2	4	2	1	1	2	2	1
2	2	1	4	1	2	2	1
2	2	2	1	1	4	4	2
1	3	1	2	1	2	2	2
1	1	1	4	2	3	2	1
2	1	2	2	2	3	2	2
2	2	1	1	1	3	2	2
2	4	1	3	2	1	2	1
2	2	1	4	1	1	2	2
1	3	2	2	1	4	1	3
2	3	1	3	1	2	1	2
1	2	1	2	2	2	2	2
2	2	2	1	1	2	1	3
2	2	1	1	1	2	1	4
2	2	1	2	2	2	1	4
1	2	2	1	2	2	1	4
2	1	2	1	1	3	2	1
1	3	1	2	2	1	2	2
2	1	2	2	1	2	2	2
1	3	2	1	1	1	1	2
1	4	1	3	2	1	2	2
1	4	1	4	2	2	2	2
1	1	1	2	2	4	2	3
1	2	1	3	1	2	1	3
1	4	2	1	2	2	1	1
1	4	2	1	1	4	2	1
2	1	1	2	2	3	2	2
1	3	1	2	1	2	1	4
1	2	2	1	2	2	2	3
2	2	2	1	1	3	2	3
2	1	2	1	2	1	2	2
1	3	2	3	1	3	2	1
1	2	2	3	1	3	1	2
1	2	1	2	1	2	2	2
2	3	1	4	1	2	1	1
1	4	2	1	1	2	2	2
2	1	1	3	2	1	1	2
1	4	2	1	1	2	2	1
2	2	2	2	1	4	2	1
2	4	1	4	2	1	2	2
1	4	2	2	2	2	2	3

a. Construct a two-way table placing gender in rows and degree in columns.
b. An employee is selected at random. Find the probability that the employee is female. Has a Ph.D. Is female and has a Ph.D.
c. Given a female employee is selected at random, find the probability that she has a Ph.D. Has a Masters or Ph.D.
d. Are the two variables, gender and degree, independent? Why or why not?

Solution

a. The variables are coded the same as on the survey.

```
MTB > NAME C1 'GENDER'   C2 'DEGREE'
MTB > READ 'GENDER' AND 'DEGREE'
DATA> 2  2
DATA> 1  4
DATA> 1  4
       .
       .
       .
DATA> 2  3
DATA> END
   200 ROWS READ
MTB > SAVE 'RDSURVEY'

Worksheet saved into file:  RDSURVEY.MTW
MTB > TABLE OF 'GENDER' AND 'DEGREE';
SUBC> TOTPERCENT.

ROWS:    GENDER      COLUMNS:   DEGREE

              1           2           3           4         ALL

    1       4.50       20.00       11.50       12.00       48.00
    2      21.00       20.50        7.00        3.50       52.00
  ALL      25.50       40.50       18.50       15.50      100.00

CELL CONTENTS -- % OF TBL

MTB > STOP
```

b. Females are coded 2 in the two-way table. We see that 52% of all employees are female. The probability that a female is randomly selected is .52.

The Ph.D. degree is coded 4 in the table. The probability that the randomly selected employee has a Ph.D. is .155.

The value at the intersection of the female row and Ph.D. column is 3.5%. The probability that the employee selected is female and has a Ph.D. is .035.

c. Since we know the employee is female, we use only the female row in the table. The probability that she has a Ph.D. is 3.50/52.00 = 0.07. The probability that she has a Masters or Ph.D. is (7.00 + 3.50)/52.00 = 0.20.

d. The two variables are not independent. From part b, the probability that a randomly selected employee has a Ph.D. is 0.155. From part c, given the employee is female, the probability she has a Ph.D. is 0.07. Knowing the gender of an employee changes the probability that the randomly selected employee has a Ph.D. The two variables are dependent.

■

EXERCISES

1. Exercise 3.18 on page 120 of the text asks you to find the probabilities associated with tossing a coin three times. Simulate the experiment by generating 100 Bernoulli trials in each of three columns. Hint: Each row can be viewed as the tossing of three coins, where a 1 represents a head and a 0 a tail. Use RSUM to count the number of heads and TALLY to find the number of times the result was 0, 1, 2 or 3 heads.

    a. Based on the simulation results, what is the probability of obtaining 0 heads? 1 head? 2 heads? 3 heads?
    b. Compare these with the respective theoretical probabilities, .125, .375, .375, and .125.

2. Simulate a die tossing experiment 1,000 times. Observe the number of times the toss results in a six. Plot the number of times you obtain a six versus the number of tosses. What proportion of outcomes do you expect to result in a six? Hint: The INTEGER subcommand with RANDOM generates random integers between any two integers, a and b.

3. An advertising agency is interested in whether a client's advertisement in the Minneapolis Tribune was noticed by the readers of the newspaper. The agency asked 50 subscribers if they read the advertisement and if they were under 30. The results are as follow.

Subscriber	Under 30?	Read Ad?	Subscriber	Under 30?	Read Ad?
1	yes	yes	26	yes	no
2	yes	no	27	no	no
3	no	no	28	yes	yes
4	no	no	29	no	no
5	no	yes	30	yes	no
6	no	no	31	yes	yes
7	yes	yes	32	yes	yes
8	yes	yes	33	no	yes
9	no	yes	34	no	no
10	yes	no	35	no	no
11	yes	yes	36	yes	no
12	yes	no	37	no	yes
13	no	no	38	no	yes
14	yes	no	39	yes	yes
15	no	yes	40	no	no
16	no	yes	41	yes	no
17	yes	yes	42	no	yes
18	no	no	43	no	yes

19	no	no	44	no	no
20	no	yes	45	yes	yes
21	yes	yes	46	yes	no
22	yes	yes	47	yes	yes
23	no	no	48	yes	yes
24	no	yes	49	no	no
25	no	no	50	no	no

a. Code the yes responses 1, and the no responses 0. Construct a two-way table with age in rows and whether the advertisement was read in columns.
b. What proportion of the sample read the advertisement?
c. What proportion read the advertisement and was under 30?
d. Is age independent of whether a subscriber read the ad? How could the advertising agency use this information?

4. Consider the class data base given in Appendix B of this supplement. Select two of the variables, for example, gender and marital status, and construct a two-way table. Write a brief report summarizing the information contained in the table.

CHAPTER 4

DISCRETE RANDOM VARIABLES

The probability distribution of a discrete random variable is described in this chapter. We illustrate commands for calculating the mean and variance, and commands for the probability distributions and cumulative probability distributions for the binomial and Poisson random variables.

NEW COMMANDS

CDF   PDF

THE PROBABILITY DISTRIBUTION OF A DISCRETE RANDOM VARIABLE

A random variable is a rule that assigns a numerical value to each simple event of an experiment. There are two types of random variables, discrete and continuous. A discrete random variable can assume a countable number of values; a continuous random variable can assume a value corresponding to any point on some interval.

The probability distribution of a discrete random variable gives all possible values of the random variable and the probability of observing each of the values. The distribution is usually presented as a table or graph. Use PRINT to obtain a table of the distribution and PLOT to graph the probability distribution.

The Mean and Variance

The mean or expected value of a discrete random variable is given by

$$\mu = \sum_{\text{All } x} xp(x).$$

The variance, a weighted average of the squared differences between the values of x and μ, is

$$\sigma^2 = \sum_{\text{All } x} (x - \mu)^2 p(x).$$

The square root of the variance is the standard deviation, denoted σ.

You can use the LET command to calculate the mean, variance and standard deviation of a discrete random variable. LET follows the usual order of precedence for evaluating algebraic equations. The results can be stored in a column or a constant location.

■ **Example 1** Consider the following probability distribution that is described in Example 4.6 on page 187 of the text. The distribution assumes that 70% of the time chemotherapy is successful in treating skin cancer. The random variable x is the number of successful cures in a sample of $n = 5$ skin cancer patients.

x	$p(x)$
0	.002
1	.029
2	.132
3	.309
4	.360
5	.168

a. Use Minitab to present the probability distribution in tabular and graphical form.
b. Calculate the mean, variance, and standard deviation of x.
c. Find the probability that x is within two standard deviations of the mean.

Solution

a. We use PRINT and PLOT to obtain a table and graph of the probability distribution.

```
MTB > NAME C1 'X' C2 'P(X)'
MTB > SET 'X'
DATA> 0:5
DATA> END
MTB > SET 'P(X)'
DATA> .002 .029 .132 .309 .360 .168
DATA> END
MTB > PRINT 'X' 'P(X)'

ROW    X    P(X)

  1    0    0.002
  2    1    0.029
  3    2    0.132
  4    3    0.309
  5    4    0.360
  6    5    0.168
```

```
MTB > PLOT 'P(X)' VS 'X';
SUBC> TITLE 'A DISCRETE PROBABILITY DISTRIBUTION';
SUBC> TITLE 'Successful Chemotherapy Treatments'.
```

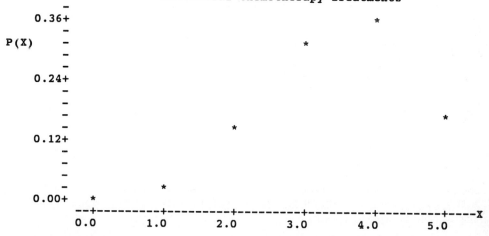

Rectangles can be drawn on this plot to obtain a graph similar to that given in Figure 4.4 on page 188 of the text.

b. We use LET to evaluate the equations for μ, σ^2, and σ. Parentheses are required with mathematical functions, such as SUM and SQRT.

```
MTB > NAME C3 'MU' C4 'VARIANCE' C5 'ST DEV'
MTB > LET 'MU' = SUM('X'*'P(X)')
MTB > LET 'VARIANCE' = SUM(('X'-'MU')**2*'P(X)')
MTB > LET 'ST DEV' = SQRT('VARIANCE')
MTB > PRINT C1-C5
 ROW      X      P(X)      MU    VARIANCE    ST DEV
  1       0     0.002     3.5     1.048      1.02372
  2       1     0.029
  3       2     0.132
  4       3     0.309
  5       4     0.360
  6       5     0.168
```

c. To find the probability that x is within two standard deviations of the mean, we find $P(\mu - 2\sigma \leq x \leq \mu + 2\sigma)$.

```
MTB > NAME C6 'MU-2SD' C7 'MU+2SD'
MTB > LET 'MU-2SD' = 'MU' - 2*'ST DEV'
MTB > LET 'MU+2SD' = 'MU' + 2*'ST DEV'
MTB > PRINT 'MU-2SD' AND 'MU+2SD'
 ROW     MU-2SD     MU+2SD
  1      1.45256    5.54744

MTB > STOP
```

The interval $\mu \pm 2\sigma$ or (1.45256, 5.54744) contains x = 2, 3, 4 and 5. The corresponding probabilities given in the table are .132, .309, .360, and .168, with sum .969. This probability is close to the value given by the Empirical Rule, which says that approximately **95%** of the measurements of a mound-shaped distribution fall within two standard deviations of the mean.

■

THE BINOMIAL RANDOM VARIABLE

The outcome of an experiment can often be described as falling in one of two categories. For example, a machine part is either defective or nondefective, a memo is either correctly typed or not, an account is either paid up or is overdue, and a product is either preferred or not preferred by a customer. The categories are usually denoted success and failure.

The binomial distribution is the probability distribution of the number of successes on n identical and independent trails. The probability of success, denoted p, is constant for each trial. The probability of failure is $q = 1 - p$. The mean or expected value of the binomial random variable is defined $\mu = np$, and the variance is $\sigma^2 = npq$. The PDF command with the BINOMIAL subcommand calculates the binomial probability distribution of x.

PDF (FOR VALUES IN E) (STORE PROBABILITIES IN E)
   BINOMIAL FOR N = K P = K

This command calculates binomial probabilities for the specified parameters, n and p. There are upper limits on the values of n, depending on the specified value of p. Release 8 limits for p = .10, .25, .50, .75 and .90 are n = 800, 300, 125, 60 and 35, respectively. Refer to HELP PDF BINOMIAL for more information. Values of x with probabilities less than .00005 or greater than .99995 are not printed on the output.

The first option calculates probabilities for specified values of the number of successes. These values are stored in a column or, if one value is desired, it may be entered as a constant. The second option stores the probabilities.

The CDF command calculates cumulative probabilities. The cumulative probability is the probability that the random variable x is less than or equal to a certain value.

CDF (FOR VALUES IN E) (STORE PROBABILITIES IN E)
   BINOMIAL FOR N = K P = K

This command calculates the cumulative binomial probability distribution function. The options and subcommands are the same as those for PDF.

Discrete Random Variables

■ **Example 2** The Heart Association's claim that only 10% of adults over 30 years old can pass the minimum requirements of the president's physical fitness test is described in Example 4.8 on page 194 of the text. Suppose four adults are given the test. Let x be the number of adults who pass the minimum requirements.

a. Construct a table and graph of the probability distribution for x.
b. Find the probability that none of the adults pass. That 1 adult passes.
c. Calculate the mean and standard deviation. Find the probability that x is within two standard deviations of the mean.

Solution The number of adults x who pass the minimum requirements is a binomial random variable with $n = 4$ and $p = .1$.

a. PDF with the BINOMIAL subcommand calculates the binomial probability distribution. To obtain the graph, we store the probabilities in a column and use PLOT.

```
MTB > NAME C1 'X' C2 'P(X)'
MTB > SET 'X'
DATA> 0:4
DATA> END
MTB > PDF 'X' STORE PROBABILITIES IN 'P(X)';
SUBC> BINOMIAL N = 4 P = .1.
MTB > PRINT 'X' 'P(X)'

  ROW     X      P(X)

    1     0     0.6561
    2     1     0.2916
    3     2     0.0486
    4     3     0.0036
    5     4     0.0001

MTB > PLOT 'P(X)' VS 'X';
SUBC> TITLE 'BINOMIAL DISTRIBUTION';
SUBC> TITLE 'Adult Physical Fitness'.
```

```
                        BINOMIAL DISTRIBUTION
                        Adult Physical Fitness
      0.75+
          -
P(X)      -    *
          -
          -
      0.50+
          -
          -
          -
          -
      0.25+
          -
          -          *
          -
          -
      0.00+
          -                        *           *         *
           --+---------+---------+---------+---------+---------+----X
           0.00      0.80      1.60      2.40      3.20      4.00
```

Chapter 4

We can draw rectangles by hand on this plot to obtain a distribution similar to that given in Figure 4.5 on page 195 of the text.

b. To calculate $P(x = 0)$ and $P(x = 1)$, we enter the constants 0 and 1 on the PDF command lines.

```
MTB > PDF X = 0;
SUBC> BINOMIAL N = 4 P = .1.
       K              P( X = K)
      0.00             0.6561
MTB > PDF X = 1;
SUBC> BINOMIAL N = 4 P = .1.
       K              P( X = K)
      1.00             0.2916
MTB > STOP
```

c. The mean is $\mu = np = .4$. The standard deviation is $\sigma = \sqrt{npq} = .6$. The probability that x is within two standard deviations of the mean is $P(\mu - 2\sigma \leq x \leq \mu + 2\sigma) = P(-.8 \leq x \leq 1.6)$. The interval contains $x = 0$ and 1; from the table, the corresponding probability is $.6561 + .2916 = .9477$.

■

■ **Example 3** Consider Example 2 of this chapter which describes the probability distribution of x, the number of four adults who can pass the president's physical fitness minimum requirements. Suppose we randomly select 10,000 samples of four adults each and observe x, the number of adults who can pass the test in each sample. Obtain the frequency and relative frequency distributions. Compare the relative frequency distribution with the binomial probability distribution given in Example 2 of this chapter. We would expect about 66% of the samples to have 0 adults who can pass, 33% to have 1 adult who can pass the test, and so on.

Solution The RANDOM command with the BINOMIAL subcommand is used to generate 10,000 samples from a binomial probability distribution with $p = .1$ and $n = 4$.

```
MTB > NAME C1 'TRIALS'
MTB > RANDOM 10000 IN 'TRIALS';
SUBC> BINOMIAL N = 4 P = .1.
MTB > HISTOGRAM 'TRIALS'

Histogram of TRIALS   N = 10000
Each * represents 135 obs.

Midpoint    Count
    0        6559    ******************************************************
    1        2926    ********************
    2         489    ****
    3          23    *
    4           3    *

MTB > NAME C2 'X' C3 'FREQ' C4 'REL FREQ'
MTB > SET 'X'
DATA> 0:4
DATA> END
MTB > SET 'FREQ'
DATA> 6559 2926 489 23 3
DATA> END
```

```
MTB > LET 'REL FREQ' = 'FREQ'/SUM('FREQ')
MTB > PRINT 'X' 'FREQ' 'REL FREQ'

  ROW     X     FREQ    REL FREQ

   1      0     6559    0.6559
   2      1     2926    0.2926
   3      2      489    0.0489
   4      3       23    0.0023
   5      4        3    0.0003

MTB > STOP
```

With 10,000 trials, the relative frequency distribution is a good approximation of the binomial probability distribution that is given in Example 2.

■

■ **Example 4** Cumulative binomial probabilities are given in Table II, Appendix A, of the text. Use Minitab to obtain the binomial probability and cumulative probability distributions for $p = .01$ and $p = .99$, with $n = 10$.

Solution The PDF and CDF commands give the probability and cumulative probability distributions for x values.

```
MTB > PDF;
SUBC> BINOMIAL N = 10 AND P = .01.

     BINOMIAL WITH N =   10   P = 0.010000
         K            P( X = K)
         0              0.9044
         1              0.0914
         2              0.0042
         3              0.0001
         4              0.0000
MTB > PDF;
SUBC> BINOMIAL N = 10 AND P = .99.

     BINOMIAL WITH N =   10   P = 0.990000
         K            P( X = K)
         6              0.0000
         7              0.0001
         8              0.0042
         9              0.0914
        10              0.9044

MTB > CDF;
SUBC> BINOMIAL N = 10 AND P = .01.

     BINOMIAL WITH N =   10   P = 0.010000
         K   P( X LESS OR = K)
         0              0.9044
         1              0.9957
         2              0.9999
         3              1.0000
```

```
MTB > CDF;
SUBC> BINOMIAL N = 10 AND P = .99.

     BINOMIAL WITH N =   10   P = 0.990000
        K   P( X LESS OR = K)
        6            0.0000
        7            0.0001
        8            0.0043
        9            0.0956
       10            1.0000

MTB > STOP
```

Notice that Minitab does not print values of x with probabilities less than .00005 or greater than .99995. You will notice some rounding differences when you compare the Minitab cumulative probability distributions with those given in Table II in Appendix A of the text. ∎

■ **Example 5** Refer to Example 4.10 on page 198 of the text. Suppose 60% of all voters in a large city are in favor of a certain candidate for mayor. A poll of 20 voters was taken to determine x, the number favoring the candidate.

a. Find the mean and standard deviation of x.
b. Find the probability that ten or fewer favor the candidate.
c. Find the probability that more than 12 favor the candidate.
d. Find the probability that 11 favor the candidate.
e. Obtain a table of the binomial distribution and of the cumulative distribution. What is the probability that x falls in the interval $\mu \pm 2\sigma$? Graph the binomial distribution.

Solution

a. The number of voters x who favor the candidate is a binomial random variable. The mean and standard deviation are computed by

$$\mu = np \qquad \sigma = \sqrt{npq}$$

Since $n = 20$ and $p = .6$, $\mu = 12$ and $\sigma = 2.19$.

b. We use CDF to find $P(x \leq 10)$.

```
MTB > CDF X = 10;
SUBC> BINOMIAL N = 20 P = .6.
       K   P( X LESS OR = K)
     10.00          0.2447
```

c. The CDF command is used to find $P(x > 12) = 1 - P(x \leq 12)$.

```
MTB > CDF X = 12;
SUBC> BINOMIAL N = 20 P = .6.
       K   P( X LESS OR = K)
     12.00          0.5841
```

The probability that more than 12 favor this candidate is $1 - .5841 = .4159$.

d. The PDF command gives $P(x = 11)$.

```
MTB > PDF X = 11;
SUBC> BINOMIAL N = 20 P = .6.
       K          P( X = K)
      11.00         0.1597
```

e. PDF calculates binomial probabilities and CDF calculates the cumulative binomial distribution.

```
MTB > NAME C1 'X' C2 'P(X)' C3 'CDF(X)'
MTB > SET 'X'
DATA> 0:20
DATA> END
MTB > PDF 'X' PUT IN 'P(X)';
SUBC> BINOMIAL N = 20 P = .6.
MTB > CDF 'X' PUT IN 'CDF(X)';
SUBC> BINOMIAL N = 20 P = .6.
MTB > PRINT 'X' 'P(X)' 'CDF(X)'

 ROW      X       P(X)       CDF(X)

   1      0     0.000000    0.00000
   2      1     0.000000    0.00000
   3      2     0.000005    0.00001
   4      3     0.000042    0.00005
   5      4     0.000270    0.00032
   6      5     0.001294    0.00161
   7      6     0.004854    0.00647
   8      7     0.014563    0.02103
   9      8     0.035497    0.05653
  10      9     0.070995    0.12752
  11     10     0.117142    0.24466
  12     11     0.159738    0.40440
  13     12     0.179706    0.58411
  14     13     0.165882    0.74999
  15     14     0.124412    0.87440
  16     15     0.074647    0.94905
  17     16     0.034991    0.98404
  18     17     0.012350    0.99639
  19     18     0.003087    0.99948
  20     19     0.000487    0.99996
  21     20     0.000037    1.00000

MTB > NAME C4 'MU-2SD' C5 'MU+2SD'
MTB > LET 'MU-2SD' = 12-2*2.19
MTB > LET 'MU+2SD' = 12+2*2.19
MTB > PRINT 'MU-2SD' 'MU+2SD'

 ROW    MU-2SD    MU+2SD

   1     7.62     16.38
```

```
MTB > PLOT 'P(X)' VS 'X';
SUBC> TITLE 'BINOMIAL PROBABILITY DISTRIBUTION';
SUBC> TITLE 'City Mayor Candidate'.
```

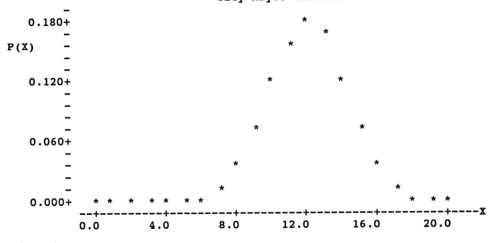

```
MTB > STOP
```

The distribution appears to be mound-shaped. The probability that x falls in the interval, $\mu \pm 2\sigma$, is $P(x = 8, 9, 10, ..., 16) = P(x \leq 16) - P(x \leq 7) = .98404 - .02103 = .96301$. Note that this probability is close to the .95 given by the Empirical Rule, which holds for symmetrical distributions. Rectangles can be drawn on the plot to obtain a graph similar to that given in Figure 4.7 on page 201 of the text.

■

THE POISSON RANDOM VARIABLE

The Poisson probability distribution is useful for describing the number of times an event occurs during a given unit of time, or other measurement. Applications include the number of computer breakdowns per month, the number of typing errors per page, the number of arrivals per ten minutes at a ticket counter, and the number of accidents per year at a production facility.

The Poisson distribution is specified by one parameter λ, the average number of events that occur in a given unit of measurement. The mean and variance of the Poisson distribution both equal λ:

$$\mu = \lambda \qquad \sigma^2 = \lambda$$

Discrete Random Variables 91

> PDF (FOR VALUES IN E) (STORE PROBABILITIES IN E)
> POISSON MU = K
>
> This command calculates the Poisson probability distribution for the specified value of the mean. The mean must be a positive value less than or equal to 50. Values of x with probabilities less than .00005 or greater than .99995 are not printed on the output.
>
> The first option calculates probabilities for specified values of x. These values are stored in a column or, if one value is desired, it may be entered as a constant. The second option stores the probabilities.
>
> CDF (FOR VALUES IN E) (STORE PROBABILITIES IN E)
> POISSON MU = K
>
> This command calculates the cumulative Poisson probabilities for the specified value of the mean. The options are the same as the options for PDF.

■ **Example 6** Example 4.11 on page 207 of the text assumes that the number of blue whale sightings reported per week has a Poisson distribution with a mean of 2.6.

a. Find the mean and standard deviation of x, the number of blue whale sightings reported per week.
b. Find the probability that fewer than two blue whales are reported during a given week.
c. Find the probability that more than five sightings are reported during a given week.
d. Find the probability that exactly five sightings are reported during a given week.
e. Use Minitab to present the probability distribution in both tabular and graphical form.

Solution

a. The mean and variance of this Poisson random variable are both equal to 2.6, the mean number of blue whale sightings.

$$\mu = 2.6 \qquad \sigma^2 = 2.6 \qquad \sigma = 1.61$$

b. The probability that fewer than two sightings are reported is $P(x \leq 1)$.

```
MTB > CDF X = 1;
SUBC> POISSON MU = 2.6.
       K    P( X LESS OR = K)
     1.00              0.2674
```

c. The probability that more than five sightings are reported is $P(x > 5) = 1 - P(x \leq 5)$.

```
MTB > CDF X = 5;
SUBC> POISSON MU = 2.6.
       K    P( X LESS OR = K)
     5.00              0.9510
```

Thus, $P(x > 5) = 1 - .9510 = .0490$.

d. PDF calculates $P(x = 5)$.

```
MTB > PDF X = 5;
SUBC> POISSON MU = 2.6.
        K           P( X = K)
      5.00            0.0735
```

e. In order to obtain the Poisson distribution, we need to specify appropriate x values. We can do this by calculating the interval $\mu \pm 3\sigma$, and taking a range slightly larger than the interval. Using $\mu = 2.6$ and $\sigma = 1.61$, the interval is (-2.23, 7.43). Since x cannot be negative, we use the values 0 to 8 for x.

```
MTB > NAME C1 'X' C2 'P(X)'
MTB > SET 'X'
DATA> 0:8
DATA> END
MTB > PDF 'X' PUT IN 'P(X)';
SUBC> POISSON MU = 2.6.
MTB > PRINT 'X' 'P(X)'

 ROW      X        P(X)

  1       0      0.074274
  2       1      0.193111
  3       2      0.251045
  4       3      0.217572
  5       4      0.141422
  6       5      0.073539
  7       6      0.031867
  8       7      0.011836
  9       8      0.003847

MTB > PLOT 'P(X)' VS 'X';
SUBC> TITLE 'POISSON DISTRIBUTION';
SUBC> TITLE 'Blue Whale Sightings'.
```

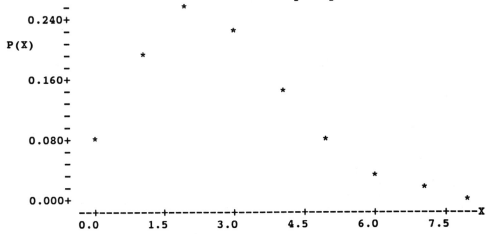

```
MTB > STOP
```

Rectangles can be added to the plot to obtain a graph as shown in Figure 4.8 on page 208 of the text.

■

■ **Example 7** Consider Example 6 of this chapter which describes the blue whale sightings. The number sighted per week has a Poisson distribution with a mean of 2.6. Suppose we randomly select 10,000 weeks, and observe x, the number of blue whale sightings. Obtain the frequency and relative frequency distributions. Compare the relative frequency distribution with the Poisson probability distribution given in Example 6 of this chapter. We would expect there to be 0 sightings in approximately 7% of the weeks, 1 sighting in 19% of the weeks, and so on.

Solution We use RANDOM with the POISSON subcommand to generate 10,000 weeks from a Poisson probability distribution with $\mu = 2.6$.

```
MTB > NAME C1 'WEEKS'
MTB > RANDOM 10000 'WEEKS';
SUBC> POISSON MU = 2.6.
MTB > HISTOGRAM 'WEEKS'

Histogram of WEEKS    N = 10000
Each * represents 50 obs.

Midpoint    Count
       0      719   ***************
       1     1980   ******************************************
       2     2426   ***************************************************
       3     2220   ********************************************
       4     1398   ****************************
       5      765   ****************
       6      319   *******
       7      115   ***
       8       46   *
       9        8   *
      10        4   *

MTB > NAME C2 'X' C3 'FREQ' C4 'REL FREQ'
MTB > SET 'X'
DATA> 0:10
DATA> END
MTB > SET 'FREQ'
DATA> 719 1980 2426 2220 1398 765 319 115 46 8 4
DATA> END
MTB > LET 'REL FREQ' = 'FREQ'/SUM('FREQ')
MTB > PRINT 'X' 'FREQ' 'REL FREQ'

    ROW     X    FREQ   REL FREQ

      1     0     719     0.0719
      2     1    1980     0.1980
      3     2    2426     0.2426
      4     3    2220     0.2220
      5     4    1398     0.1398
      6     5     765     0.0765
      7     6     319     0.0319
```

94 Chapter 4

```
     8     7    115    0.0115
     9     8     46    0.0046
    10     9      8    0.0008
    11    10      4    0.0004

MTB > STOP
```

With 10,000 weeks, the relative frequency distribution is a good approximation of the Poisson probability distribution given in Example 6. ■

Comment There are subcommands for PDF and CDF to calculate other discrete probability distributions. The subcommands are the same as those listed with RANDOM on page 70 in Chapter 3. Use the HELP facility for specific information.

EXERCISES

1. Consider the following hypothetical probability distribution given in Exercise 4.14 on page 184 of the text. Suppose x is the number of football games that Coach Bear Bryant of the University of Alabama would win in the first half of his season.

x	$p(x)$
0	.001
1	.010
2	.060
3	.185
4	.324
5	.302
6	.118

   a. Obtain a graph of the probability distribution.
   b. Find the expected value and standard deviation of x. Interpret.
   c. Obtain the intervals $\mu \pm \sigma$, $\mu \pm 2\sigma$, and $\mu \pm 3\sigma$. Find the probability that x falls in each of the intervals. Compare these probabilities with the Empirical Rule.

2. In his book *100% American*, Daniel Evan Weiss reported that 40% of Americans do not think a college education is important to succeed in the business world. Consider a random sample of ten Americans, and let the random variable x be the number of Americans who do not think a college education is important to succeed in the business world.

   a. Obtain a table and graph of the probability distribution.
   b. Find the expected value and standard deviation of x.
   c. Find the probability that four Americans in the sample of ten believe a college education is not important for business success.
   d. Find the probability that more than five Americans in the sample believe a college education

is not important for business success.

3. A 30% defective rate in gas tanks installed in an automobile manufacturer's 1988 compact model is reported in Exercise 4.40 on page 204 of the text. Let x equal the number of defective gas tanks in a random sample of 15 cars.

   a. Obtain the probability distribution. Calculate the mean and standard deviation of x.
   b. Obtain the intervals $\mu \pm \sigma$, $\mu \pm 2\sigma$, and $\mu \pm 3\sigma$. Find the probability that x falls in each of the intervals. Compare these probabilities with the Empirical Rule.
   c. Graph the probability distribution. Label μ, and the intervals obtained in part b on the graph.
   d. Find the probability of finding more than ten defective tanks.
   e. Consider 1000 samples of 15 cars. Simulate the probability distribution of x. Compare this with the probability distribution of part a.

4. A new drug that has been synthesized to reduce a person's blood pressure is discussed in Exercise 4.47 on page 205 of the text. Twenty randomly selected hypertensive patients receive the new drug. Suppose the blood pressure drops in 18 or more patients.

   a. What is the probability of observing a blood pressure drop in 18 or more out of 20 patients if the new drug is ineffective? Assume that the probability of a blood pressure drop in untreated patients is .5. What do you conclude about the drug?
   b. Assume the probability of a blood pressure drop in untreated patients is .3. Would you be more or less likely to conclude the drug is effective than in part a?

5. Suppose x is a binomial random variable with $n = 10$.

   a. For $p = .01, .1, .5, .8,$ and $.9$, obtain the binomial probability distribution and a graph of each distribution.
   b. How does the shape of the binomial probability distribution change as p increases?
   c. For what values of p are the binomial probability distributions symmetric or nearly symmetric?

6. Suppose x is a binomial random variable with $p = .2$.

   a. Obtain the probability distribution and graph of each distribution for the following values of n: 2, 5, 10, 20, and 40.
   b. What happens to the shape of the binomial probability distribution as n increases?
   c. For $n = 40$, calculate the probability that x is within one standard deviation of the mean. Within two standard deviations of the mean. How do these probabilities compare with the Empirical Rule?

7. Suppose a company samples 200 parts from a large shipment. The company accepts the shipment if there are 10 or fewer defectives in the sample, and rejects it otherwise.

   a. What is the probability of accepting the shipment if the shipment is 3% ($p = .03$) defective?
   b. What is the probability of rejecting the shipment if the shipment is 10% defective?

8. Exercise 4.56 on page 211 of the text describes the Environmental Protection Agency pollution standards. Suppose the mean number of vinyl chloride and similar compounds emissions from a particular factory is 4 parts per million. Assume that x, the parts per million of these compounds in air samples, has a Poisson probability distribution.

   a. Find the mean and standard deviation of x.
   b. The EPA limits a factory's emissions of these compounds to 10 parts per million. What is the probability that an air sample from this particular factory exceeds this limit?
   c. Obtain a table and graph of the probability distribution of x.

9. Statistics from the Internal Revenue Service disclose that the chance of a tax return audit in 1992 was about .009 for taxpayers with income less than $50,000. Let x be the number of tax return audits in a random sample of 100 tax returns.

   a. Calculate the mean and standard deviation of x.
   b. What is the probability that more than three tax returns with income less than $50,000 are audited?
   c. Obtain the intervals $\mu \pm \sigma$, $\mu \pm 2\sigma$, and $\mu \pm 3\sigma$. Find the probability that x falls in each of the intervals.
   d. Find the probability that no one will be audited, assuming the IRS's disclosure is accurate.

10. The *Sourcebook of Criminal Justice Statistics* published by the Bureau of Justice Statistics gives statistics on cigarette use among college students. According to the statistics, 23.2% of college students used cigarettes in 1991. Suppose this accurately represents the percentage of all college students who use cigarettes. Let x be the number who use cigarettes in a random sample of 20 college students.

    a. Obtain the probability distribution of x. Graph the probability distribution.
    b. Calculate the mean and standard deviation of x.
    c. Obtain the intervals $\mu \pm \sigma$, $\mu \pm 2\sigma$, and $\mu \pm 3\sigma$. Find the probability that x falls in each of the intervals. Compare these probabilities with the Empirical Rule.
    d. Find the probability that no one in the sample of 20 students uses cigarettes. That more than five use cigarettes.

11. Consider Exercise 4.59 on page 211. The safety supervisor at a large manufacturing plant believes that the expected number of industrial accidents per month is 3.4. Assume that the number of accidents has an approximate Poisson probability distribution.

    a. Find the probability that there are exactly two accidents occurring next month.
    b. Find the probability that three or more accidents will occur next month.
    c. Use Minitab to present the probability distribution in both tabular and graphical form.

CHAPTER 5
CONTINUOUS RANDOM VARIABLES

Probability distributions associated with continuous random variables are discussed in this chapter. We demonstrate the commands for the uniform continuous, normal and exponential distributions.

NEW COMMANDS

INVCDF   MPLOT

CONTINUOUS PROBABILITY DISTRIBUTIONS

A continuous random variable x can assume any value within an interval. Examples include the length of time for a drug to be effective, the daily percent changes in stock prices, the strength of steel beams, and the starting salaries of business college graduates.

The graph of a continuous probability distribution is a smooth curve called the probability density function, denoted $f(x)$. The function is a rule that determines how the probability is distributed over the range of the possible values of x. To describe a probability distribution, the total area under the smooth curve must equal 1. The probability of a continuous random variable x over an interval is represented by the corresponding area under the curve. Since the area over any point is zero, the probability at a point is zero; thus, $P(z \leq a) = P(z < a)$.

The PDF and CDF commands determine the probability density function values and cumulative probabilities for values of x. We illustrate the subcommand for the uniform, normal, and exponential distribution. Subcommands for some other continuous probability distributions are the same as those listed for RANDOM in Chapter 3 on page 70. Use the HELP facility to obtain specific information.

THE UNIFORM DISTRIBUTION

One of the simplest continuous probability distributions is the uniform distribution. The probability density function value is the same for all values of the random variable between two points, c and d. Examples include the time of a ride at an amusement park and the distance from a refinery to a break in a mile long pipeline.

Chapter 5

The probability density function is given by

$$f(x) = 1/(d - c),$$

for $c \leq x \leq d$. The expected value or mean is

$$\mu = (c + d)/2$$

and the variance is

$$\sigma^2 = (d - c)^2/12.$$

The PDF and CDF commands with the UNIFORM subcommand calculate densities and cumulative probabilities of the continuous uniform distribution.

PDF FOR VALUES IN E (STORE DENSITIES IN E)
    UNIFORM C = K D = K

PDF calculates the probability density function for values of x between the points c and d. The values of x are stored in a column, or if one value is desired, it may be entered as a constant on the command line. The option stores the density function values.

CDF FOR VALUES IN E (STORE PROBABILITIES IN E)
    UNIFORM C = K D = K

CDF calculates the cumulative probability for each specified value of x between c and d. The option stores the probabilities.

■ **Example 1** Consider Example 5.1 on page 223 of the text. A used car dealer sells a car that will have a breakdown in the next six months. The dealer provides a warranty of 45 days on all cars sold. The length of time until the breakdown occurs is a uniform random variable with values between 0 and 6 months.

a. What is the probability density function of x, the length of time before a breakdown? Calculate the mean and standard deviation.
b. Graph the probability distribution of x.
c. A car with a breakdown in less than 45 days or 1.5 months is under warranty. Calculate the probability that a breakdown will occur while the car is under warranty.

Solution The values for c and d are 0 and 6 months.

a. The probability density function is $f(x) = 1/(d - c) = 1/(6 - 0) = 1/6 = .167$.

   The mean $\mu = (c + d)/2 = (0 + 6)/2 = 3$ months.

The variance $\sigma^2 = (d - c)^2/12 = (6 - 0)^2/12 = 3$, and the standard deviation $\sigma = \sqrt{3} = 1.73$ months.

b. To graph the continuous uniform distribution, we evaluate the probability density function for some x values: 0, .5, 1, ..., 6.

```
MTB > NAME C1 'X' C2 'F(X)'
MTB > SET 'X'
DATA> 0:6/.5
DATA> END
MTB > PDF 'X' STORE DENSITIES IN 'F(X)';
SUBC> UNIFORM C = 0 D = 6.
MTB > PLOT 'F(X)' VS 'X';
SUBC> TITLE 'CONTINUOUS UNIFORM DISTRIBUTION';
SUBC> TITLE 'Major Car Breakdown';
SUBC> YINCREMENT .1.
```

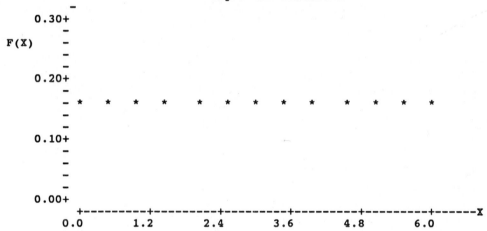

The graph of the continuous uniform distribution as shown in Figure 5.3 on page 223 of the text can be drawn by hand on the Minitab plot. The graph has a rectangular shape with constant height $f(x) = .167$ at all points on the interval, 0 to 6 months.

c. The probability that a breakdown occurs while the car is under warranty is given by $P(x < 1.5)$.

```
MTB > CDF X = 1.5;
SUBC> UNIFORM C = 0 D = 6.
    1.5000      0.2500
MTB > STOP
```

The probability that a breakdown occurs while the car is under warranty is .25.

THE NORMAL DISTRIBUTION

The normal probability distribution is one of the most useful distributions in statistics. This mound-shaped distribution is completely specified by the mean μ and standard deviation σ. Some examples include the monthly rate of return of a stock, the cargo weights on a carrier, the weekly sales of a lumber corporation, and the time required to tune an engine.

The standard normal random variable, typically denoted z, has a mean $\mu = 0$ and a standard deviation $\sigma = 1$. Any normal random variable x can be standardized by calculating the z-score:

$$z = (x - \mu)/\sigma.$$

The z-score gives the number of standard deviations between the observation x and the mean μ.

The PDF and CDF commands with the NORMAL subcommand determine the densities and cumulative probabilities of the normal probability distribution. The PDF command gives the value of $f(x)$ for each specified x. The value of $f(x)$ represents the height of the normal curve at the specified x.

Use CDF to calculate probabilities of events associated with a continuous random variable. For the normal distribution, the CDF command gives the area under the probability density function from the left tail to the value specified for the statistic. This differs from Table IV in Appendix A of the text which gives the area from the center of the normal distribution to the positive value of the z-score. We suggest that you always sketch the normal curve and shade the area corresponding to the probability that you are calculating.

PDF FOR E (STORE DENSITY FUNCTION VALUES IN E)
   NORMAL WITH MU = K AND SIGMA = K

PDF gives the probability density function for each value specified on the command line. These values are stored in a column, or if one value is desired, it may be entered as a constant on the command line. The option stores the density function values.

If you do not give a subcommand, Minitab calculates the probability density function for a standard normal distribution with $\mu = 0$ and $\sigma = 1$.

CDF FOR E (STORE PROBABILITIES IN E)
   NORMAL WITH MU = K AND SIGMA = K

CDF calculates the area under the probability density function from the left tail to each specified value of x. The options are the same as those for PDF.

The default distribution is the standard normal distribution with $\mu = 0$ and $\sigma = 1$.

The INVCDF command is the inverse of CDF. INVCDF determines the value of z associated with a specific cumulative probability.

> INVCDF FOR PROBABILITIES IN E (STORE RESULTS IN E)
> NORMAL WITH MU = K AND SD = K
>
> This command calculates the value a associated with a specified probability p such that $P(x \le a) = p$. If the results are not stored, Minitab prints the probability first, and then the value a.
>
> The standard normal distribution with $\mu = 0$ and $\sigma = 1$ is the default distribution.

■ **Example 2** Graph the standard normal probability distribution.

Solution The default distribution for PDF is the standard normal distribution. According to the Empirical Rule, almost all of the observations fall within three standard deviations of the mean. Using a slightly larger interval, we graph the distribution for some z values: -4.0, -3.5, -3.0, ..., 4.0.

```
MTB > NAME C1 'Z' C2 'F(Z)'
MTB > SET 'Z'
DATA> -4:4/.5
DATA> END
MTB > PDF 'Z' STORE DENSITIES IN 'F(Z)'
MTB > PLOT 'F(Z)' VS 'Z';
SUBC> TITLE 'STANDARD NORMAL DISTRIBUTION'.
```

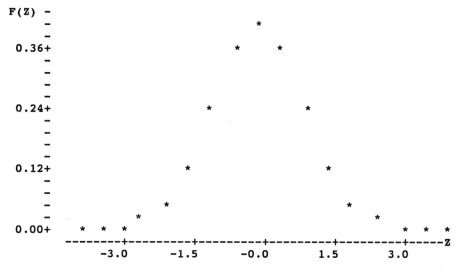

```
MTB > STOP
```

102   *Chapter 5*

A smooth curve can be drawn through the asterisks to obtain the graph of the standard normal distribution as shown in Table 5.1 on page 228 of the text.

■

■ **Example 3**   The text, beginning with Example 5.2 on page 229, gives some problems using the standard normal distribution. Show the corresponding Minitab work. Refer to the graphs in the text which illustrate the areas corresponding to the probabilities.

a. Example 5.2, page 229.   Find the probability that the standard normal random variable falls between -1.33 and 1.33.

Solution   Since CDF gives the area from the left tail of the normal curve to the specified z value, we rewrite $P(-1.33 < z < 1.33) = P(z < 1.33) - P(z \leq -1.33)$. We do not need a subcommand to calculate cumulative probabilities of the standard normal distribution.

```
MTB > CDF 1.33
     1.3300      0.9082
MTB > CDF -1.33
    -1.3300      0.0918
MTB > LET K1 = .9082 - .0918
MTB > PRINT K1
K1         0.816400
```

b. Example 5.3, page 230.   Find the probability that a standard normal random variable exceeds 1.64.

Solution   The probability that z exceeds 1.64 is $P(z > 1.64) = 1 - P(z \leq 1.64)$.

```
MTB > CDF 1.64
     1.6400      0.9495
MTB > LET K2 = 1 - 0.9495
MTB > PRINT K2
K2         0.0505000
```

c. Example 5.4, page 231.   Find the probability that a standard normal random variable lies to the right of a point -.74 standard deviation from the mean.

Solution   The probability that z exceeds -.74 is given by $P(z > -.74) = 1 - P(z \leq -.74)$.

```
MTB > CDF -.74
    -0.7400      0.2296
MTB > LET K3 = 1 - .2296
MTB > PRINT K3
K3         0.770400
```

d. Example 5.5, page 231.   Find the probability that a normally distributed random variable lies beyond 1.96 standard deviations from its mean.

Solution   We need to find the probability that z is less than -1.96 or greater than 1.96: $P(z < -1.96) + P(z > 1.96) = P(z < -1.96) + [1 - P(z \leq 1.96)]$.

```
MTB > CDF -1.96
   -1.9600      0.0250
MTB > CDF 1.96
    1.9600      0.9750
MTB > LET K4 = .0250 - (1 - .9750)
MTB > PRINT K4
K4       0.0500000
```

e. Example 5.8, page 235. Find the value of z_0 that will be exceeded 10% of the time.

Solution In this problem, we are looking for the value of z_0 such that the area to the right of that value is .10. The area to the left of z_0 is .90. INVCDF .90 gives the value of z_0 such that $P(z \leq z_0)$ = .90.

```
MTB > INVCDF .90
    0.9000      1.2816
```

About 10% of the standard normal distribution exceeds $z = 1.28$.

f. Example 5.9, page 236. Find the value of z_0 such that 95% of the standard normal values falls within $-z_0$ and $+z_0$.

Solution We need to find z_0 such that $P(-z_0 \leq z \leq z_0) = .95$. Because the normal distribution is symmetrical, the area below $-z_0$ is .025 and below $+z_0$ is .975.

```
MTB > INVCDF .025
    0.0250     -1.9600
MTB > INVCDF .975
    0.9750      1.9600
MTB > STOP
```

Approximately 95% of the values of a normal random variable fall within 1.96 standard deviations of the mean.

■

■ **Example 4** Refer to Example 5.7 on page 233 of the text. The in-city mileage of a new automobile model is approximately normally distributed with $\mu = 27$ and $\sigma = 3$ miles per gallon.

a. Use Minitab to obtain a graph of the normal distribution.
b. What proportion of this model car averages less than 20 miles per gallon? What should you conclude if you purchased this model and it averaged less than 20 miles per gallon?
c. Ten percent of this model averages less than what mileage?

Solution

a. To approximate the values of x for the PDF command line, we calculate the interval, $\mu \pm 3\sigma$, or (18, 36). According to the Empirical Rule, almost all values of x fall within this interval.

```
MTB > NAME C1 'X' C2 'F(X)'
MTB > SET 'X'
DATA> 18:36
DATA> END
```

104 Chapter 5

```
MTB > PDF 'X' STORE DENSITIES IN 'F(X)';
SUBC> NORMAL MU = 27 AND SIGMA = 3.
MTB > PLOT 'F(X)' VS 'X';
SUBC> TITLE 'NORMAL DISTRIBUTION';
SUBC> TITLE 'In-city Mileage'.
```

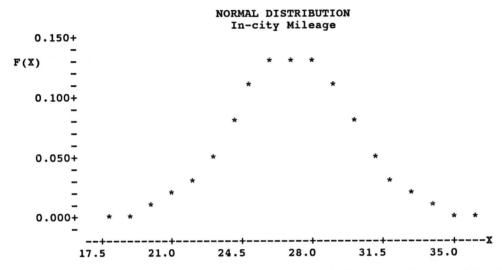

A smooth curve can be drawn through the asterisks to obtain the graph of the normal distribution as shown in Figure 5.13 on page 234 of the text.

b. We use CDF to find $P(x < 20)$.

```
MTB > CDF 20;
SUBC> NORMAL MU = 27 AND SIGMA = 3.
   20.0000    0.0098
```

Less than 1% of all cars of this type of car should average less than 20 miles per gallon. If your car averages less than 20 miles per gallon, you may have been unfortunate to have purchased one of the few cars that averages less than 20 miles per gallon, or the probability distribution of mileage is not normal with $\mu = 27$ and $\sigma = 3$.

c. We use INVCDF to find the mileage x_0 such that $P(x < x_0) = .1$.

```
MTB > INVCDF .1;
SUBC> NORMAL MU = 27 SIGMA = 3.
    0.1000   23.1553
MTB > STOP
```

Ten percent of all cars of this model average less than 23.2 miles per gallon. ■

■ **Example 5** Consider the normal probability distribution for the in-city mileage given in Example 4. Suppose we want to sample 1,000 cars.

a. Simulate the distribution of in-city mileages for the 1,000 cars.

b. How many cars average less than 20 miles per gallon? How does this compare with the theoretical probability calculated in Example 4?

Solution

a. We use RANDOM with the NORMAL subcommand to simulate the normal probability distribution.

```
MTB > NAME C1 'CARS'
MTB > RANDOM 1000 'CARS';
SUBC> NORMAL MU = 27 SIGMA = 3.
MTB > HISTOGRAM 'CARS'

Histogram of CARS    N = 1000
Each * represents 10 obs.

Midpoint     Count
      18         5  *
      20        20  **
      22        62  *******
      24       135  **************
      26       262  ***************************
      28       253  **************************
      30       173  ******************
      32        63  *******
      34        25  ***
      36         2  *
```

The simulated distribution of the sample of 1,000 cars is approximately normal.

b. We use LET to select the cars with mileage less than 20 miles per gallon. If the comparison ('CARS' < 20) given on the command line is true, the result is set in C2 to 1; if the comparison is not true ('CARS' ≥ 20), the result is set to 0. TALLY summarizes the results.

```
MTB > LET C2 = ('CARS' < 20)
MTB > TALLY C2

       C2      COUNT
        0        986
        1         14
       N=       1000
MTB > STOP
```

In our simulated sample, 14 of the 1,000 cars or 1.4% have mileages less than 20 miles per gallon. This simulated result is approximately equal to the exact probability of 1% that was calculated in Example 4. ∎

■ **Example 6** Consider Example 5.10 on page 236 of the text. The scores on a college entrance examination are assumed to be approximately normal with $\mu = 550$ and $\sigma = 100$. Suppose a certain university considers those applicants whose scores exceed the 90th percentile for admission. What is the minimum score an applicant must receive in order to be considered for admission?

106   *Chapter 5*

Solution   We use the INVCDF command to find the 90th percentile of the distribution.

```
MTB > INVCDF .90;
SUBC> NORMAL MU = 550 AND SIGMA = 100.
    0.9000    678.1552
MTB > STOP
```

The 90th percentile of the test score distribution is 678. An applicant must score at least 678 to be considered for admission.

■

THE NORMAL APPROXIMATION OF THE BINOMIAL DISTRIBUTION

The normal distribution is a good approximation of the binomial distribution if n is large. To demonstrate this graphically, we superimpose a plot of the normal distribution onto a plot of the binomial distribution using MPLOT. The MPLOT command constructs two or more plots on the same axes.

MPLOT OF C VS C, C VS C,..., C VS C
  YINCREMENT = K
  YSTART AT K (END AT K)
  XINCREMENT = K
  XSTART AT K (END AT K)
  TITLE = 'TEXT'
  FOOTNOTE = 'TEXT'
  XLABEL = 'TEXT'
  YLABEL = 'TEXT'

MPLOT puts several plots on the same axes. The first column of each pair is put on the vertical axis, and the second column on the horizontal axis. The first pair of columns is plotted with an A, the second with a B, and so on. If two or more points fall at the same location, the number of points is displayed on the plot. If there is no footnote, a legend describing the symbols is printed on the output.

The subcommands control the scales and labels. INCREMENT sets the distances between the tick (+) marks on the plot. START specifies the first points and END, the last points plotted on the axes.

A maximum of three titles and two footnotes may be added to the plot with TITLE and FOOTNOTE. XLABEL and YLABEL name the two axes. Any characters may be used as text. The maximum number of characters printed for TITLE, FOOTNOTE, and XLABEL equals the width of the plot (57 by default). The maximum number of characters for YLABEL equals the height of the plot (17 by default).

Continuous Random Variables

■ **Example 7** Refer to Example 5.11 on page 243 of the text. Suppose that 6% of the circuit boards for a brand of pocket calculator are defective. Let the random variable x be the number of defective circuit boards in a sample of 200 boards. Then x is a binomial random variable with $n = 200$ and $p = .06$.

a. Illustrate the relationship between the binomial and the normal distribution in tabular and graphical form using Minitab.
b. Use the normal approximation to the binomial distribution to approximate $P(x \geq 20)$. Compare it to the exact probability using the binomial distribution.

Solution

a. The mean of the binomial distribution is $\mu = np$ or 12, and the standard deviation is $\sigma = \sqrt{npq}$ or 3.36. The values of x within the interval, $\mu \pm 3\sigma$, range from 2 to 22.

```
MTB > NAME C1 'X' C2 'BINOMIAL' C3 'NORMAL'
MTB > SET 'X'
DATA> 2:22
DATA> END
MTB > PDF 'X' STORE PROBABILITIES IN 'BINOMIAL';
SUBC> BINOMIAL N = 200 P = .06.
MTB > PDF 'X' STORE DENSITIES IN 'NORMAL';
SUBC> NORMAL MU = 12 AND SIGMA = 3.36.
MTB > PRINT 'X' 'BINOMIAL' 'NORMAL'

  ROW     X     BINOMIAL      NORMAL

    1     2     0.000342     0.001416
    2     3     0.001442     0.003285
    3     4     0.004534     0.006976
    4     5     0.011344     0.013555
    5     6     0.023533     0.024107
    6     7     0.041630     0.039239
    7     8     0.064106     0.058455
    8     9     0.087294     0.079701
    9    10     0.106424     0.099457
   10    11     0.117334     0.113589
   11    12     0.117958     0.118733
   12    13     0.108885     0.113589
   13    14     0.092833     0.099457
   14    15     0.073477     0.079701
   15    16     0.054228     0.058455
   16    17     0.037464     0.039239
   17    18     0.024312     0.024107
   18    19     0.014865     0.013555
   19    20     0.008587     0.006976
   20    21     0.004698     0.003285
   21    22     0.002440     0.001416
```

```
MTB > MPLOT 'BINOMIAL' VS 'X' AND 'NORMAL' VS 'X';
SUBC> TITLE 'NORMAL APPROXIMATION TO THE BINOMIAL DISTRIBUTION'.
```

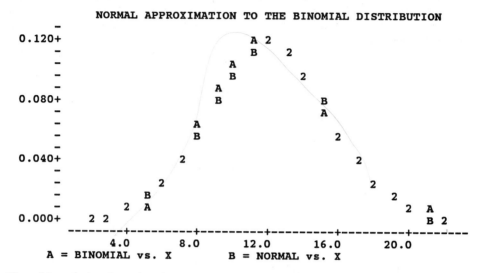

The table and plot show that the normal distribution is a good approximation of the binomial distribution for $n = 200$ and $p = .06$. The number 2 on the plot indicates that the corresponding binomial and normal densities are about the same.

b. For the normal approximation, we subtract .5, a correction for continuity, from 20 and calculate $P(x \geq 19.5)$. We apply the correction because a continuous probability distribution is used to approximate a discrete probability distribution.

```
MTB > # APPROXIMATE PROBABILITY USING NORMAL DISTRIBUTION
MTB > CDF 19.5;
SUBC> NORMAL MU = 12 AND SIGMA = 3.36.
     19.5000      0.9872
MTB > LET K1 = 1 - .9872
MTB > PRINT K1
K1         0.0128000
MTB > # EXACT BINOMIAL PROBABILITY
MTB > CDF 19;
SUBC> BINOMIAL N = 200 AND P = .06.
       K   P( X LESS OR = K)
     19.00            0.9821
MTB > LET K2 = 1 - .9821
MTB > PRINT K2
K2         0.0179000
MTB > STOP
```

The probability using the normal approximation of .0128 is close to the exact binomial probability of 0.0179. Although p is only .06, the normal distribution is a good approximation because of the large n.

THE EXPONENTIAL DISTRIBUTION

The exponential distribution is useful for describing the length of time between arrivals or other events. For example, the length of time between arrivals of an airport shuttle bus and the length of time between breakdowns of a mainframe computer may be approximated by the exponential distribution.

The exponential distribution is specified by one parameter θ, the average length of time between occurrences of random events. The probability density function is

$$f(x) = 1/\theta \, e^{-x/\theta}$$

for $x > 0$ and $\theta > 0$. The mean and standard deviation are given by

$$\mu = \theta \text{ and } \sigma = \theta$$

PDF and CDF calculate densities and cumulative probabilities for the exponential probability distribution.

PDF FOR E (STORE DENSITY FUNCTION VALUES IN E)
EXPONENTIAL WITH MU = K

This command calculates the exponential probability distribution density function for the specified mean. The values of x may be stored in a column, or if one value is desired, it may be entered as a constant on the command line. The option stores the density function values.

CDF FOR E (STORE PROBABILITIES IN E)
EXPONENTIAL MU = K

The cumulative probability is given for each specified value of x. The option stores the probabilities.

■ **Example 8** Refer to Example 5.12 on page 251 of the text concerning the length of time between emergency arrivals at a certain hospital. Suppose the length of time is approximately an exponential random variable with $\theta = 2$ hours.

a. Graph the probability distribution. Find the mean and standard deviation.
b. What is the probability that more than five hours go by without an emergency arrival?
c. Simulate the lengths of time between 1,000 emergency arrivals. Obtain a histogram of the distribution. Estimate the probability that more than five hours go by without an emergency.

Solution

a. The mean and standard deviation of an exponential random variable are both equal to $\theta = 2$. To approximate the values of x for the PDF command line, we calculate the interval, $\mu \pm 3\sigma$, or (-4, 8), and use a slightly larger interval. Since x must be greater than 0, we use values of x from 0 to 10.

```
MTB > NAME C1 'X' C2 'F(X)'
MTB > SET 'X'
DATA> 0:10
DATA> END
MTB > PDF 'X' STORE DENSITIES IN 'F(X)';
SUBC> EXPONENTIAL MU = 2.
MTB > PLOT 'F(X)' VS 'X';
SUBC> TITLE 'EXPONENTIAL DISTRIBUTION';
SUBC> TITLE 'Days between Automobile Sales'.
```

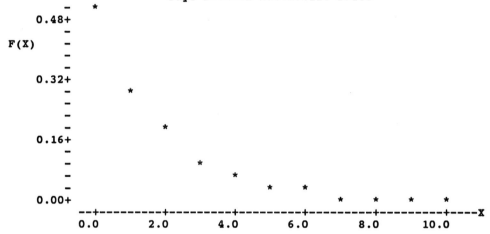

The graph is similar to that given in Figure 5.24 on page 251 of the text. The graph crosses the $f(x)$ axis at $1/\theta = .5$.

b. We use CDF to find $P(x > 5) = 1 - P(x \leq 5)$.

```
MTB > CDF X = 5;
SUBC> EXPONENTIAL MU = 2.
    5.0000    0.9179
MTB > LET K1 = 1 - .9179
MTB > PRINT K1
K1        0.0821000
```

The probability of more than five hours between emergency arrivals is .082.

c. RANDOM with the EXPONENTIAL subcommand generates observations from an exponential distribution.

```
MTB > NAME C3 'DAYS'
MTB > RANDOM 1000 'DAYS';
SUBC> EXPONENTIAL MU = 2.
MTB > HISTOGRAM 'DAYS';
SUBC> START .5;
SUBC> INCREMENT 1.

Histogram of DAYS   N = 1000
Each * represents 10 obs.

Midpoint    Count
    0.50     384    ****************************************
    1.50     244    *************************
    2.50     160    ****************
    3.50      89    *********
    4.50      51    ******
    5.50      23    ***
    6.50      21    ***
    7.50      15    **
    8.50       3    *
    9.50       4    *
   10.50       4    *
   11.50       0
   12.50       2    *

MTB > DESCRIBE 'DAYS'

                N       MEAN     MEDIAN     TRMEAN      STDEV     SEMEAN
DAYS         1000     1.9803     1.4475     1.7692     1.9063     0.0603

              MIN        MAX         Q1         Q3
DAYS       0.0054    12.8427     0.6065     2.7715

MTB > LET C2 = ('DAYS' > 5)
MTB > TALLY C2

      C2    COUNT
       0      928
       1       72
      N=     1000

MTB > STOP
```

The shape of the histogram approximates the graph of the exponential probability distribution given in part a. The mean and standard deviation are slightly less than $\mu = \sigma = 2$. In our simulated sample, about 7% of the time there is more than 5 hours between emergency arrivals. ∎

■ **Example 9** Waiting line or queuing theory is concerned with modeling the characteristics of a service system. Both the time between arrivals of customers (interarrival time) and the time to service a customer (service time) are reasonably approximated by an exponential distribution. Suppose the length of time between the arrival of customers at a hair stylist salon can be described by an exponential distribution, with $\theta = 6$ minutes between customers.

a. Use Minitab to graph the exponential distribution.
b. Find the probability that the interarrival time of two customers is greater than 10 minutes.
c. Find the probability that the interarrival time is within two standard deviations of the mean.

112 Chapter 5

Solution The mean interarrival time is $\mu = 6$ minutes, and $\sigma = 6$ minutes.

a. We use the interval $\mu \pm 3\sigma$ or (-12, 24) to approximate the range of x values for PDF. Since interarrival time must be greater than 0, we use the range (0, 24).

```
MTB > NAME C1 'X' C2 'F(X)'
MTB > SET 'X'
DATA> 0:24
DATA> END
MTB > PDF 'X' STORE THE DENSITIES IN 'F(X)';
SUBC> EXPONENTIAL MU = 6.
MTB > PLOT 'F(X)' VS 'X';
SUBC> TITLE 'EXPONENTIAL DISTRIBUTION'.
```

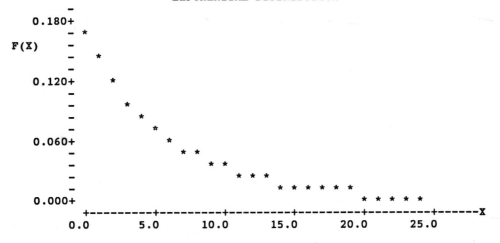

A smooth curve can be drawn through the asterisks to obtain the graph of the exponential distribution. The curve meets the $f(x)$ axis at the point $f(x) = .167$ customers per minute.

b. To find the probability that the interarrival time between two customers is greater than 10 minutes, we find $P(x > 10) = 1 - P(x \leq 10)$.

```
MTB > CDF X = 10;
SUBC> EXPONENTIAL MU = 6.
    10.0000    0.8111
MTB > LET K1 = 1 - .8111
MTB > PRINT K1
  K1       .1889
```

The probability that the interarrival time of two customers is greater than 10 minutes is about .19.

c. The probability that the interarrival time is within two standard deviations of the mean is given by $P(\mu - 2\sigma \leq x \leq \mu + 2\sigma) = P(-6 \leq x \leq 18)$. Since interarrival time must be greater than 0, we find $P(0 < x \leq 18) = P(x \leq 18)$.

```
MTB > CDF X = 18;
SUBC> EXPONENTIAL MU = 6.
```

```
          18.0000       0.9502
      MTB > STOP
```

The probability that the interarrival time is between $\mu \pm 2\sigma$ is .9502.

■

EXERCISES

1. Compute and plot the probability density function for a uniform random variable x with $c = 100$ and $d = 130$. Find the mean and standard deviation. What is the probability that x is within two standard deviations of the mean?

2. Refer to Exercise 5.9 on page 225 of the text. The bottling company manager believes that a soft drink machine is equally likely to dispense any amount between 6.5 and 7.5 ounces.

    a. What is the probability density function of x, the amount of soft drink that is dispensed for the machine? Calculate the mean and standard deviation.
    b. Graph the probability distribution of x.
    c. Suppose that amounts less than 6.75 ounces have to be refilled. Calculate the fraction that must be refilled.

3. Compute and plot the probability density function for a normal random variable x with $\mu = 100$ and $\sigma = 10$. Use values of x from three standard deviations below the mean to three above the mean.

4. The Cloverhills Dairy uses a filling machine to fill one quart (32 ounces) bottles of milk. The net weight of milk placed in the bottles is normally distributed with a standard deviation of 0.5 ounces. The average amount of milk placed in the bottles is controlled by adjusting a setting on the filling machine.

    a. Assume the filling machine is adjusted to fill bottles with an average of 32.2 ounces. What proportion of the bottles are filled with less than 32 ounces?
    b. Companies must meet certain standards. Assume Cloverhills Dairy is required to fill no more than 1% of the bottles with less than 32 ounces. At what level should the machine be set so the requirement is met with the least amount of average fill?

5. Refer to Exercise 5.35 on page 240 of the text. The pulse rate per minute of United States adult males between 18 and 25 years of age is known to be normally distributed with $\mu = 72$ beats per minute and $\sigma = 9.7$.

    a. Use Minitab to graph the normal distribution curve.
    b. What proportion of the adult males have a pulse rate less than 60 beats per minute? Shade the corresponding area under the normal curve obtained in part a.
    c. If the military service requires that anyone with a pulse rate over 100 is medically unsuitable for service, what proportion of adult males between 18 and 25 years old would be declared unsuitable because their pulse rates are too high?

114   Chapter 5

d. Find the pulse rate such that 80% of the adult males exceed this rate?

6. Suppose that x is a binomial random variable with $n = 100$ and $p = 0.1$. Compare the binomial distribution and normal approximation in tabular and graphical form. Discuss the results.

7. Suppose that a company samples 200 parts from a large shipment. The company accepts the shipment if there are 10 or fewer defective in the sample, and rejects it otherwise.

   a. Use the normal approximation of the binomial distribution to estimate the probability of accepting the shipment if the shipment is 3% ($p = .03$) defective? Compare this with the exact probability using the binomial distribution.
   b. What is the approximate probability of rejecting the shipment if the shipment is 10% defective? Compare this with the exact probability using the binomial distribution.

8. Market research reported in Exercise 5.55 on page 248 of the text reveals that an estimated 30% of United States households own one or more personal computers. Suppose that in a sample of 200 households in a high-tech community, 72 own personal computers.

   a. What are the mean and standard deviation of x, the number of households in a sample of 200 that own personal computers.
   b. Graphically compare the binomial distribution and normal approximation. Use values of x from about three standard deviations below the mean to three above the mean.
   c. Use the normal approximation to the binomial distribution to estimate the probability that 72 or more in the sample of 200 households would own personal computers if the research is correct. What would you conclude about this community?

9. The following data are the recorded minutes between the arrivals of visitors at the state park.

   Park Visits

.2	.4	.5	.9	1.2	1.9	3.2	1.2	.2
8.0	10.5	7.1	13.5	.5	.5	5.5	2.2	.7
9.0	1.4	5.7	7.2	2.2	3.2	.5	5.0	.5
.4	6.4	.1	3.8	18.2	5.8	.4	.3	.7
13.2	.3	14.9	5.3	15.5	2.1	1.9	8.1	1.0
7.2	2.4	.3	.6	1.3	8.3	1.9	3.1	.3
.7	6.6	4.4	10.8	4.3	3.5	1.0	11.4	1.1

   a. Graph the data. Does the exponential distribution reasonably describe the data?
   b. If the exponential distribution describes the data, estimate μ and σ.

10. Refer to Exercise 5.70 on page 255 of the text. Suppose the length of time that an individual has to wait in line to be served at a fast food franchise is exponentially distributed with $\theta = 1$ minute.

    a. Graph the exponential distribution.
    b. Find the probability that a customer will wait more than 2 minutes before being served.
    c. What is the probability that an individual will be served within 30 seconds?
    d. What is the probability that the time to be served is within $\mu \pm \sigma$? Within $\mu \pm 2\sigma$?
    e. Find the length of time x such that 90% of all customers wait longer than x.

CHAPTER 6
SAMPLING DISTRIBUTIONS

A primary objective of statistics is to make inferences about population parameters based on sample data. Sample statistics, such as the mean and standard deviation, are used to estimate population parameters. Since a sample statistic is computed from random variables, it is random, and has a probability distribution called a sampling distribution. This chapter illustrates some sampling techniques and the sampling distribution of the sample mean.

NEW COMMANDS

ECHO   EXECUTE   NOECHO   SAMPLE   STORE

STATISTICAL SAMPLING TECHNIQUES

Oftentimes, decisions must be made on the basis of a subset of the population of data, called a sample. Time and cost constraints, and the unavailability of data, preclude a study of the entire population. The statistical technique of selecting a sample of data from the population is called sampling.

The most fundamental sampling technique is simple random sampling, which is a method of selecting a sample from a population such that every sample of size n has an equal chance of being selected. We can use a random number generator in Minitab to select a simple random sample from a finite or infinite population.

Sampling from a Finite Population

A finite population consists of a countable number of observations. Sometimes we are interested in randomly selecting a sample from a finite population. For example, suppose we have collected daily sales data over a period of several years, and want a preliminary study of a subset of the data. Or we may want to randomly select a sample of starting salaries of 1993 college graduates. The SAMPLE command generates a simple random sample with or without replacement from a finite population.

116   *Chapter 6*

> SAMPLE K ROWS FROM C,...,C, PUT IN C,...,C
> REPLACE
>
> This command randomly selects a sample of size K without replacement from one or more finite populations. If several columns are specified, the same rows are selected from each column. The REPLACE subcommand samples with replacement.

■ **Example 1**   Appendix B of this supplement contains a data set compiled by the Minnesota Real Estate Research Center on homes sold in St. Cloud, Minnesota in 1988. Consider the selling prices of homes. Select three random samples of size five from the 197 selling prices. Calculate the mean of each sample.

Solution   The selling prices of 197 homes are saved in the file HOMES. We use SAMPLE to select and store the three random sample, and LET to calculate the mean of each sample. The number in parentheses following 'MEANS' is the row number corresponding to the number of the sample. For example, the mean of the first sample is entered in MEANS(1).

```
MTB > RETRIEVE 'HOMES'
 WORKSHEET SAVED   1/10/1994

Worksheet retrieved from file: HOMES.MTW
MTB > INFORMATION

COLUMN      NAME         COUNT     MISSING
C1          AREA          197
C2          BEDROOMS      197
C3          LIST PR       197
C4          SOLD PR       197
C5          FINANCE       197
C6          DAYS          197          2
C7          MTH SOLD      197
C8          DAY SOLD      197

CONSTANTS USED: NONE

MTB > NAME C9 'SAMPLE' C10 'MEANS'
MTB > SAMPLE 5 OBS FROM 'SOLD PR' PUT IN 'SAMPLE'
MTB > PRINT 'SAMPLE'

SAMPLE
   45000     68900     67200     38000     41000

MTB > LET 'MEANS'(1) = MEAN('SAMPLE')    # FIRST SAMPLE MEAN
MTB > SAMPLE 5 OBS FROM 'SOLD PR' PUT IN 'SAMPLE'
MTB > PRINT 'SAMPLE'

SAMPLE
   69200    116000     88900    116000     37000

MTB > LET 'MEANS'(2) = MEAN('SAMPLE')    # SECOND SAMPLE MEAN
MTB > SAMPLE 5 OBS FROM 'SOLD PR' PUT IN 'SAMPLE'
MTB > PRINT 'SAMPLE'
```

```
    SAMPLE
       56000    88000    56500    61000    92000

MTB > LET 'MEANS'(3) = MEAN('SAMPLE')   # THIRD SAMPLE MEAN
MTB > PRINT 'MEANS'

    MEANS
       52020    85420    70700

MTB > STOP
```

The sample means are estimates of the mean selling price of homes in St. Cloud, Minnesota in 1988. Notice the sample mean varies from sample to sample. The second sample mean is high because of the unusually high selling prices in the sample.

∎

The sampling process, as illustrated in the previous example, is repetitive. You can reduce the time and the amount of typing by using STORE and EXECUTE to repeat a set of Minitab commands. STORE creates a command file called a macro.

STORE THE FOLLOWING COMMANDS ('FILENAME')

Minitab responds to this command with a STOR> prompt. Following the prompt, enter the set of commands that you want to repeat. Use END to indicate the end of the set of stored commands. The option stores the commands in a file. The default file extension is MTB. If the option is not used, the default name is MINITAB.MTB.

EXECUTE THE STORED COMMANDS ('FILENAME') (K TIMES)

This command repeats the stored commands K number of times. If the number K is not specified, the set of commands is executed once.

When the STORE and EXECUTE commands are used in Release 8, the set of commands is printed on the output each time the set is executed. NOECHO prints only the results; the printing of the commands resumes with ECHO. In Release 9, NOECHO is the default for executing a macro. You need to enter an ECHO to print the commands.

NOECHO THE COMMANDS THAT FOLLOW

Following NOECHO, Minitab prints only the output of the commands.

ECHO THE COMMANDS THAT FOLLOW

Following ECHO, Minitab resumes printing the commands with the output.

Chapter 6

Comment   We recommend that you ECHO all commands until you are familiar with the STORE and EXECUTE commands. If the commands are printed each time they are executed, you can directly observe the execution of the stored commands. Usually this makes it easier to locate errors.

■ **Example 2**   Use STORE and EXECUTE to repeat the sampling process that is described in Example 1 on page 116 of this chapter. Select three random samples of size five from the 1988 selling prices of homes in St. Cloud, Minnesota. Calculate the means. Repeat the sampling a second time using NOECHO.

Solution   We use a constant K1 as a pointer to specify the sample number and the row of MEANS in which we enter the corresponding sample mean   K1 is initially set at 1, and then increased by 1 each time the stored commands are executed.

```
MTB > RETRIEVE 'HOMES'
 WORKSHEET SAVED   1/10/1994

Worksheet retrieved from file: HOMES.MTW
MTB > NAME C9 'SAMPLE' C10 'MEANS'
MTB > LET K1=1
MTB > STORE 'SAMPLING'
STOR> SAMPLE 5 OBS FROM 'SOLD PR' PUT IN 'SAMPLE'
STOR> PRINT K1 'SAMPLE'
STOR> LET 'MEANS'(K1) = MEAN('SAMPLE')
STOR> LET K1=K1+1
STOR> END
MTB > ECHO
MTB > EXECUTE 'SAMPLING' 3 TIMES   # THE STORED COMMANDS ARE NOW EXECUTED
MTB > SAMPLE 5 OBS FROM 'SOLD PR' PUT IN 'SAMPLE'
MTB > PRINT K1 'SAMPLE'
K1        1.00000

SAMPLE
    48300      64475     41900     34900     50000

MTB > LET 'MEANS'(K1) = MEAN('SAMPLE')
MTB > LET K1=K1+1
MTB > END
MTB > SAMPLE 5 OBS FROM 'SOLD PR' PUT IN 'SAMPLE'
MTB > PRINT K1 'SAMPLE'
K1        2.00000

SAMPLE
    79900      55500     46900     69000     59900

MTB > LET 'MEANS'(K1) = MEAN('SAMPLE')
MTB > LET K1=K1+1
MTB > END
MTB > SAMPLE 5 OBS FROM 'SOLD PR' PUT IN 'SAMPLE'
MTB > PRINT K1 'SAMPLE'
K1        3.00000

SAMPLE
    85500      26000     70000     42500     47900

MTB > LET 'MEANS'(K1) = MEAN('SAMPLE')
MTB > LET K1=K1+1
MTB > END
```

```
MTB > PRINT 'MEANS'
MEANS
    47915    62240    54380
```

The stored commands and results are printed each time they are executed. The NOECHO command suppresses the printing of the commands. To resume all printing, we enter an ECHO command. The entire program is repeated below.

```
MTB > LET K1=1
MTB > STORE 'SAMPLING'
STOR> SAMPLE 5 OBS FROM 'SOLD PR' PUT IN 'SAMPLE'
STOR> PRINT K1 'SAMPLE'
STOR> LET 'MEANS'(K1) = MEAN('SAMPLE')
STOR> LET K1=K1+1
STOR> END
MTB > NOECHO
MTB > EXECUTE 'SAMPLING' 3 TIMES

K1        1.00000

SAMPLE
   69000    17900    41000    70000    49900

K1        2.00000

SAMPLE
   63000    37800    55500    53900    47500

K1        3.00000

SAMPLE
   16000    47900    47500    72000    45800

MTB > ECHO
MTB > PRINT 'MEANS'

MEANS
    49560    51540    45840

MTB > STOP
```

The NOECHO command prints only the results, saving time and storage space. Notice some variability in the sample means produced by the three sampling processes.

■

Sampling from an Infinite Population

An infinite population consists of an undefined number of observations. If a production process continues indefinitely under the same conditions, the process can be considered an infinite population. For example, the manufacturing of computer chips is an infinite population.

The RANDOM command described in Chapter 3 generates random samples of n observations from an infinite population with a specified probability distribution. The subcommand specifies the distribution.

120   Chapter 6

■ **Example 3** A Growth Share Account, an aggressive investment account in a retirement plan, focuses on stocks. The account's monthly mean rate of return for the five year period ending June, 1993, was .016 with standard deviation .0550. Assume the distribution of monthly rates is normal.

a. Simulate the monthly rates of return for a five year time period. Describe the sample of monthly rates.
b. Find the probability that x, the monthly rate of return, is negative.
c. Refer to the sample generated in part a. Find the number of months in which the rate of return was negative. How does this compare with the probability found in part b?

Solution

a. We use RANDOM with the NORMAL subcommand to generate the 60 monthly rates of return.

```
MTB > NAME C1 'RATES'
MTB > RANDOM 60 MONTHS IN 'RATES';
SUBC> NORMAL MU = .016 SIGMA = .055.
MTB > PRINT 'RATES'

RATES
   0.052414  -0.061694   0.018518   0.000527   0.022996   0.030647  -0.043021
  -0.005917   0.113645   0.002246  -0.089594   0.016101   0.029568   0.026611
  -0.196774  -0.009025   0.016639   0.084604   0.009441  -0.032295   0.011439
   0.067244   0.072741   0.091710  -0.077976   0.024079  -0.044139   0.014710
  -0.041893   0.000383   0.089314   0.004443   0.015004  -0.004938  -0.016876
   0.041042  -0.020224  -0.007768   0.012669   0.056126  -0.044992   0.014218
   0.053284   0.047870  -0.039461   0.053560   0.096867   0.039201   0.094779
   0.044412   0.023712   0.071429   0.107423  -0.036141  -0.017341   0.055469
   0.043547   0.055912  -0.015529  -0.039989

MTB > HISTOGRAM 'RATES'

Histogram of RATES    N = 60

Midpoint    Count
   -0.20      1    *
   -0.16      0
   -0.12      0
   -0.08      3    ***
   -0.04      9    *********
    0.00     20    ********************
    0.04     17    *****************
    0.08      8    ********
    0.12      2    **

MTB > DESCRIBE 'RATES'

                N      MEAN    MEDIAN    TRMEAN     STDEV    SEMEAN
RATES          60   0.01468   0.01555   0.01717   0.05412   0.00699

              MIN       MAX        Q1        Q3
RATES    -0.19677   0.11365  -0.01654   0.05307
```

The histogram shows that the simulated distribution of 60 monthly rates is approximately normal with an extremely low rate for one month. The sample mean, $\bar{x} = .015$, is close to $\mu = .016$.

b. We use CDF to find $P(x < 0)$.

```
MTB > CDF X < 0;
SUBC> NORMAL MU = .016 SIGMA = .0550.
    0.0000   0.3856
```

The probability that the monthly rate of return is negative is about .39.

c. To count the number of months that the simulated rates are negative, we CODE the negative rates -1, and the positive rates 1.

```
MTB > NAME C2 'CODE'
MTB > CODE (-1:0)-1 AND (0:1)1 'RATES' PUT IN 'CODE'
MTB > TALLY 'CODE';
SUBC> COUNT;
SUBC> PERCENT.

    CODE   COUNT  PERCENT
     -1     20     33.33
      1     40     66.67
     N=     60

MTB > STOP
```

In 20 out of 60 months, or about 33% of the time, the simulated account has a negative rate of return. This is somewhat lower than the probability of .39 found in part b.   ∎

THE SAMPLING DISTRIBUTION OF THE SAMPLE MEAN

A sample statistic is used to estimate a population parameter. For example, the sample mean $\bar{x}$ is a point estimator of μ. Since $\bar{x}$ is computed from random variables, it is random, and has a probability distribution called the sampling distribution of $\bar{x}$.

The exact sampling distribution of $\bar{x}$ is the probability distribution of means of all possible samples of n observations from a population. If the number of possible samples is small, you can list all the samples and calculate the sample means. For example, consider a population of five stocks purchased in equal amounts by an investor. Define the population mean μ as the mean rate of return of the stocks. If we consider samples of $n = 3$ stocks, there are 10 possible samples (the combinations of 5 stocks taken 3 at a time). The 10 samples can be listed and the sample mean $\bar{x}$ calculated for each sample.

Most of the time, however, the possible number of samples is too large to list. For example, if the investor had purchased 10 stocks and is interested in samples of $n = 5$ stocks, there would be 252 possible samples.

When the number of possible samples is large, the sampling distribution of $\bar{x}$ can be approximated by generating many simple random samples of size n and computing the mean of each sample. The probability distribution of the sample means is the sampling distribution of $\bar{x}$, with mean, $\mu_{\bar{x}} = \mu$, and standard deviation, $\sigma_{\bar{x}} = \sigma/\sqrt{n}$.

Chapter 6

■ **Example 4** Find the approximate sampling distribution of the mean of a random sample of $n = 5$ observations randomly selected from the population of 1988 selling prices of homes in St. Cloud, Minnesota given in Appendix B of this supplement.

Solution The 197 selling prices are in the file HOMES. We use the SAMPLE command to generate 100 random samples, and STORE and EXECUTE to repeat the sampling. The constant K1 specifies the sample number and the row of the column MEANS used to store the sample mean.

```
MTB > RETRIEVE 'HOMES'
 WORKSHEET SAVED   1/10/1994

Worksheet retrieved from file: HOMES.MTW
MTB > INFORMATION

COLUMN    NAME         COUNT    MISSING
C1        AREA         197
C2        BEDROOMS     197
C3        LIST PR      197
C4        SOLD PR      197
C5        FINANCE      197
C6        DAYS         197         2
C7        MTH SOLD     197
C8        DAY SOLD     197

CONSTANTS USED: NONE

MTB > NAME C9 'SAMPLE' C10 'MEANS'
MTB > LET K1=1
MTB > STORE 'SAMPLING'
STOR> SAMPLE 5 OBS FROM 'SOLD PR' PUT IN 'SAMPLE'
STOR> LET 'MEANS'(K1) = MEAN('SAMPLE')
STOR> LET K1=K1+1
STOR> END
MTB > NOECHO
MTB > EXECUTE 'SAMPLING' 100 TIMES
MTB > ECHO
MTB > PRINT 'MEANS'

MEANS
    60580    72197    56145    68607    94240    66400    68210    50940
    75700    55980    52980    46285    66380    73080    58420    65175
    65820    57980   100280    81580    54960    42440    50980    81537
    70780    68720    81680    74080    61600    48655    58255    68060
    85197    61755    56460    61630    63790    71500    70760    59620
    55500    61260    51990    61405    74580    84780    58300    52540
    56590    58900    69720    53180    83200    50970    64000    55280
    50795    54660    63980    66590    69000    73560    65195    51930
    57700    39680    74640    44360    52380    75690    54580    56400
    47315    82900    62170    39040    55020    61200    73957    44597
    58280    56150    50120    80640    72980    47780    78640    62460
    60180    56760    57350    58200    53840    54960    78535    90680
    59680    60560    76580    49600
```

Sampling Distributions 123

```
MTB > DOTPLOT 'SOLD PR' 'MEANS';
SUBC> SAME.
```

```
                    :   :
                    :   :
                    :   :
                  : :   :
                 :. :::::   :
                 ::.:::::   :
                .:::::::: :   :
               ::::::::::::. .::
             .:::::::::::::::::.        :
            ::..:::::::::::::::::: . :: :      . ..          . .
         -------+---------+---------+---------+---------+---------SOLD PR

                      :
                      :.
                     .::
                    ::::
                    :::::
                   ::::::.:
                   .::::::.
                  .::::::::::
                  :.:::::::::::  .. .
         -------+---------+---------+---------+---------+---------MEANS
             35000     70000    105000    140000    175000    210000
```

The first graph is the population distribution of selling prices. The next is the simulated sampling distribution of $\bar{x}$. The population of selling prices is more variable than the sample means. DESCRIBE gives more information.

```
MTB > DESCRIBE 'SOLD PR' 'MEANS'

              N      MEAN    MEDIAN    TRMEAN     STDEV    SEMEAN
SOLD PR     197     61534     57500     59231     26736      1905
MEANS       100     63029     60890     62641     12100      1210

            MIN       MAX        Q1        Q3
SOLD PR   16000    191000     45400     70400
MEANS     39040    100280     54960     71320
MTB > STOP
```

The mean, $63,029, of the approximate sampling distribution of $\bar{x}$ is close to the population mean μ = $61,534. The standard deviation $s_{\bar{x}}$ = $12,100 is an estimate of $\sigma_{\bar{x}}$. ∎

■ **Example 5** Use Minitab to repeat the sampling described in Example 6.2 on page 268 of the text. Generate 1,000 random samples of 11 measurements from a continuous uniform distribution, with lower limit 0 and upper limit 1. Calculate the sample mean $\bar{x}$ and the sample median m for each sample. Obtain graphs of the sampling distributions. Discuss.

Solution We use RANDOM with the UNIFORM subcommand described in Chapter 3 to generate samples with n = 11 measurements. We enter the samples across 1,000 rows rather than columns for two reasons. The first is that the number of arguments allowed on a command line is limited with some Minitab systems. For example, you cannot generate 50 observations in 1,000 columns with Minitab on a microcomputer. A second reason is the use of the row commands. For instance, RMEAN and RMEDIAN compute means and medians across rows and store the results in columns.

124 *Chapter 6*

We can then construct dot plots or other graphs of the means and medians.

> **Comment** Because of the large number of data points, you cannot replicate this experiment if you are using the student version of Minitab or a microcomputer with limited memory.

```
MTB > RANDOM 1000 MEASUREMENTS IN C1-C11;
SUBC> UNIFORM C = 0 D = 1.
MTB > NAME C12 'MEANS' C13 'MEDIANS'
MTB > RMEANS C1-C11 PUT SAMPLE MEANS IN 'MEANS'
MTB > DESCRIBE 'MEANS'

                 N      MEAN    MEDIAN    TRMEAN     STDEV    SEMEAN
MEANS         1000   0.49950   0.50020   0.49956   0.08686   0.00275

               MIN       MAX        Q1        Q3
MEANS      0.18609   0.82155   0.44055   0.55922

MTB > RMEDIAN C1-C11 PUT SAMPLE MEDIANS IN 'MEDIANS'
MTB > DESCRIBE 'MEDIANS'

                 N      MEAN    MEDIAN    TRMEAN     STDEV    SEMEAN
MEDIANS       1000   0.49895   0.49628   0.49875   0.13644   0.00431

               MIN       MAX        Q1        Q3
MEDIANS    0.14131   0.93488   0.40029   0.59273

MTB > DOTPLOT 'MEANS' 'MEDIANS';
SUBC> SAME.

Each dot represents 5 points
                              :  ..
                           :.:!..
                         ..:::::::
                       :::::::::::.. .
                    .:::::::::::::::.
                    ::::::::::::::::::.
                  :::::::::::::::::::::::
            . . ..:.::::::::::::::::::::..:...   .
      -+---------+---------+---------+---------+---------+-----MEANS

Each dot represents 3 points
                              .
                            :  ..
                     ::    ::  :::
                  :::::::.:::::::.
                 ::::::::::::::::::::.
              .:  :::::::::::::::::::  .:.:
         . . :::::::::::::::::::::::::::::::...
         :.::::::::::::::::::::::::::::::::::..
       ..:.:::::::::::::::::::::::::::::::::::..:.  . .
      -+---------+---------+---------+---------+---------+-----MEDIANS
     0.15      0.30      0.45      0.60      0.75      0.90

MTB > STOP
```

DESCRIBE gives the mean of the sample means, 0.49950, and the mean of the sample medians, 0.49895. The respective standard deviations are 0.08686 and 0.13644. The dot plots show the sampling distributions of $\bar{x}$ and m are approximately normal, and that the sampling distribution of $\bar{x}$ is less variable than that of m. For these samples from the continuous uniform distribution, we conclude that the sample mean $\bar{x}$ contains more information about μ than the sample median m.

■

The Central Limit Theorem

Suppose random samples are selected from a population distribution. The Central Limit Theorem states that the sampling distribution of $\bar{x}$ is approximately normal if n is sufficiently large. The normal approximation is better for larger sample sizes. If a random sample is selected from a normal population distribution, the sampling distribution of $\bar{x}$ is a normal distribution regardless of sample size.

■ **Example 6** Consider the population distributions listed below in parts a, b, and c. For each population, graph the distribution and simulate the sampling distribution of the sample mean by generating 1,000 samples of sizes 2, 5, 30, and 50. Describe the sampling distributions and tell how they illustrate the Central Limit Theorem. Compare with the graphs in Figure 6.10 on page 281 of the text.

a. Uniform distribution with lower limit 0 and upper limit 1.
b. Exponential distribution with $\theta = 1$.
c. Standard normal distribution with $\mu = 0$ and $\sigma = 1$.

Solution

a. The samples are generated in 1,000 rows of 50 columns using RANDOM with the UNIFORM subcommand. To calculate the sample means, we use the number of columns corresponding to each desired sample size.

```
MTB > NAME C1 'X' C2 'F(X)'
MTB > SET 'X'
DATA> 0:1/.1
DATA> END
MTB > PDF 'X' STORE IN 'F(X)';
SUBC> UNIFORM C = 0 D = 1.
MTB > PRINT 'X' 'F(X)'

 ROW      X     F(X)

  1      0.0     1
  2      0.1     1
  3      0.2     1
  4      0.3     1
  5      0.4     1
  6      0.5     1
  7      0.6     1
  8      0.7     1
  9      0.8     1
 10      0.9     1
 11      1.0     1
```

```
MTB > PLOT 'F(X)' VS 'X';
SUBC> TITLE 'UNIFORM PROBABILITY DISTRIBUTION';
SUBC> YINCREMENT 1.
```

```
                        UNIFORM PROBABILITY DISTRIBUTION
         3.0+
            -
F(X)        -
            -
            -
         2.0+
            -
            -
            -
            -
         1.0+   *    *    *    *    *    *    *    *    *    *
            -
            -
            -
            -
         0.0+
            -
            --+---------+---------+---------+---------+---------+----X
            0.00      0.20      0.40      0.60      0.80      1.00
```

The uniform probability distribution has a rectangular shape. The mean is $\mu = (c + d)/2 = .5$ and $\sigma^2 = (d - c)^2/12 = .0833$ and $\sigma = .2887$.

```
MTB > RANDOM 1000 OBSERVATIONS IN C1-C50;
SUBC> UNIFORM C = 0 D = 1.
MTB > NAME C51 'N=2' C52 'N=5' C53 'N=30' C54 'N=50'
MTB > RMEAN C1-C2 PUT IN 'N=2'      # MEANS FOR SAMPLE SIZE 2
MTB > RMEAN C1-C5 PUT IN 'N=5'      # MEANS FOR SAMPLE SIZE 5
MTB > RMEAN C1-C30 PUT IN 'N=30'    # MEANS FOR SAMPLE SIZE 30
MTB > RMEAN C1-C50 PUT IN 'N=50'    # MEANS FOR SAMPLE SIZE 50
```

```
MTB > DOTPLOT 'N=2'-'N=50';
SUBC> SAME.
```

Each dot represents 3 points

```
                          . .
               :       :  ::.
               :    . ::  :::::  .   .
          .:  :   ::.: ::  :::::  :.  :
          .:: ::  ::::::::.:::::.:::::.    :
          .::::::::::::::::::::::::::::::. .  .
         .:...::::::::::::::::::::::::::::::: .:
         .:.:::::::::::::::::::::::::::::::::::.
        -+---------+---------+---------+---------+---------+-----N=2
```

Each dot represents 4 points

```
                               o
                          : :.:
                          :::::::::. . .
                          ::::::::: : :
                          :::::::::::::
                     ....  :::::::::::::
                     ::::::::::::::::::::.
                     ..:::::::::::::::::::.
                    . :.:::::::::::::::::::::.... ..
        -+---------+---------+---------+---------+---------+-----N=5
```

Each dot represents 12 points

```
                              .
                              :
                           ..:.:
                           :::::
                           :::::
                          .:::::::
                          :::::::::.
                       ...:::::::::::.
        -+---------+---------+---------+---------+---------+-----N=30
```

Each dot represents 14 points

```
                            ::
                            :::
                           .:::.
                           :::::
                          .::::::
                          .:::::::
                         ..:::::::...
        -+---------+---------+---------+---------+---------+-----N=50
        0.00      0.20      0.40      0.60      0.80      1.00
```

The dot plots illustrate the sampling distributions of $\bar{x}$ when sampling from a uniform distribution. The distributions are approximately normal even for small sample sizes. This is generally the case with symmetrical population distributions. The graphs are similar to those given in Figure 6.10 on page 281 of the text.

Each sampling distribution is centered around the population mean, $\mu = .5$. MEAN calculates the mean of each sampling distribution.

128   Chapter 6

```
MTB > MEAN 'N=2'
   MEAN    =      0.50046
MTB > MEAN 'N=5'
   MEAN    =      0.49492
MTB > MEAN 'N=30'
   MEAN    =      0.49868
MTB > MEAN 'N=50'
   MEAN    =      0.50009
```

The dot plots show a decrease in the variation of the sampling distributions of the sample mean as the sample size is increased. The standard deviation of the sampling distribution of $\bar{x}$ is defined as $\sigma/\sqrt{n}$. For the actual sampling distributions of $\bar{x}$, the standard deviations for $n = 2$, 5, 30, and 50 are equal to .2041, .1291, .0527, and .0408, respectively. The following standard deviations of the simulated sampling distributions are close approximations of the actual results.

```
MTB > STDEV 'N=2'
   ST.DEV. =      0.20960
MTB > STDEV 'N=5'
   ST.DEV. =      0.13034
MTB > STDEV 'N=30'
   ST.DEV. =      0.052563
MTB > STDEV 'N=50'
   ST.DEV. =      0.040209
```

b. We repeat the Minitab program of part a using the EXPONENTIAL subcommand with RANDOM.

```
MTB > SET 'X'
DATA> 0:4/.2
DATA> END
MTB > PDF 'X' STORE IN 'F(X)';
SUBC> EXPONENTIAL MU = 1.
MTB > PLOT 'F(X)' VS 'X';
SUBC> TITLE 'EXPONENTIAL PROBABILITY DISTRIBUTION'.
```

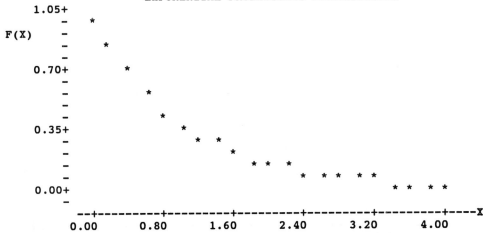

The exponential distribution, with mean $\mu = 1$ and standard deviation $\sigma = 1$, is skewed to the right.

```
MTB > RANDOM 1000 OBSERVATIONS IN C1-C50;
SUBC> EXPONENTIAL MU = 1.
MTB > RMEAN C1-C2 PUT IN 'N=2'
MTB > RMEAN C1-C5 PUT IN 'N=5'
MTB > RMEAN C1-C30 PUT IN 'N=30'
MTB > RMEAN C1-C50 PUT IN 'N=50'
MTB > DOTPLOT 'N=2'-'N=50';
SUBC> SAME.
```

Each dot represents 5 points

```
                 :..:
                .::::.
                ::::::.
              ::::::::::::   .
             .:::::::::::::.:
             ::::::::::::::::. :.
             :::::::::::::::::::.
            .::::::::::::::::::::::...........  .   ...  .
    -+---------+---------+---------+---------+---------+-----N=2
```

Each dot represents 6 points

```
                    :...
                   .:::::
                   ::::::.
                  .:::::::
                  ::::::::.
                 .::::::::::..
                 :::::::::::::.
               ..::::::::::::::::....  ..    . .
    -+---------+---------+---------+---------+---------+-----N=5
```

Each dot represents 15 points

```
                    ..
                    ::
                    :::
                   .:::
                   :::::
                   :::::
                   ::::::.
                  .::::::::... .
    -+---------+---------+---------+---------+---------+-----N=30
```

Each dot represents 19 points

```
                    .
                   .:
                   ::.
                   :::
                  .:::
                  ::::.
                  :::::
                 .:::::::...
    -+---------+---------+---------+---------+---------+-----N=50
    0.0       1.0       2.0       3.0       4.0       5.0
```

When sampling from an exponential distribution which is skewed, the sample size must be large to obtain an approximately normal sampling distribution of $\bar{x}$.

130 Chapter 6

The means of the simulated sampling distributions approximate the population mean $\mu = 1$.

```
MTB > MEAN 'N=2'
   MEAN    =         1.0187
MTB > MEAN 'N=5'
   MEAN    =         1.0215
MTB > MEAN 'N=30'
   MEAN    =         0.99661
MTB > MEAN 'N=50'
   MEAN    =         0.99657
```

The standard deviations of the sampling distributions of $\bar{x}$ equal $\sigma/\sqrt{n} = 1/\sqrt{n}$, or .7071, .4472, .1826 and .1414 for $n = 2, 5, 30,$ and 50, respectively. The following are close approximations.

```
MTB > STDEV 'N=2'
   ST.DEV. =         0.71304
MTB > STDEV 'N=5'
   ST.DEV. =         0.45389
MTB > STDEV 'N=30'
   ST.DEV. =         0.18449
MTB > STDEV 'N=50'
   ST.DEV. =         0.14443
```

c. To simulate the sampling distributions of $\bar{x}$ using a standard normal distribution, we repeat the Minitab program of part a. By default, the PDF and RANDOM commands use a standard normal distribution if a subcommand is not specified.

```
MTB > SET 'X'
DATA> -3.5:3.5/.2
DATA> END
MTB > PDF 'X' STORE IN 'F(X)'
MTB > PLOT 'F(X)' VS 'X';
SUBC> TITLE 'NORMAL PROBABILITY DISTRIBUTION'.
```

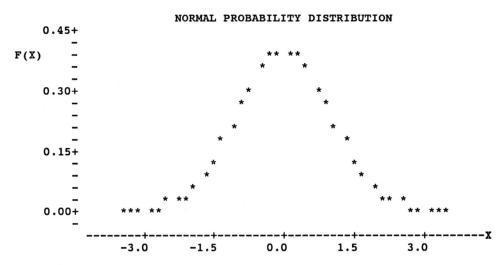

The standard normal distribution has $\mu = 0$ and $\sigma = 1$.

```
MTB > RANDOM 1000 OBSERVATIONS IN C1-C50
MTB > RMEAN C1-C2 PUT IN 'N=2'
MTB > RMEAN C1-C5 PUT IN 'N=5'
MTB > RMEAN C1-C30 PUT IN 'N=30'
MTB > RMEAN C1-C50 PUT IN 'N=50'
MTB > DOTPLOT 'N=2'-'N=50';
SUBC> SAME.
```

Each dot represents 3 points

```
                              ..
                       :     ::.. :
                       :: .:.::::   :.
                   :  . ::::::::: ::
                  : :. ::::::::::::::.:  .
                :.::::::::::::::::::::::.::
              . :::::::::::::::::::::::::::
              ...:::::::::::::::::::::::::.:.
      .    ...::::::::::::::::::::::::::::::: ::  . .
      -------+---------+---------+---------+---------+---------N=2
```

Each dot represents 5 points

```
                            .:::
                           ..:::: .
                           ::::::::
                          ..:::::::..
                          ::::::::::::
                        .::::::::::::::. .
                        ::::::::::::::::::.: .
                   .. .::::.:::::::::::::::::  . .
      -------+---------+---------+---------+---------+---------N=5
```

Each dot represents 13 points

```
                             .
                            .:
                           ::.
                          .::::.
                          :::::
                         ::::::::
                         :::::::::.
                       ....:::::::::...
      -------+---------+---------+---------+---------+---------N=30
```

Each dot represents 16 points

```
                           ::
                          :::
                          :::
                         .::::
                         :::::
                         ::::::.
                        .:::::::..
      -------+---------+---------+---------+---------+---------N=50
          -1.60     -0.80      0.00      0.80      1.60      2.40
```

When sampling from a normal population, the sampling distribution of $\bar{x}$ is normal for any sample size.

The means of the simulated sampling distributions approximate the population mean $\mu = 0$.

```
MTB > MEAN 'N=2'
   MEAN    =   -0.0010128
MTB > MEAN 'N=5'
   MEAN    =   -0.020197
MTB > MEAN 'N=30'
   MEAN    =   -0.0038487
MTB > MEAN 'N=50'
   MEAN    =   -0.0057162
```

The standard deviations of the sampling distributions of $\bar{x}$ equal $\sigma/\sqrt{n} = 1/\sqrt{n}$, or .7071, .4472, .1826 and .1414 for $n = 2, 5, 30,$ and 50, respectively. The following are approximations.

```
MTB > STDEV 'N=2'
   ST.DEV. =      0.69584
MTB > STDEV 'N=5'
   ST.DEV. =      0.44132
MTB > STDEV 'N=30'
   ST.DEV. =      0.18574
MTB > STDEV 'N=50'
   ST.DEV. =      0.13923
MTB > STOP
```

In summary, these simulated sampling distributions of $\bar{x}$ illustrate the Central Limit Theorem. If n is sufficiently large, the sampling distribution of the sample mean is approximately normal. The normal approximation is better as n is increased.

The mean of each simulated sampling distribution approximates the mean of the population from which the sample was selected. The standard deviation approximates $\sigma/\sqrt{n}$. As the sample size increases, the standard deviation decreases; that is, the sampling distribution becomes more concentrated around the population mean.

■

■ **Example 7** Example 6.9 on page 289 describes a Bernoulli random variable. If $n = 1$, a binomial random variable x is a Bernoulli random variable, where $x = 1$ with probability p, and $x = 0$ with probability $q = 1 - p$. Success is generally associated with 1, and failure, with 0.

a. Simulate the sampling distribution of the mean by generating 1,000 samples of sizes 1, 10, 25, and 100 from a Bernoulli distribution with $p = .8$. Construct a histogram of each distribution. Summarize.
b. What is the approximate mean and standard deviation of each sampling distribution of $\bar{x}$? Compare these values with the exact values.

Comment *Because of the large number of data points, you cannot replicate this experiment if you are using the student version of Minitab or a microcomputer with limited memory.*

Solution

a. The samples are generated in 1,000 rows of 100 columns using RANDOM with the BERNOULLI subcommand. Because of computer system limitations, these samples are generated in two parts. To calculate the sample means, we use the number of columns corresponding to each desired sample size.

```
MTB > RANDOM 1000 OBS C1-C50;
SUBC> BERNOULLI p = .8.
MTB > RANDOM 1000 OBS C51-C100;
SUBC> BERNOULLI p = .8.
MTB > NAME C101 'N=1'  C102 'N=10'  C103 'N=25'  C104 'N=100'
MTB > RMEAN C1 PUT IN 'N=1'
MTB > RMEAN C1-C10 PUT IN 'N=10'
MTB > RMEAN C1-C25 PUT IN 'N=25'
MTB > RMEAN C1-C100 PUT IN 'N=100'
MTB > HISTOGRAM 'N=1'-'N=100'

Histogram of N=1   N = 1000
Each * represents 20 obs.

Midpoint    Count
       0      213   ***********
       1      787   ****************************************

Histogram of N=10   N = 1000
Each * represents 10 obs.

Midpoint    Count
    0.40       3   *
    0.45       0
    0.50      32   ****
    0.55       0
    0.60      93   **********
    0.65       0
    0.70     177   ******************
    0.75       0
    0.80     311   ********************************
    0.85       0
    0.90     252   **************************
    0.95       0
    1.00     132   **************

Histogram of N=25   N = 1000
Each * represents 5 obs.

Midpoint    Count
    0.50       2   *
    0.55       5   *
    0.60      11   ***
    0.65      26   ******
    0.70     181   *************************************
    0.75     182   *************************************
    0.80     182   *************************************
    0.85     160   ********************************
    0.90     222   *********************************************
    0.95      24   *****
    1.00       5   *
```

```
Histogram of N=100   N = 1000
Each * represents 5 obs.

Midpoint   Count
  0.68       1    *
  0.70       6    **
  0.72      22    *****
  0.74      63    *************
  0.76      99    ********************
  0.78     163    *********************************
  0.80     179    ************************************
  0.82     185    *************************************
  0.84     138    ****************************
  0.86      90    ******************
  0.88      43    *********
  0.90      10    **
  0.92       1    *
```

These histograms resemble the frequency histograms given in Figure 6.15 on page 291 of the text. As n increases, the sampling distribution of $\bar{x}$ more closely approximates the normal distribution.

b. The mean and standard deviation of the sampling distribution of $\bar{x}$ are given on page 290 of the text: $\mu_{\bar{x}} = .8$, and $\sigma_{\bar{x}} = .4/\sqrt{n}$.

```
MTB > MEAN 'N=1'
    MEAN    =       0.78700
MTB > MEAN 'N=10'
    MEAN    =       0.80450
MTB > MEAN 'N=25'
    MEAN    =       0.79996
MTB > MEAN 'N=100'
    MEAN    =       0.80067
MTB > STDEV 'N=1'       # EXACT VALUE IS .4000
    ST.DEV. =       0.40963
MTB > STDEV 'N=10'      # EXACT VALUE IS .1265
    ST.DEV. =       0.12902
MTB > STDEV 'N=25'      # EXACT VALUE IS .0800
    ST.DEV. =       0.082474
MTB > STDEV 'N=100'     # EXACT VALUE IS .0400
    ST.DEV. =       0.040093
MTB > STOP
```

The mean and standard deviation of each simulated sampling distribution closely approximate the exact values. ■

EXERCISES

1. The managerial assertiveness scores for $n = 101$ managers of a consulting firm are given in the table. We expect the characteristics of a sample to be similar to the corresponding characteristics of the underlying population. Sample 30 scores from the following population. Graphically and numerically summarize the population and the sample. Compare and discuss the characteristics.

Managerial Assertiveness Scores

98	76	89	65	58	77	90	92	83	83	96	78
94	93	84	78	74	70	68	54	68	79	77	75
72	83	89	90	93	78	96	82	87	86	94	59
71	89	90	99	74	77	78	65	68	98	73	85
89	69	72	78	65	97	94	89	67	58	40	84
78	77	78	89	66	70	72	89	73	75	84	86
89	74	95	83	74	55	77	82	98	89	65	78
77	90	93	99	52	89	85	89	76	74	89	67
66	89	76	89	93							

2. Estimate the probability that two or more people in a group of 25 have the same birthday. Use RANDOM with the INTEGER subcommand to simulate 50 samples of $n = 25$ birthdays. Assume the 365 days of the year are equally likely to be birthdays. How many samples of 25 had matching birthdays? How does this compare with what you may expect?

3. Consider the exponential distribution with $\mu = 2$. Obtain a graph of the probability distribution. Generate a random sample of 200 observations from the distribution and construct a histogram of the 200 observations. Compare the simulated distribution with the exponential distribution.

4. Suppose the random variable x has a continuous uniform distribution with a lower boundary of 0 and an upper boundary of 10.

   a. Generate 100 samples of size $n = 9$. Calculate the means of the 100 samples and construct a histogram of the sample means. This is an approximation of the sampling distribution of the sample mean. Comment on the shape of the distribution.
   b. Calculate the mean and standard deviation of the 100 sample means. Compare with $\mu = 5$ and $\sigma_{\bar{x}} = .962$.
   c. Repeat parts a and b for $n = 18$; $\mu = 5$ and $\sigma_{\bar{x}} = .680$. Summarize.

5. If a random sample is selected from a normal population distribution, the sampling distribution of $\bar{x}$ is normal for any sample size. Generate 100 samples from a normal probability distribution with $\mu = 300$ and $\sigma = 20$ for sample sizes 4, 8 and 16. For each sample size, construct a histogram of the sample means, and calculate the mean and the standard deviation of the 100 sample means. Summarize the results.

6. Consider the 1988 data on homes sold in St. Cloud, Minnesota that are given in Appendix B of this supplement. Use 100 random samples of $n = 5$ to find the approximate sampling distribution of the mean. Find the estimates of μ and $\sigma_{\bar{x}}$.

136   Chapter 6

7. Assume that the distribution of the number of violent crimes per day in a certain city described in Exercise 6.27 on page 287 of the text is normal with $\mu = 1.3$ and $\sigma = 1.7$.

   a. Use 100 samples to approximate the sampling distribution of $\bar{x}$ for $n = 30$. What is the approximate mean and standard deviation of $\bar{x}$? How do these compare with the exact μ and $\sigma_{\bar{x}}$?
   b. Use 100 samples to approximate the sampling distribution of $\bar{x}$ for $n = 50$. Compare with the distribution in part a.
   c. Calculate $P(1 < \bar{x} < 1.9)$ for $n = 30$ and $n = 50$. Which probability is greater? Why?

8. Refer to Exercise 6.29 on page 287 of the text. A bottler's quality control program monitors a vendor's production process to verify the claim that the mean internal strength of its bottles is 157 psi and standard deviation is 3 psi. The bottler randomly selected 40 bottles.

   a. Simulate 50 samples of $n = 40$ bottles. Assume the production process is normally distributed with $\mu = 157$ psi and $\sigma = 3$ psi. Numerically and graphically describe the sampling distribution of $\bar{x}$. Estimate the probability that the process mean is less than 155.
   b. Repeat part a for each of the following combinations of μ and σ.

   Random 50 c1-c40;
   NORM 157 3,

   1. $\mu = 156, \sigma = 3$
   2. $\mu = 158, \sigma = 3$
   3. $\mu = 157, \sigma = 2$
   4. $\mu = 157, \sigma = 6$

   c. Compare the results for parts a and b.

9. Refer to the class data set given in Appendix B. Generate a random sample of 40 students.

   a. Numerically and graphically describe the grade point averages of the 40 students. Compare with the set of 200 students.
   b. Repeat part a for the ages of students.

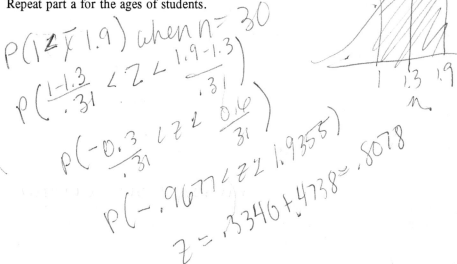

CHAPTER 7

INFERENCES BASED ON A SINGLE SAMPLE: ESTIMATION

Statistical estimation consists of a point estimator of the population parameter and a reliability measure based on the sampling distribution of the point estimator. Minitab constructs confidence intervals of a population mean from the sample observations. If we know only the sample statistics, such as the mean and standard deviation, it is easier to compute confidence intervals by hand. Confidence interval estimation of the population proportion is presented in the final section.

NEW COMMANDS

TINTERVAL   ZINTERVAL

ESTIMATION OF A POPULATION MEAN

One of the most important parameters characterizing a population is the mean. The estimation of the mean assertiveness training score, the estimation of the mean number of airline ticket cancellations per flight, and a test of the claim that a radial tire has a mean life of 50,000 miles are situations in which we want to make an inference about the population mean. This section is concerned with the estimation of the population mean, μ.

To estimate μ, we select a sample from the population, and calculate the sample mean. The sample mean $\bar{x}$ is a point estimator of μ. The previous chapter illustrated that, although the sample mean varies from sample to sample, it has certain characteristics. Specifically, the sampling distribution of $\bar{x}$ is approximately normal if the sample size is large, or if the population distribution is approximately normal regardless of sample size. The mean of the sampling distribution is the same as the population mean, $\mu_{\bar{x}} = \mu$, and the standard deviation is $\sigma_{\bar{x}} = \sigma/\sqrt{n}$.

Different Minitab procedures are used to make inferences about the population mean, depending on whether we have a large or small sample. Both ZINTERVAL and TINTERVAL construct confidence intervals for μ. If you have a large sample and know σ, use ZINTERVAL. The formula is

$$\bar{x} \pm z_{\alpha/2} \sigma/\sqrt{n}$$

If you have a large sample but do not know σ, use either ZINTERVAL or TINTERVAL; use the sample standard deviation to approximate σ with ZINTERVAL. If you have a small sample, use ZINTERVAL if σ is known, and TINTERVAL if σ is unknown. In either case, the sampled population must have an approximately normal distribution.

ZINTERVAL (K PERCENT CONFIDENCE) SIGMA = K, FOR C,...,C

This command calculates and prints a confidence interval for μ on the data in each column. The value for z is from the standard normal distribution corresponding to a K% confidence level. The default is a 95% confidence level. Use the sample standard deviation s to estimate an unknown σ for large samples.

Use TINTERVAL for small samples, and for large samples if σ is unknown. Small sample estimation assumes that the population distribution is approximately normal. The formula is

$$\bar{x} \pm t_{\alpha/2} s/\sqrt{n}$$

TINTERVAL (K PERCENT CONFIDENCE) FOR C,...,C

This command calculates and prints a confidence interval for μ on the data in each column. The t value is from the t distribution corresponding to $(n - 1)$ degrees of freedom and K% confidence level. The default confidence level is 95%.

■ **Example 1** In Chapter 2 of this supplement, we analyzed the EPA mileage ratings of a random sample of 100 new model cars. The mileage ratings are repeated below.

EPA Mileage Ratings

36.3	41.0	36.9	37.1	44.9	36.8	30.0	37.2	42.1	36.7
32.7	37.3	41.2	36.6	32.9	36.5	33.2	37.4	37.5	33.6
40.5	36.5	37.6	33.9	40.2	36.4	37.7	37.7	40.0	34.2
36.2	37.9	36.0	37.9	35.9	38.2	38.3	35.7	35.6	35.1
38.5	39.0	35.5	34.8	38.6	39.4	35.3	34.4	38.8	39.7
36.3	36.8	32.5	36.4	40.5	36.6	36.1	38.2	38.4	39.3
41.0	31.8	37.3	33.1	37.0	37.6	37.0	38.7	39.0	35.8
37.0	37.2	40.7	37.4	37.1	37.8	35.9	35.6	36.7	34.5
37.1	40.3	36.7	37.0	33.9	40.1	38.0	35.2	34.8	39.5
39.9	36.9	32.9	33.8	39.8	34.0	36.8	35.0	38.1	36.9

a. Construct a 95% confidence interval for μ, the population mean mileage rating. Interpret the confidence interval.
b. Construct a 90% confidence interval for μ. Compare the widths of the two intervals.

Solution

a. The mileage ratings are saved in a file named EPA.

```
MTB > RETRIEVE 'EPA"
  WORKSHEET SAVED   1/10/1994

Worksheet retrieved from file: EPA.MTW
MTB > INFORMATION

COLUMN      NAME       COUNT
C1          MPG         100

CONSTANTS USED: NONE

MTB > DESCRIBE 'MPG'

              N        MEAN     MEDIAN    TRMEAN     STDEV    SEMEAN
MPG         100      36.994     37.000    36.992     2.418     0.242

              MIN        MAX         Q1        Q3
MPG        30.000     44.900     35.625    38.375
```

The point estimator of the population mean mileage rating is $\bar{x} = 36.99$ miles per gallon. For a sample of $n = 100$, we can use either ZINTERVAL or TINTERVAL to construct confidence intervals. The sample standard deviation, $s = 2.418$, estimates σ.

```
MTB > ZINTERVAL SIGMA = 2.418 'MPG'

THE ASSUMED SIGMA =2.42

              N        MEAN      STDEV    SE MEAN     95.0 PERCENT C.I.
MPG         100      36.994      2.418      0.242   ( 36.519,  37.469)
```

The Minitab output for ZINTERVAL includes the assumed σ, the sample size, sample mean and standard deviation, standard error of the mean, and the confidence interval. We are 95% confident that μ, the mean mileage rating is between 36.5 and 37.5 miles per gallon.

The TINTERVAL command does not require an estimate of σ. With the large sample size, the output is very similar to that of ZINTERVAL.

```
MTB > TINTERVAL OF 'MPG'

              N        MEAN      STDEV    SE MEAN     95.0 PERCENT C.I.
MPG         100      36.994      2.418      0.242   ( 36.514,  37.474)
```

b. Confidence coefficients other than 95% must be specified on the command line.

```
MTB > ZINTERVAL 90% CONFIDENCE SIGMA = 2.418 'MPG'

THE ASSUMED SIGMA =2.42

              N        MEAN      STDEV    SE MEAN     90.0 PERCENT C.I.
MPG         100      36.994      2.418      0.242   ( 36.596,  37.392)
MTB > STOP
```

Chapter 7

The width of the 90% confidence interval is less than the width of the 95% confidence interval. In general, the lower the confidence coefficient, the narrower the confidence interval.

■

■ **Example 2** Different random samples from a population give different confidence intervals. To illustrate, consider the 1988 data on selling prices of homes in St. Cloud, Minnesota provided in Appendix B of this supplement. Select two random samples of 30 selling prices. Obtain plots of the population and the two samples of selling prices. Use each sample to construct a 90% confidence interval to estimate μ, the mean selling price of all homes. Interpret the output. Do both intervals include the population mean?

Solution The data for home prices are saved in HOMES. DOTPLOT with the SAME subcommand plots the population and sample data on the same scale. We use TINTERVAL with a 90% confidence level to estimate μ.

```
MTB > RETRIEVE 'HOMES'
  WORKSHEET SAVED   1/10/1994

Worksheet retrieved from file: HOMES.MTW
MTB > INFORMATION

COLUMN      NAME         COUNT     MISSING
C1          AREA         197
C2          BEDROOMS     197
C3          LIST PR      197
C4          SOLD PR      197
C5          FINANCE      197
C6          DAYS         197          2
C7          MTH SOLD     197
C8          DAY SOLD     197

CONSTANTS USED: NONE

MTB > NAME C9 'SAMPLE 1' C10 'SAMPLE 2'
MTB > SAMPLE 30 OBS FROM 'SOLD PR' PUT IN 'SAMPLE 1'
MTB > PRINT 'SAMPLE 1'

SAMPLE 1
   82500    38500    54000    88900    76000    53000    84900    38400    47500
   41900    79000    69000    42000    19500    40875    22000    22000    66000
   16000    52250    50900    69000    47500    17000    57500    55000    96900
   54000    61000    85500

MTB > SAMPLE 30 OBS FROM 'SOLD PR' PUT IN 'SAMPLE 2'
MTB > PRINT 'SAMPLE 2'

SAMPLE 2
   27000    55000    53000    55500    63500    34900   117000    17000
   32500    37800    70000    43000    61900    70000    79900    40000
   60000    88000    61000    63900    43200    75900    46000    73650
   29000    47900   107000    69000    71000    55600
```

```
MTB > DOTPLOT 'SOLD PR' 'SAMPLE 1' 'SAMPLE 2';
SUBC> SAME.
```

<pre>
                              : :
                              : :
                              : :
                              : : .
                           :. ::::: :
                           ::.::::: :
                          .:::::::: :   :
                          ::::::::::. .::
                       .: :::::::::::::::.   :
                       ::.:::::::::::::::: .:: :   . ..   . .
                   ----+---------+---------+---------+---------+---------+------SOLD PR

                                .
                            .   :   .
                         :: :: :::. .: ...:. .
                   ----+---------+---------+---------+---------+---------+------SAMPLE 1

                              . . :
                         . :..:...::: :...   .   . .
                   ----+---------+---------+---------+---------+---------+------SAMPLE 2
                     35000   70000   105000   140000   175000   210000
</pre>

```
MTB > DESCRIBE 'SOLD PR' 'SAMPLE 1' 'SAMPLE 2'

                  N      MEAN    MEDIAN    TRMEAN    STDEV    SEMEAN
SOLD PR         197     61534     57500     59231    26736      1905
SAMPLE 1         30     54284     53500     54220    22496      4107
SAMPLE 2         30     58305     57800     56967    22461      4101

                MIN       MAX        Q1        Q3
SOLD PR       16000    191000     45400     70400
SAMPLE 1      16000     96900     40281     70750
SAMPLE 2      17000    117000     42250     70250
```

Neither sample contains the unusually high selling prices that are found in the population. As a result, the means and standard deviations of these samples are lower than the population parameters.

```
MTB > TINTERVAL 90% CONFIDENCE FOR 'SAMPLE 1' AND 'SAMPLE 2'

              N      MEAN     STDEV    SE MEAN    90.0 PERCENT C.I.
SAMPLE 1     30   54284.2   22496.2     4107.2   ( 47303.9, 61264.4)
SAMPLE 2     30   58305.0   22461.0     4100.8   ( 51335.6, 65274.4)

MTB > STOP
```

From the first sample, we are 90% confident that the true mean selling price is between $47,304 and $61,264. From the second sample, we are 90% confident that the true mean selling price is between $51,336 and $65,274. The first sample confidence interval fails to include the population mean, μ = $61,534. ∎

> **Comment** *This example ignores the finite population correction factor. The standard error should be multiplied by the factor if the sample size is large relative to the size of the finite population. TINTERVAL assumes the sample size represents a small percentage of the total number in the population.*

Sample Size to Estimate μ

Increasing the sample size provides more information about the population, resulting in more reliable estimates of population parameters. For large sample sizes, the width W of a confidence interval for μ is given by

$$W = 2z_{\alpha/2}\sigma/\sqrt{n}$$

where $z_{\alpha/2}$ is the z-score corresponding to a $(1 - \alpha)100\%$ confidence coefficient and σ is the known or estimated population standard deviation. Since n is in the denominator, the width decreases as the sample size increases. The next example illustrates the relationship between the sample size and width of a confidence interval.

■ **Example 3** Suppose the large hospital discussed on page 312 of the text wants to use a 95% confidence interval to estimate the mean length of stay for patients. The sample standard deviation $s = 4.9$ is an estimate of σ.

a. Calculate the approximate widths of confidence intervals for μ for samples of sizes $n = 100, 200, 300, \ldots, 2{,}000$ accounts. Graph the width versus the sample size. What is the relationship between the sample size and width? What happens to the marginal benefit of 100 additional observations as the sample size increases?
b. Use the results of part a to to determine the factor by which a sample size must be increased to reduce the width by 50%?
c. Use the table of widths calculated in part a to estimate the sample size that will provide a width of about $10.

Solution The width of a 95% confidence interval with $\sigma = 4.9$ is given by

$$W = 2(1.96)(4.9)/\sqrt{n}$$

a. We set the sample sizes 100, 200, 300, ..., 2,000 in the column named SIZE, and use LET to calculate the widths in the column named WIDTH.

```
MTB > NAME C1 'SIZE' C2 'WIDTH'
MTB > SET 'SIZE'
DATA> 100:2000/100
DATA> END
MTB > LET 'WIDTH'=2*1.96*4.9/SQRT('SIZE')
MTB > PRINT 'SIZE' 'WIDTH'

ROW     SIZE     WIDTH

  1      100    1.92080
  2      200    1.35821
  3      300    1.10897
  4      400    0.96040
  5      500    0.85901
  6      600    0.78416
  7      700    0.72599
  8      800    0.67911
  9      900    0.64027
 10     1000    0.60741
 11     1100    0.57914
```

12	1200	0.55449
13	1300	0.53273
14	1400	0.51336
15	1500	0.49595
16	1600	0.48020
17	1700	0.46586
18	1800	0.45274
19	1900	0.44066
20	2000	0.42950

```
MTB > PLOT 'WIDTH' VS 'SIZE';
SUBC> TITLE '95% CONFIDENCE INTERVALS';
SUBC> TITLE 'Width as a Function of Sample Size'.
```

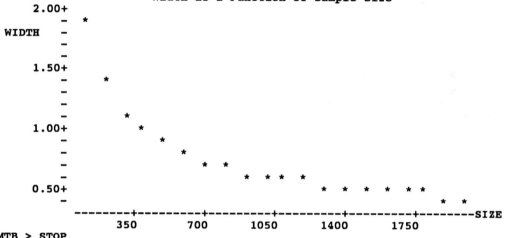

```
MTB > STOP
```

The graph illustrates a nonlinear relationship between width and sample size. As the sample size increases, the width decreases rapidly with smaller sample sizes, and at a slower rate with larger sizes. There is a diminishing benefit in terms of interval width gained by increasing the sample size by 100 observations for larger sample sizes. For example, as n is increased from 100 to 200, the width decreases by about $10; as n is increased from 1,900 to 2,000, the width decreases by about $0.20.

b. To reduce the width by 50%, the sample size has to be increased by a factor of four. For example, as we increase n from 10 to 40, the width decreases from $35.28 to $17.64.

c. To obtain a width of about $10, we use the table to find that $n = 1,200$ gives a width of about $10.20 and $n = 1,300$ gives a width of about $9.80. We estimate that about $n = 1,250$ will provide a 95% confidence interval with a width of about $10.

■

SIMULATING CONFIDENCE INTERVALS

Suppose that we select many random samples from a population with mean μ and standard deviation σ, and construct a confidence interval for μ using each sample. We expect that the percent of intervals containing μ would approximately equal the confidence level. For example, if the confidence level is 90%, we expect that about 90% of the confidence intervals would contain μ and about 10% would not contain μ.

■ **Example 4** Simulate 30 samples of size 20 from a normal distribution with $\mu = 70$ and $\sigma = 10$. Use TINTERVAL to construct 90% confidence intervals for the 30 samples. How many of the intervals contain the population mean μ? How does this compare with what you would expect?

Solution Because Minitab constructs confidence intervals on column data, we generate the samples in 30 columns.

```
MTB > RANDOM 20 OBSERVATIONS IN COLUMNS C1-C30;
SUBC> NORMAL WITH MU = 70 AND SIGMA = 10.
MTB > TINTERVAL 90% CONFIDENCE ON C1-C30

            N      MEAN    STDEV   SE MEAN      90.0 PERCENT C.I.
C1         20     68.19    10.79     2.41    (   64.02,    72.36)
C2         20     70.26     9.99     2.23    (   66.39,    74.12)
C3         20     71.08     8.62     1.93    (   67.74,    74.41)
C4         20     72.27     9.83     2.20    (   68.47,    76.07)
C5         20     69.03    10.93     2.45    (   64.80,    73.26)
C6         20     67.72    10.42     2.33    (   63.69,    71.75)
C7         20     72.29     9.24     2.07    (   68.71,    75.86)
C8         20     69.44     8.78     1.96    (   66.05,    72.84)
C9         20     70.08    12.80     2.86    (   65.13,    75.03)
C10        20     73.09     9.03     2.02    (   69.60,    76.58)
C11        20     68.51    11.04     2.47    (   64.24,    72.78)
C12        20     74.70     9.47     2.12    (   71.04,    78.36)
C13        20     68.65    10.90     2.44    (   64.44,    72.87)
C14        20     71.75    13.87     3.10    (   66.39,    77.12)
C15        20     67.30    10.17     2.27    (   63.37,    71.23)
C16        20     68.87    12.20     2.73    (   64.15,    73.59)
C17        20     76.39    10.22     2.29    (   72.44,    80.35)
C18        20     70.65    12.56     2.81    (   65.79,    75.50)
C19        20     70.34    10.37     2.32    (   66.33,    74.35)
C20        20     72.37     8.57     1.92    (   69.05,    75.68)
C21        20     72.81     9.76     2.18    (   69.03,    76.58)
C22        20     70.57    10.25     2.29    (   66.60,    74.53)
C23        20     69.77     7.98     1.78    (   66.69,    72.86)
C24        20     69.73     8.32     1.86    (   66.51,    72.95)
C25        20     68.26    10.58     2.37    (   64.17,    72.35)
C26        20     66.44     9.44     2.11    (   62.79,    70.09)
C27        20     67.53     7.26     1.62    (   64.72,    70.34)
C28        20     69.56     8.26     1.85    (   66.36,    72.75)
C29        20     68.48     9.78     2.19    (   64.70,    72.26)
C30        20     71.43    11.96     2.67    (   66.80,    76.05)

MTB > STOP
```

We expect about 27 (90% of 30) confidence intervals would contain μ. Twenty-eight intervals contain $\mu = 70$; the samples in C12 and C17 fail to include the true mean. ■

Inferences Based on a Single Sample: Estimation 145

LARGE SAMPLE ESTIMATION OF A BINOMIAL PROBABILITY

The binomial probability distribution, characterized by two outcomes, termed success and failure, is described in Chapter 4. The probability of success is p and the probability of failure is $q = 1 - p$. This section describes statistical estimation of p. The point estimator is $\hat{p}$, the proportion of successes in a random sample of n trials. If x is the number of successes, $\hat{p} = x/n$.

From the Central Limit Theorem, the sampling distribution of $\hat{p}$ is approximately normal for large sample sizes. As discussed on page 327 of the text, a sample size is large if 0 or 1 is not included in the interval $\hat{p} \pm 3\sigma_{\hat{p}}$. The mean of the sampling distribution of $\hat{p}$ is p and the standard deviation is $\sqrt{pq/n}$. There is no Minitab command to construct confidence intervals for the binomial probability. Use LET to calculate the upper and lower limits of the confidence intervals.

■ **Example 5**   Consider Example 7.5 on page 328 of the text. An objective of the Equal Employment Opportunity Commission (EEOC) is to monitor the treatment of minorities in the workplace. Suppose the EEOC sampled 135 recent hires of a large company and found 12 minorities.

a. Use Minitab to determine if the sample size is sufficiently large to be able to use the normal approximation to estimate p, the proportion of minority hires for this company.
b. Construct and interpret a 90% confidence interval for p.

Solution

a. The sample size is sufficiently large if 0 and 1 are not within three standard deviations of $\hat{p}$. We use LET to calculate $\hat{p} \pm 3\sigma_{\hat{p}}$.

```
MTB > LET K1 = 12/135
MTB > LET K2=K1-3*SQRT(K1*(1-K1)/135)
MTB > LET K3=K1+3*SQRT(K1*(1-K1)/135)
MTB > PRINT K2 K3
K2        0.0154795
K3        0.162521
```

The confidence interval does not include 0 or 1. We can assume the sampling distribution of $\hat{p}$ is approximately normal.

b. The z value for a 90% confidence interval is 1.645.

```
MTB > LET K2=K1-1.645*SQRT(K1*(1-K1)/135)
MTB > LET K3=K1+1.645*SQRT(K1*(1-K1)/135)
MTB > PRINT K2 K3
K2        0.0486862
K3        0.129314
MTB > STOP
```

We can be 90% confident that the proportion of minority hires with this company is between .049 and .129.

■

146   Chapter 7

■ **Example 6**   The population proportion p can range from 0 to 1. This example illustrates the relationship between p and the sample size required to obtain a confidence interval of a desired width W. If you do not have an estimate for p, the most conservative sample size for a given confidence level and a desired width is obtained using $p = .5$. Determine the sample sizes required to obtain 90% confidence intervals for p having a width $W = .02$. Use values of $p = .05, .10, .15, ..., .95$. Graph the sample size versus p. Summarize the results.

Solution   We use LET to calculate n given by

$$n = 4z_{\alpha/2}^2 pq/W^2$$

The z-score for 90% confidence intervals is 1.645.

```
MTB > NAME C1 'p' C2 'SIZE'
MTB > SET 'p'
DATA> .05:.95/.05
DATA> END
MTB > LET 'SIZE' = 4*(1.645)**2*'p'*(1-'p')/(.02)**2
MTB > PRINT 'p' 'SIZE'

  ROW        p        SIZE

    1     0.05     1285.36
    2     0.10     2435.42
    3     0.15     3450.18
    4     0.20     4329.64
    5     0.25     5073.80
    6     0.30     5682.65
    7     0.35     6156.21
    8     0.40     6494.46
    9     0.45     6697.41
   10     0.50     6765.06
   11     0.55     6697.41
   12     0.60     6494.46
   13     0.65     6156.21
   14     0.70     5682.65
   15     0.75     5073.80
   16     0.80     4329.64
   17     0.85     3450.18
   18     0.90     2435.42
   19     0.95     1285.36
```

```
MTB > PLOT 'SIZE' VS 'p';
SUBC> TITLE '90% CONFIDENCE INTERVALS';
SUBC> TITLE 'Sample Size as a Function of p'.
```

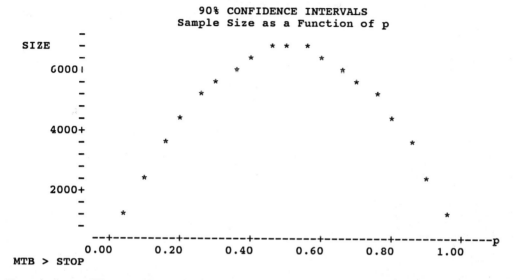

```
MTB > STOP
```

To estimate p with a specified confidence level and width, the table and graph indicate that we need the largest sample sizes when p is close to .5. This result is consistent with the fact that you obtain conservatively large samples by using $p = .5$. This demonstrates the importance of having an approximate value of p; the farther p is from .5, the smaller the necessary sample size. ∎

EXERCISES

1. Simulate 50 samples of size 15 from a normal distribution with $\mu = 300$ and $\sigma = 50$. Use the TINTERVAL command to construct 90% confidence intervals for the 50 samples. How many of the intervals included $\mu = 300$? How many would you expect to include μ?

2. The width of confidence intervals can be decreased by increasing the sample size. If the sample size is changed by a factor of k, the width decreases by a factor of approximately $1/\sqrt{k}$. For example, if the sample size is increased by a factor of $k = 4$, the width is about one half of what it was before.

   a. Simulate a sample of 30 observations from a normal distribution with $\mu = 300$ and $\sigma = 50$. Use the TINTERVAL command to construct a 95% confidence interval for each sample.
   b. Repeat part a using samples of 120 and 480 observations.
   c. Calculate the width of each interval. Do the results agree with the relationship between sample size and the width of a confidence interval stated above? Discuss.

3. A large grocery store has a packaging machine to weigh and package ground beef. An adjustment on the machine allows the operator to fill packages of different weights. Suppose the

148   Chapter 7

filling machine is adjusted to fill one pound packages. All packages do not contain exactly one pound because of either random variability in the weighing process, or an incorrect adjustment setting. To check if the machine is adjusted correctly, the following weights of a random sample of 40 packages were recorded. The packaging process has a standard deviation of .05 pounds.

Packaging Weights

0.97	1.05	0.96	1.00	1.03	0.98	1.03	0.98	1.09	1.01
1.05	1.05	1.05	1.00	0.96	1.02	1.02	1.07	1.01	1.06
1.09	0.94	1.08	1.07	1.03	1.00	1.05	1.05	0.98	1.10
1.05	0.98	1.12	1.04	1.05	1.06	1.07	1.09	1.00	1.09

a. Construct a stem and leaf display. What does the display imply about the machine setting?
b. Construct a 95% confidence interval to estimate the mean weight of all packages filled by the machine. Interpret the confidence interval.

4. A study concerning employee participation in 401(k) plans is reported in Exercise 7.16 on page 310 of the text. A company, concerned that its employee participation rate was low, sampled 30 other companies with similar 401(k) plans and obtained the following rates.

401(k) Participation Rates

80	76	81	77	82	80	85	60	80	79
82	70	88	85	80	79	83	75	87	78
80	84	72	75	90	84	82	77	75	86

a. Construct a stem and leaf display. Describe the distribution of participation rates.
b. Construct a 95% confidence interval to estimate the mean participation rate for all companies with 401(k) plans. Interpret the confidence interval. What assumptions are necessary for this confidence interval to be valid?

5. Refer to the HOMES data base given in Appendix B. The variable DAYS is the number of days a home is on the market before it is sold.

a. Use DOTPLOT and DESCRIBE to summarize the distribution of the number of days.
b. Randomly generate a sample of size $n = 30$ from the population of DAYS. Graphically and numerically summarize the sample. Construct a 90% confidence interval of the mean number of days a home is on the market before it is sold.
c. Compare the population and sample results.

6. Consider Exercise 7.42 on page 324 of the text. A large corporation obtained the following randomly selected costs of hiring an entry level secretary.

Hiring Costs

$2,100	$1,650	$1,315	$2,035
2,245	1,980	1,700	2,190

a. Construct a 90% confidence interval for μ, the mean cost of hiring an entry level secretary. Interpret.
b. Construct a 95% confidence interval for μ. Compare the widths of the two intervals.

7. In Exercise 7.53 on page 331 of the text, a random sample of 122 Illinois law firms was selected to determine computer usage. The survey showed that 76 used microcomputers.

   a. Use Minitab to determine if the sample size is sufficiently large for assuming an approximately normal distribution to estimate p, the proportion of all Illinois law firms who use microcomputers. Explain why this is important.
   b. Use a 95% confidence interval to estimate the proportion of all Illinois law firms who used microcomputers. Interpret.

8. Consider the class data set given in Appendix B of this supplement.

   a. The administrator of the business program wants to estimate μ, the mean grade point average. Construct and interpret 90% and 95% confidence intervals on μ. Do the confidence intervals suggest that the mean grade point average differs significantly from 2.75?
   b. Construct and interpret a 90% confidence interval to estimate p, the proportion of male students taking business statistics.

CHAPTER 8

INFERENCES BASED ON A
SINGLE SAMPLE: TESTS OF HYPOTHESES

Hypothesis testing is a statistical procedure used to make an inference about the value of a population parameter. This chapter illustrates Minitab commands to test hypotheses about the population mean, binomial probability, and variance based on a single sample.

NEW COMMANDS

TTEST   ZTEST

HYPOTHESIS TESTS ABOUT A POPULATION MEAN

A hypothesis test uses sample information to test a claim or make a decision about the value of a population mean. For example, a bank manager may want to test the claim that the mean time spent waiting in line for banking services is less than five minutes, or the Environmental Protection Agency may want to test the claim that an automobile obtains 35 miles per gallon in city driving. In either situation, a sample provides information to test a hypothesis about μ.

The null and alternative hypotheses for a two-tailed test are

$H_0: \mu = \mu_0$
$H_a: \mu \neq \mu_0$

For a large sample test, the formula for the test statistic is

$$z = \frac{\bar{x} - \mu_0}{\sigma/\sqrt{n}},$$

For small sample tests, the sampled population is assumed to have an approximately normal distribution. The value of the test statistic is

$$t = \frac{\bar{x} - \mu_0}{s/\sqrt{n}}$$

The Minitab commands to test a hypothesis about the population mean are ZTEST and TTEST. If you have a large sample, you can use ZTEST if σ is known, and either ZTEST or TTEST if σ is unknown. If you have a small sample, use ZTEST if σ is known and TTEST if σ is unknown. Both commands require the sample observations.

ZTEST (OF MU = K) SIGMA = K ON C,...,C
    ALTERNATIVE = K

This command calculates and prints the z test statistic and the p-value for the hypothesis test H_0: $\mu = K$ on the data in each column, where $K = \mu_0$. The default value of K is 0. You must specify a value for σ. For large samples, use the sample standard deviation to estimate an unknown σ.

TTEST (OF MU = K) ON C,...,C
    ALTERNATIVE = K

This command calculates and prints the t test statistic and p-value for the hypothesis test on the data in each column.

For both ZTEST and TTEST, use the ALTERNATIVE subcommand for a one-tailed test:

   ALTERNATIVE = +1 to test H_a: $\mu > K$
   ALTERNATIVE = -1 to test H_a: $\mu < K$

You do not need a subcommand for a two-tailed test of H_a: $\mu \neq K$.

■ **Example 1** Refer to the EPA mileage ratings for a certain new model car that are given in Chapter 2. Suppose that a car dealer believes that the mean mileage of this model car is less than 38 miles per gallon. Do the sample data provide sufficient evidence to support this belief? Use $\alpha = .01$.

Solution The mileage ratings are saved in a file named EPA. To test whether the mean mileage is less than 38, we test

   H_0: $\mu = 38$
   H_a: $\mu < 38$ $M > 55$

```
MTB > RETRIEVE 'EPA'
  WORKSHEET SAVED  1/10/1994

Worksheet retrieved from file: EPA.MTW
MTB > INFORMATION

COLUMN      NAME       COUNT
C1          MPG         100

CONSTANTS USED: NONE
```

```
MTB > PRINT 'MPG'

MPG
  36.3   32.7   40.5   36.2   38.5   36.3   41.0   37.0   37.1   39.9   41.0
  37.3   36.5   37.9   39.0   36.8   31.8   37.2   40.3   36.9   36.9   41.2
  37.6   36.0   35.5   32.5   37.3   40.7   36.7   32.9   37.1   36.6   33.9
  37.9   34.8   36.4   33.1   37.4   37.0   33.8   44.9   32.9   40.2   35.9
  38.6   40.5   37.0   37.1   33.9   39.8   36.8   36.5   36.4   38.2   39.4
  36.6   37.6   37.8   40.1   34.0   30.0   33.2   37.7   38.3   35.3   36.1
  37.0   35.9   38.0   36.8   37.2   37.4   37.7   35.7   34.4   38.2   38.7
  35.6   35.2   35.0   42.1   37.5   40.0   35.6   38.8   38.4   39.0   36.7
  34.8   38.1   36.7   33.6   34.2   35.1   39.7   39.3   35.8   34.5   39.5
  36.9

MTB > DESCRIBE 'MPG'

              N       MEAN    MEDIAN   TRMEAN    STDEV   SEMEAN
MPG         100     36.994   37.000   36.992    2.418    0.242

              MIN       MAX        Q1       Q3
MPG        30.000   44.900    35.625   38.375

MTB > ZTEST MU = 38 SIGMA = 2.418 OF 'MPG';
SUBC>  ALTERNATIVE = -1.

TEST OF MU = 38.000 VS MU L.T. 38.000
THE ASSUMED SIGMA = 2.42

              N      MEAN     STDEV   SE MEAN       Z    P VALUE
MPG         100    36.994    2.418    0.242     -4.16    0.0000
```

The null and alternative hypotheses are stated on the output. The assumed σ is the sample standard deviation that we entered on the command line. The sample size, sample mean and standard deviation, and standard error of the mean are provided on the output.

The test statistic, $z = -4.16$, is the number of standard deviations that the sample mean 36.994 is from the hypothesized mean, $\mu = 38$. We determine the rejection region for $\alpha = .01$ with INVCDF, which calculates the value of z for a specified area. We do not need a subcommand for the standard normal distribution.

```
MTB > INVCDF OF .01
     0.0100    -2.3263
MTB > STOP
```

The rejection region consists of all values of z less than -2.33; that is, if the sample mean is more than 2.33 standard deviations below $\mu = 38$, we reject H_0. Since $z = -4.16$, there is sufficient evidence in the sample to conclude that the mean mileage rating for the new model car is less than 38 miles per gallon.

■

The Observed Significance Level: p-Value

In the previous example, we made the decision to reject the null hypothesis if the value of the test statistic falls in the rejection region. An alternative way to make a decision about the value of a population mean is to calculate the *p*-value, or the observed significance level. The *p*-value is the probability of obtaining a value of the test statistic as extreme as that which is observed, assuming the null hypothesis is true. The *p*-value is very useful if the researcher prefers not to select an α level for the test. The reader can then determine whether to reject the null hypothesis based on the *p*-value. Usually the null hypothesis is rejected if the *p*-value is less than the selected α level.

■ **Example 2** The changes in housing prices over short time periods are in part determined by supply and demand. The real estate board in a Minnesota community projected an increase in selling prices of homes in 1989 over the mean 1988 selling price of $61,534. The reason for the projection was an increase in demand due to some expansions of several area businesses and the subsequent increase in labor. To test the accuracy of the projection, a random sample of 16 homes sold in 1989 was selected, and the following selling prices recorded.

Housing Prices

$42,000	$39,900	$51,300	$71,000	$50,900	$93,500
57,000	62,200	67,800	41,000	85,000	38,400
70,950	84,500	77,900	64,000		

a. Use a box plot to summarize the 1989 selling prices. Are there any serious departures from normality?
b. Test whether the average home price increased from 1988 to 1989. Use $\alpha = .01$. Interpret the *p*-value.

Solution

a. A small sample hypothesis test assumes the population of selling prices is approximately normal.

```
MTB > NAME C1 'PRICE'
MTB > SET 'PRICE'
DATA> 42000  39900  51300  71000  50900  93500  57000  82200
DATA> 67800  41000  85000  38400  70950  84500  77900  64000
DATA> END
MTB > BOXPLOT 'PRICE'
```

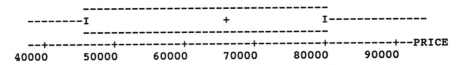

The box plot indicates a median selling price of about $66,000. The distribution is slightly skewed to the left, but does not have any outliers. There does not seem to be a serious departure from normality.

b. TTEST with the ALTERNATIVE subcommand tests whether the 1989 average home price has increased from the 1988 average price. The hypotheses are

$H_0: \mu = \$61,534$
$H_a: \mu > \$61,534$

```
MTB > TTEST MU = 61534 DATA IN 'PRICE';
SUBC> ALTERNATIVE = 1.

TEST OF MU = 61534.000 VS MU G.T. 61534.000

              N        MEAN      STDEV    SE MEAN        T     P VALUE
PRICE        16    63584.375  18250.111   4562.528     0.45      0.33

MTB > STOP
```

The p-value of 0.33 is the probability of obtaining a t test statistic as large as 0.45, assuming that H_0: μ = \$61,534, is true. Since the p-value is greater than the significance level, α = 0.01, there is insufficient evidence in the sample to conclude that the 1989 mean selling price is greater than the 1988 mean selling price; that is, there is not enough sample evidence to support the real estate board's projection.

■

LARGE SAMPLE TEST OF HYPOTHESIS ABOUT A BINOMIAL PROBABILITY

This section describes tests of hypotheses for the binomial probability. The point estimator of p is $\hat{p}$, the proportion of successes in a random sample of n trials. The null and alternative hypotheses for a two-tailed test are

H_0: $p = p_0$
H_a: $p \neq p_0$

For a large sample test, the formula for the test statistic is

$$z = \frac{\hat{p} - p_0}{\sigma_{\hat{p}}},$$

where p_0 is the hypothesized binomial probability with standard deviation $\sigma_{\hat{p}} = \sqrt{p_0 q_0 / n}$. There is no Minitab command to do this hypothesis test. You can use LET to calculate the z test statistic and CDF to determine the p-value.

■ **Example 3** Consider Example 8.6 on page 374 of the text. The manufacturer of alkaline batteries wants a defective rate of less than 5%. In a random sample of 300 batteries from a large shipment, 10 batteries are found to be defective.

   a. Is the sample size large enough to use the normal approximation for the sampling distribution of $\hat{p}$?
   b. Does this sample provide sufficient evidence for the manufacturer to conclude that the proportion of defective batteries is less than .05? Test using α = .01. Interpret the p-value.

Solution

a. The sample size is sufficiently large if 0 or 1 are not within three standard deviations of p_0. To check this, we determine the interval, $p_0 \pm 3 \sigma_{\hat{p}}$, with p_0 = .05.

```
MTB > LET K1=.05
MTB > LET K2=K1-3*SQRT(K1*(1-K1)/300)
MTB > LET K3=K1+3*SQRT(K1*(1-K1)/300)

MTB > PRINT K2 K3
K2        0.0122508
K3        0.0877492
```

The interval, $.01 \leq p_0 \leq .09$, does not include 0 or 1. We can assume the sampling distribution of $\hat{p}$ is approximately normal.

b. The null and alternative hypotheses to test whether the proportion of defectives is less than .05 are

H_0: $p = .05$
H_a: $p < .05$

```
MTB > LET K1=10      # 10 DEFECTIVE BATTERIES
MTB > LET K2=.05     # ASSUMED BINOMIAL PROBABILITY
MTB > LET K3=(K1/300-K2)/SQRT(K2*(1-K2)/300)    # Z TEST STATISTIC
MTB > PRINT K3
K3      -1.32453
MTB > # CALCULATE THE P-VALUE
MTB > CDF K3
   -1.3245    0.0927
MTB > STOP
```

The p-value = .093 is greater than $\alpha = .01$. There is not sufficient evidence in the sample to conclude that the defective rate is less than 5%. The manager cannot be reasonably certain that fewer than 5% of the batteries are defective. ∎

SIMULATING HYPOTHESES TESTS: TYPE I AND TYPE II ERRORS

When testing a hypothesis, α is the probability of making a Type I error; that is, α is the probability of rejecting a true H_0. If we select many random samples from a population, we would expect to reject a true H_0 in about $(\alpha)100\%$ of the tests, and fail to reject a true H_0 in about $(1 - \alpha)100\%$. For example, if $\alpha = .05$, the probability of rejecting a true H_0 is 0.05. This means that the test procedure in repeated sampling leads to the rejection of a true H_0 about 5% of the time. The interpretation is similar to the repeated sampling interpretation of a confidence interval.

■ **Example 4** Simulate 30 samples of size $n = 20$ from a normal distribution with $\mu = 70$ and $\sigma = 10$. For each sample, use TTEST to test the null hypothesis that the population mean is 70. Use $\alpha = .10$. How many times did you reject the true null hypothesis? How many times would you expect to reject the true null hypothesis?

Solution The samples of 20 observations are randomly generated in 30 columns. We test

H_0: $\mu = 70$
H_a: $\mu \neq 70$

```
MTB > RANDOM 20 OBSERVATIONS IN COLUMNS C1-C30;
SUBC> NORMAL MU = 70 AND SIGMA = 10.
MTB > TTEST OF MU = 70 ON THE SAMPLES IN C1-C30
```

	N	MEANS	STDEV	SEMEAN	T	P VALUE
C1	20	72.191	11.570	2.587	0.85	0.41
C2	20	68.638	9.223	2.062	-0.66	0.52
C3	20	66.244	7.249	1.621	-2.32	0.032
C4	20	72.922	9.751	2.180	1.34	0.20
C5	20	69.113	6.242	1.396	-0.64	0.53
C6	20	68.339	12.617	2.821	-0.59	0.56
C7	20	71.567	8.792	1.966	0.80	0.44
C8	20	71.609	11.747	2.627	0.61	0.55
C9	20	70.411	10.599	2.370	0.17	0.86
C10	20	66.583	15.412	3.446	-0.99	0.33
C11	20	69.645	9.740	2.178	-0.16	0.87
C12	20	72.970	8.187	1.831	1.62	0.12
C13	20	69.281	10.374	2.320	-0.31	0.76
C14	20	67.530	7.972	1.782	-1.39	0.18
C15	20	68.594	8.840	1.977	-0.71	0.49
C16	20	69.286	10.493	2.346	-0.30	0.76
C17	20	72.423	8.393	1.877	1.29	0.21
C18	20	71.861	9.071	2.028	0.92	0.37
C19	20	69.281	7.392	1.653	-0.44	0.67
C20	20	70.621	10.479	2.343	0.27	0.79
C21	20	67.683	8.585	1.920	-1.21	0.24
C22	20	63.153	8.708	1.947	-3.52	0.0023
C23	20	71.099	11.891	2.659	0.41	0.68
C24	20	68.887	11.831	2.646	-0.42	0.68
C25	20	71.840	9.717	2.173	0.85	0.41
C26	20	69.698	10.335	2.311	-0.13	0.90
C27	20	68.912	14.667	3.280	-0.33	0.74
C28	20	68.891	9.857	2.204	-0.50	0.62
C29	20	69.741	10.152	2.270	-0.11	0.91
C30	20	71.343	9.377	2.097	0.64	0.53

```
MTB > STOP
```

Since the samples were randomly selected from a normal population distribution with a mean of 70, we know the null hypothesis is true. In two of the 30 samples (C3 and C22), the true null hypothesis was rejected. Using $\alpha = .10$, we would expect to reject the null hypothesis incorrectly in 3 (10% of 30) samples.

■

The probability of making a Type II error or the probability of accepting a false H_0 is called ß. In any given test, ß depends on the true value of μ and is difficult to specify. For this reason, we generally conclude that we fail to reject H_0 rather than risk a Type II error.

Suppose we select many random samples from a population and test a false H_0. Then we should accept a false H_0 in about (ß)100% of the tests, and reject the false H_0 in about (1 - ß)100% of the tests. For example, if ß = .05, in repeated sampling we should accept the false H_0 about 5% of the time.

158 Chapter 8

■ **Example 5** Simulate 30 samples of size $n = 20$ from a normal distribution with $\mu = 73$ and $\sigma = 10$. For each sample, use TTEST to test the null hypothesis that $\mu = 70$. Use $\alpha = .10$. How many tests result in the acceptance of the false null hypothesis? Estimate ß.

Solution For each of the 30 samples, we test

$$H_0: \mu = 70$$
$$H_a: \mu \neq 70$$

```
MTB > RANDOM 20 OBSERVATIONS IN C1-C30;
SUBC> NORMAL WITH MU = 73 AND SIGMA = 10.
MTB > TTEST OF MU = 70, DATA IN C1-C30

TEST OF MU = 70.000 VS MU N.E. 70.000
```

	N	MEAN	STDEV	SE MEAN	T	P VALUE
C1	20	70.251	10.880	2.433	0.10	0.92
C2	20	74.742	9.997	2.235	2.12	0.047
C3	20	67.661	9.999	2.236	-1.05	0.31
C4	20	75.035	10.590	2.368	2.13	0.047
C5	20	74.972	9.606	2.148	2.31	0.032
C6	20	70.788	10.248	2.292	0.34	0.73
C7	20	75.840	8.747	1.956	2.99	0.0076
C8	20	77.721	10.682	2.389	3.23	0.0044
C9	20	70.495	10.307	2.305	0.21	0.83
C10	20	75.851	10.824	2.420	2.42	0.026
C11	20	73.773	10.438	2.334	1.62	0.12
C12	20	73.813	9.594	2.145	1.78	0.091
C13	20	70.221	10.837	2.423	0.09	0.93
C14	20	68.968	10.235	2.289	-0.45	0.66
C15	20	73.414	6.898	1.542	2.21	0.039
C16	20	68.323	8.079	1.807	-0.93	0.37
C17	20	76.656	12.420	2.777	2.40	0.027
C18	20	72.500	10.679	2.388	1.05	0.31
C19	20	74.252	11.366	2.542	1.67	0.11
C20	20	75.056	10.205	2.282	2.22	0.039
C21	20	74.972	9.831	2.198	2.26	0.036
C22	20	71.985	9.306	2.081	0.95	0.35
C23	20	72.982	6.050	1.353	2.20	0.040
C24	20	72.927	10.470	2.341	1.25	0.23
C25	20	69.583	8.198	1.833	-0.23	0.82
C26	20	72.094	9.592	2.145	0.98	0.34
C27	20	76.259	11.228	2.511	2.49	0.022
C28	20	71.701	9.045	2.022	0.84	0.41
C29	20	74.444	12.263	2.742	1.62	0.12
C30	20	73.673	8.336	1.864	1.97	0.064

```
MTB > STOP
```

Since the samples were randomly selected from a normal distribution with a mean of 73, we know the null hypothesis is false. We select the false H_0 if the *p*-value is greater than $\alpha = .10$. In this example, the false H_0 is accepted in 16 of the 30 samples. Thus we estimate ß to be about 16/30 or .53.

■

INFERENCES ABOUT A POPULATION VARIANCE

The point estimate of a population variance is the sample variance s^2. If the sampled population has a normal distribution, the sampling distribution of s^2 has an approximate chi-square distribution with $(n - 1)$ degrees of freedom. There is not a Minitab command to construct a confidence interval or to test a hypothesis about a population variance. For a $(1 - \alpha)100\%$ confidence interval, use INVCDF and LET to evaluate the formula,

$$\frac{(n-1)s^2}{\chi^2_{1-\alpha/2}} \leq \sigma^2 \leq \frac{(n-1)s^2}{\chi^2_{\alpha/2}}$$

where $\chi^2_{1-\alpha/2}$ and $\chi^2_{\alpha/2}$ are the respective right and left tail values from a chi-square distribution with $(n - 1)$ degrees of freedom. This notation differs from the text, but is consistent with the Minitab command structure.

The null and alternative hypothesis for a two-tailed test are

H_0: $\sigma^2 = \sigma_0^2$
H_a: $\sigma^2 \neq \sigma_0^2$

where σ_0^2 is the hypothesized value for σ^2. The test statistic is

$$\chi^2 = \frac{(n-1)s^2}{\sigma_0^2}$$

Use CDF with the CHISQUARE subcommand to determine the *p*-value.

■ **Example 6** Refer to the EPA mileage ratings for a certain new model car that are given in Chapter 2. Suppose the car dealer believes that the standard deviation of the miles per gallon for this model car should be less than 3 miles per gallon.

a. Numerically and graphically describe the miles per gallon.
b. Test whether the standard deviation is less than 3 mpg. Use $\alpha = .05$.
c. Construct and interpret a 95% confidence interval for the true variance of miles per gallon for this model.
d. Construct and interpret a 95% confidence interval for the true standard deviation of miles per gallon for this model.

Solution The mileage ratings are saved in a file named EPA.

a. We use DESCRIBE and DOTPLOT to describe the data.

```
MTB > RETRIEVE 'EPA'
   WORKSHEET SAVED  1/10/1994

Worksheet retrieved from file: EPA.MTW
```

```
MTB > INFORMATION

COLUMN   NAME      COUNT
C1       MPG       100

CONSTANTS USED: NONE

MTB > PRINT 'MPG'

MPG
  36.3   32.7   40.5   36.2   38.5   36.3   41.0   37.0   37.1   39.9   41.0
  37.3   36.5   37.9   39.0   36.8   31.8   37.2   40.3   36.9   36.9   41.2
  37.6   36.0   35.5   32.5   37.3   40.7   36.7   32.9   37.1   36.6   33.9
  37.9   34.8   36.4   33.1   37.4   37.0   33.8   44.9   32.9   40.2   35.9
  38.6   40.5   37.0   37.1   33.9   39.8   36.8   36.5   36.4   38.2   39.4
  36.6   37.6   37.8   40.1   34.0   30.0   33.2   37.7   38.3   35.3   36.1
  37.0   35.9   38.0   36.8   37.2   37.4   37.7   35.7   34.4   38.2   38.7
  35.6   35.2   35.0   42.1   37.5   40.0   35.6   38.8   38.4   39.0   36.7
  34.8   38.1   36.7   33.6   34.2   35.1   39.7   39.3   35.8   34.5   39.5
  36.9

MTB > DESCRIBE 'MPG'

               N       MEAN     MEDIAN     TRMEAN      STDEV     SEMEAN
MPG          100     36.994     37.000     36.992      2.418      0.242

             MIN        MAX         Q1         Q3
MPG       30.000     44.900     35.625     38.375

MTB > DOTPLOT 'MPG'
```

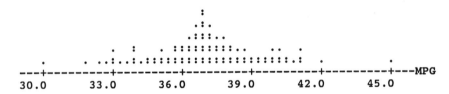

The point estimate of the population standard deviation is 2.418 mpg. The dot plot indicates an unusually high mileage car and an unusually low mileage car when compared with the others.

b. To test whether the standard deviation is less than 3 mpg, the hypothesis test is stated in terms of the variance. The sample size is $n = 100$ and $\sigma_0^2 = 9$.

H_0: $\sigma^2 = 9$
H_a: $\sigma^2 < 9$

```
MTB > # CALCULATE THE TEST STATISTIC
MTB > LET K1=(100-1)*STDEV('MPG')**2/9
MTB > PRINT K1
K1        64.3085
MTB > CDF K1;
SUBC> CHISQUARE 99 DF.
    64.3085    0.0027
```

The *p*-value of .0027 is less than $\alpha = .05$. There is sufficient evidence that the variance is less than 9. The car dealer can conclude that the standard deviation is less than 3 mpg.

c. INVCDF with the CHISQUARE subcommand gives the χ^2 values for the confidence interval.

```
MTB > INVCDF .975;
SUBC> CHISQUARE 99 DF.
      0.9750   128.4209
MTB > INVCDF .025;
SUBC> CHISQUARE 99 DF,
      0.0250    73.3611
MTB > LET K1=(100-1)*STDEV('MPG')**2/128.4209
MTB > LET K2=(100-1)*STDEV('MPG')**2/73.3611
MTB > PRINT K1 K2
K1         4.50687
K2         7.88942
```

We are 95% confident that the population variance of miles per gallon for this model is between 4.51 to 7.89 mpg.

d. A confidence interval for σ is obtained by taking the positive square root of the end points of the σ^2 confidence interval.

```
MTB > LET K3=SQRT(K1)
MTB > LET K4=SQRT(K2)
MTB > PRINT K3 K4
K3         2.12294
K4         2.80881
MTB > STOP
```

We are 95% confident that the population standard deviation of miles per gallon for this model is between 2.12 and 2.81 mpg.

■

EXERCISES

1. Simulate 50 samples of $n = 15$ from a normal distribution with $\mu = 100$ and $\sigma = 10$ in order to illustrate the meaning of α and β.

   a. Use TTEST and $\alpha = .10$ to test the true null hypothesis, $\mu = 100$, for the 50 samples. How many times did you reject the true null hypothesis? How many times would you expect to reject the true null hypothesis?

   b. Use the TTEST command to test the false null hypothesis, $\mu = 105$, for the 50 samples, at $\alpha = .10$. How many times did you accept the false null hypothesis? Estimate β.

2. A large international corporation customarily rents several homes for relocating employees in Phoenix. The Personnel Division of the corporation is interested in studying the monthly rents of homes within the city. The following prices of 40 three bedroom homes were randomly selected from Phoenix in August, 1989.

Chapter 8

Rental Prices

$625	$795	$365	$ 595	$600
600	600	575	540	850
475	450	625	595	650
550	465	445	735	620
750	465	625	975	425
475	720	500	800	985
535	575	545	1,175	650
525	565	850	625	770

  a. Graphically summarize the rental prices.
  b. Does the sample of rental prices provide sufficient evidence to conclude that mean rental price for three bedroom homes in Phoenix exceeds $550? Use $\alpha = .05$.

3. A manufacturer claims that its laser-based inspection equipment for printed circuit boards (PCBs) can inspect on average at least 10 solder joints per second when the joints are spaced .1 inch apart. In order to check this claim, a potential buyer tested the equipment on 48 different PCBs. The equipment was operated for exactly one second on each PCB. The following number of solder joints inspected on each run are given in Exercise 8.20 on page 359 of the text.

Solder Joints Inspected

10	9	10	10	11	9	12	8	8	9	6	10
7	10	11	9	9	13	9	10	11	10	12	8
9	9	9	7	12	6	9	10	10	8	7	9
11	12	10	0	10	11	12	9	7	9	9	10

  a. Obtain a box plot of the number of solder joints inspected per second by this laser-based inspection equipment. Are there any unusual observations?
  b. Do the data provide sufficient evidence to refute the manufacturer's claim? Use a significance level of $\alpha = .05$.
  c. Construct and interpret a 95% confidence interval to estimate the true mean number of solder joints inspected per second.

4. A large grocery store has a packaging machine to weigh and package ground beef. An adjustment on the machine allows the operator to fill packages of different weights. Suppose the filling machine is adjusted to fill one pound packages. The meat department manager of the grocery store wants to test if the packaging machine is operating correctly. Specifically, the manager would like to determine if the average fill in the packages differs from one pound. If the weight is less than one pound, customers are dissatisfied. If the weight is greater than one pound, the store loses potential earnings.

Packaging Weights

0.97	1.05	0.96	1.00	1.03	0.98	1.03	0.98	1.09	1.01	
1.05	1.05	1.05	1.00	0.96	1.02	1.02	1.07	1.01	1.06	
1.09	0.94	1.08	1.07	1.03	1.00	1.05	1.05	0.98	1.10	
1.05	0.98	1.12	1.04	1.05	1.06	1.07	1.09	1.00	1.09	

a. Obtain a dot plot of the packaging weights. Do the weights appear to be normally distributed?
b. Do the data provide sufficient evidence to conclude that the population mean differs from 1 pound? Use a significance level of $\alpha = .05$.

5. Refer to Exercise 8.49 on page 371 of the text. A new fumigant for the control of nematodes has been developed for strawberry growers. With previous fumigation, the yield was eight pounds of marketable fruit for a standard plot. With the application of the new fumigant, the yields from six standard plots of strawberries were 9, 9, 13, 9 10, and 8 pounds.

   a. Does the sample provide sufficient evidence to conclude that the new fumigant increases the average yield? Use $\alpha = .05$. Interpret the *p*-value.
   b. What assumptions are necessary for this hypothesis test?

6. Child care benefits are discussed in Exercise 8.62 on page 378 of the text. A union claims that at least 90% of manufacturing firms do not offer child care benefits to workers. To test this claim, a random sample of 350 manufacturing firms is selected. Only 28 firms offer child care benefits.

   a. Construct a confidence interval to determine whether the sample size is large enough to use the normal approximation for the sampling distribution of $\hat{p}$.
   b. Does this sample provide sufficient evidence to support the union's claim? Use $\alpha = .10$.
   c. Calculate and interpret the *p*-value of the test.

7. Consider the class data set given in Appendix B of this supplement. The administrator of the business program claims that the mean grade point average for the majors is different than the required 2.75. Does the sample evidence suggest that the mean grade point average differs significantly from 2.75? Use a significance level of $\alpha = .05$.

CHAPTER 9

INFERENCES BASED ON TWO SAMPLES: ESTIMATION AND TESTS OF HYPOTHESES

This chapter includes some statistical methods for comparing two populations. We describe estimation and hypothesis tests about the difference between population means for two different sampling procedures. In one procedure, random samples from two populations are obtained independently of each other. In the other, observations from one sample are paired or matched with corresponding observations from another sample. The last sections give confidence intervals and hypothesis tests for the differences between two population proportions and two population variances.

NEW COMMANDS

TWOSAMPLE-T   TWOT

ESTIMATION AND HYPOTHESIS TESTING OF $\mu_1 - \mu_2$: INDEPENDENT SAMPLING

The objective is to make inferences about the difference between population means. For example, experimenters may want to compare the mean yields in a chemical laboratory using two different processes. They record samples of daily yield for each process, and make an inference on the difference between mean yields. Or a product manager may want to assess the difference between mean lengths of time required to package a product using two different types of packages.

Suppose one population has mean μ_1 and variance σ_1^2, and the other has mean μ_2 and variance σ_2^2. Independent samples are obtained if the random selection of n_1 measurements from the first population is unrelated to the random selection of n_2 measurements from the other population. The sample statistic, $(\bar{x}_1 - \bar{x}_2)$, is a point estimate of the population parameter, $(\mu_1 - \mu_2)$.

The $(1 - \alpha)100\%$ confidence interval for $(\mu_1 - \mu_2)$ is defined

$$(\bar{x}_1 - \bar{x}_2) \pm t_{\alpha/2}\sqrt{s_1^2/n_1 + s_2^2/n_2}$$

The null and alternative hypotheses for a two-tailed test are

H_0: $\mu_1 - \mu_2 = D_0$
H_a: $\mu_1 - \mu_2 \neq D_0$

where D_0 is the hypothesized difference between the means. Often $D_0 = 0$. The test statistic is defined

$$t = \frac{(\bar{x}_1 - \bar{x}_2) - D_0}{\sqrt{s_1^2/n_1 + s_2^2/n_2}}$$

If the two populations have approximately equal variances, the variances in the test statistic calculation are replaced by the pooled variance s_p^2, which is a weighted average of the sample variances. The formula for the pooled variance is

$$s_p^2 = \frac{(n_1 - 1)s_1^2 + (n_2 - 1)s_2^2}{n_1 + n_2 - 2}$$

The Minitab commands for making inferences about $(\mu_1 - \mu_2)$ for independent samples are TWOSAMPLE-T or TWOT. Both commands use the t test statistic for large and small samples. Use TWOSAMPLE-T if the data are organized as unstacked data; that is, if each set of sample data is in a separate column. Use TWOT if the data are stacked in the same column, and a code to identify each set is in another column.

We recommend that you obtain descriptive statistics and plots in addition to confidence intervals and hypothesis tests. HISTOGRAM, STEM AND LEAF, BOXPLOT and DOTPLOT, introduced in Chapter 2, are useful graphical procedures. Use DESCRIBE for comparative descriptive statistics.

TWOSAMPLE-T (K PERCENT CONFIDENCE) DATA IN C, C
    POOLED
    ALTERNATIVE

This command gives the confidence interval and t test for two independent samples stored in the specified columns. The default confidence level is 95%. If the two populations have approximately equal variances, the POOLED subcommand is used to obtain an estimate of the pooled variance, s_p^2.

Use the ALTERNATIVE subcommand for one-tailed tests:

    ALTERNATIVE = +1 for H_a: $\mu_1 - \mu_2 > 0$
    ALTERNATIVE = -1 for H_a: $\mu_1 - \mu_2 < 0$

TWOT (K PERCENT CONFIDENCE) DATA IN C, CODE IN C
    POOLED
    ALTERNATIVE

Stack both samples in one column, and the code to identify the groups in a second column. The subcommands and output with TWOT are similar to those with TWOSAMPLE-T.

Inferences Based on Two Samples: Estimation and Tests of Hypotheses 167

Comment An assumption of TWOSAMPLE-T and TWOT for small independent samples is that both sampled populations are normally distributed. If there is serious departure from normality, a nonparametric test such as one described in Chapter 11 should be used to make inferences about the difference between means.

■ **Example 1** A large international corporation rents several apartments for relocating employees in Phoenix, Arizona. The Personnel Division of the corporation is interested in comparing the mean monthly rents of apartments within the city and the surrounding suburbs. The following random samples of apartment rentals were independently selected from each locality.

City			Suburb		
$515	$995	$345	$575	$700	$850
525	600	575	640	650	495
475	550	725	595	650	750
850	465	435	725	600	1075
750	465	625	975	625	650
495	750	500	700	685	100
535	575	549	1175	650	545
515	515	495	850	825	750
525	525	550	50	575	550
475	1150	650	1100	585	1000

a. Graphically and numerically describe the samples of rents for city and suburb.
b. Construct 95% confidence intervals for the true difference in the mean monthly rents for the two localities with and without the POOLED subcommand. Compare the outputs.
c. Is there evidence in the samples that the mean monthly rent in the city is less than in the suburbs? Test using $\alpha = .01$.

Solution

a. We use DOTPLOT and DESCRIBE to compare the sample data. The SAME subcommand is used with DOTPLOT so that both graphs have the same scale.

```
MTB > NAME C1 'CITY' C2 'SUBURB'
MTB > SET 'CITY'
DATA> 515 525 475 850 750 495 535 515 525 475 995 600 550
DATA> 465 465 750 575 515 525 1150 345 575 725 435 625 500
DATA> 549 495 550 650
DATA> END
MTB > SET 'SUBURB'
DATA> 575 640 595 725 975 700 1175 850 650 1100 700
DATA> 650 650 600 625 685 650 825 575 585 850 495 750
DATA> 1075 650 1100 545 750 550 1000
DATA> END
MTB > SAVE 'RENTS'

Worksheet saved into file:  RENTS.MTW
```

168   Chapter 9

```
MTB > DOTPLOT 'CITY' AND 'SUBURB';
SUBC> SAME.
```

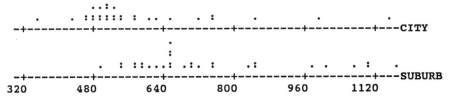

The dot plots indicate that the distribution of city rents is shifted to the left of the distribution of suburb rents. At first glance, it appears that apartment rents tend to be lower in the city.

```
MTB > DESCRIBE 'CITY' AND 'SUBURB'

                N       MEAN    MEDIAN   TRMEAN    STDEV   SEMEAN
CITY           30      590.0     530.0    568.2    168.7     30.8
SUBURB         30      743.2     667.5    730.0    189.2     34.5

              MIN        MAX        Q1        Q3
CITY        345.0     1150.0     495.0     631.3
SUBURB      495.0     1175.0     598.7     850.0
```

The sample of apartment rents in the city of Phoenix has a lower mean, lower quartiles (Q1, median, and Q3), and a lower standard deviation than the sample of rents in the suburbs. The point estimator for the difference in mean monthly rents is 590.0 - 742.3 = -137.7; that is, the mean city rent is $137.70 lower than the mean suburb rent.

b. Since the sample data are organized as unstacked data, we use TWOSAMPLE-T to obtain a 95% confidence interval for the true difference between means. The POOLED calculation assumes the population variances are equal.

```
MTB > TWOSAMPLE-T 'CITY' 'SUBURB'

TWOSAMPLE T FOR CITY VS SUBURB
            N       MEAN    STDEV   SE MEAN
CITY       30        590      169        31
SUBURB     30        743      189        35

95 PCT CI FOR MU CITY - MU SUBURB: (-246, -60)

TTEST MU CITY = MU SUBURB (VS NE): T= -3.31  P=0.0016  DF=  57
```

We are 95% confident that the difference between the mean monthly rents in the two localities falls between -246 and -60; that is, the mean monthly rent in the city is from $60 to $246 lower than that in the suburbs.

```
MTB > TWOSAMPLE-T 'CITY' 'SUBURB';
SUBC> POOLED.

TWOSAMPLE T FOR CITY VS SUBURB
            N       MEAN    STDEV   SE MEAN
CITY       30        590      169        31
SUBURB     30        743      189        35
```

```
95 PCT CI FOR MU CITY - MU SUBURB: (-246, -61)

TTEST MU CITY = MU SUBURB (VS NE): T= -3.31  P=0.0016  DF= 58

POOLED STDEV =            179
```

Since the sample standard deviations are about the same and $n_1 = n_2$, the POOLED output is similar to the first output. The degrees of freedom and confidence intervals differ slightly. Both outputs include a two-tailed hypothesis test of the difference between population means.

c. To test whether the mean rent in the city is less than in the suburbs, we do a one-tailed hypothesis test of

H_0: $\mu_1 - \mu_2 = 0$
H_a: $\mu_1 - \mu_2 < 0$

```
MTB > TWOSAMPLE-T 'CITY' AND 'SUBURB';
SUBC> ALTERNATIVE = -1.

TWOSAMPLE T FOR CITY VS SUBURB
           N      MEAN      STDEV     SE MEAN
CITY      30       590        169        31
SUBURB    30       743        189        35

95 PCT CI FOR MU CITY - MU SUBURB: (-246, -60)

TTEST MU CITY = MU SUBURB (VS LT): T= -3.31  P=0.0008  DF= 57

MTB > STOP
```

The test statistic $t = -3.31$ has a p-value of .0008. Since the p-value is less than $\alpha = .01$, we have sufficient evidence in the sample to conclude that the mean monthly rent in the city is less than that of the suburbs.

Comment *The p-value for this one-tailed test equals one-half of the reported p-value for the two-tailed test of part b.*

■

INFERENCES ABOUT THE DIFFERENCE BETWEEN TWO POPULATION MEANS: PAIRED DIFFERENCE EXPERIMENTS

A paired difference experiment is one in which the samples consist of matched or paired observations randomly selected from a population of paired observations. For example, each pair may correspond to the same city, the same week, the same person, or the same business. Frequently, a paired difference experiment provides more information about the difference between two means than independent sampling procedures. By pairing the observations when possible, some sampling variability is removed.

For example, suppose we want to compare the number of visitors to two city museums. If we collect admissions data to both museums for 16 random weeks of the year, there may be a similar pattern of variation in the mean admissions at both museums. This week-to-week variability in admissions is

Chapter 9

removed by calculating the weekly differences in admissions to the two museums, and making an inference on the mean difference.

The parameter of interest is the mean, μ_D, of the population differences. The point estimate of μ_D is $\bar{x}_D$, the mean of the sample differences. If the population distribution of differences is approximately normal, the t statistic is used to make an inference on μ_D.

To construct a confidence interval and test a hypothesis for paired data with Minitab, compute the differences between corresponding observations and then use TINTERVAL on the sample of differences. You can compare the paired samples graphically by plotting the paired observations on the same graph with MPLOT.

■ **Example 2** Refer to Example 9.8 on page 433 of the text. Use Minitab to compare the starting salaries of male and female college graduates for pairs of graduates with the same major and similar grade point averages.

Pair	Male	Female
1	$24,300	$23,800
2	26,500	26,600
3	25,400	24,800
4	23,500	23,500
5	28,500	27,600
6	22,800	23,000
7	24,500	24,200
8	26,200	25,100
9	23,400	23,200
10	24,200	23,500

a. Graphically compare the starting salaries of male and female graduates.
b. Construct a 95% confidence interval for μ_D. Interpret.
c. Do the data provide sufficient evidence that the mean starting salary for males exceeds the mean starting salary for females? Test using $\alpha = .05$.

Solution

a. We use MPLOT to graphically compare the paired samples.

```
MTB > NAME C1 'PAIR' C2 'M SALARY' C3 'F SALARY'
MTB > SET 'PAIR'
DATA> 1:10
DATA> END
MTB > SET 'M SALARY'
DATA> 24300   26500   25400   23500   28500   22800   24500   26200   23400   24200
DATA> END
MTB > SET 'F SALARY'
DATA> 23800   26600   24800   23500   27600   23000   24200   25100   23200   23500
DATA> END
MTB > SAVE 'SALARIES'

Worksheet saved into file:   SALARIES.MTW
```

```
MTB > MPLOT 'M SALARY' VS 'PAIR' AND 'F SALARY' VS 'PAIR';
SUBC> TITLE 'MALE AND FEMALE STARTING SALARIES'.
```

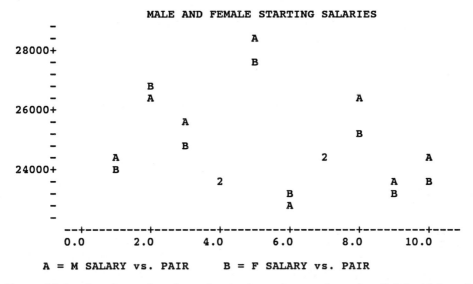

```
       A = M SALARY vs. PAIR     B = F SALARY vs. PAIR
```

The multiple plot shows that the male starting salary tends to be slightly higher than the corresponding female salary in the matched pair. We also see that both salaries within pairs tend to increase or decrease from one pair to the next; that is, the two samples are not independent and should not be analyzed with procedures for independent samples.

b. To construct a 95% confidence interval, we calculate the differences in male and female starting salaries. The population differences are assumed to be normal.

```
MTB > NAME C4 'DIFF'
MTB > LET 'DIFF' = 'M SALARY' - 'F SALARY'
MTB > PRINT 'PAIR' 'M SALARY' 'F SALARY' 'DIFF'

ROW    PAIR   M SALARY   F SALARY   DIFF

 1      1      24300      23800      500
 2      2      26500      26600     -100
 3      3      25400      24800      600
 4      4      23500      23500        0
 5      5      28500      27600      900
 6      6      22800      23000     -200
 7      7      24500      24200      300
 8      8      26200      25100     1100
 9      9      23400      23200      200
10     10      24200      23500      700

MTB > DESCRIBE 'DIFF'

              N      MEAN    MEDIAN   TRMEAN   STDEV   SEMEAN
DIFF         10       400       400      387     435      137

             MIN       MAX        Q1       Q3
DIFF        -200      1100       -25      750
```

172   Chapter 9

```
MTB > DOTPLOT 'DIFF'
```

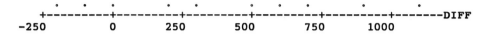

```
MTB > TINTERVAL 'DIFF'

              N      MEAN     STDEV    SE MEAN    95.0 PERCENT C.I.
DIFF         10   400.000   434.613    137.437   ( 89.013, 710.987)
```

The dot plot does not indicate any extreme values. Male graduates average $400 more starting salary than female graduates for pairs of students with the same major and grade point average. We are 95% confident the mean difference between salaries is between $89 and $711; that is, salaries average between $89 and $711 higher for males than for females.

c.   To test whether the mean difference between starting salaries is greater than 0, the hypotheses are

$H_0: \mu_D = 0$
$H_a: \mu_D > 0$

```
MTB > TTEST 'DIFF';
SUBC> ALTERNATIVE = 1.

TEST OF MU =   0.000 VS MU G.T.   0.000

              N      MEAN     STDEV    SE MEAN        T     P VALUE
DIFF         10   400.000   434.613    137.437     2.91     0.0087

MTB > STOP
```

Since the observed significance level or p-value of .0087 is less than $\alpha = .05$, we reject the null hypothesis. We have sufficient evidence to conclude the mean starting salary for males exceeds the mean starting salary for females.   ∎

Comment   *If there is doubt about the normal population of differences, a nonparametric procedure that is described in Chapter 11 should be used.*

INFERENCES ABOUT THE DIFFERENCES BETWEEN POPULATION PROPORTIONS: INDEPENDENT BINOMIAL EXPERIMENTS

Suppose the proportion of successes in one population is p_1, and p_2 in another population. To make an inference about the difference between these proportions, the point estimator is the difference between corresponding sample proportions $(\hat{p}_1 - \hat{p}_2)$, where $\hat{p}_1 = x_1/n_1$ and $\hat{p}_2 = x_2/n_2$. With large sample sizes, the sampling distribution of $(\hat{p}_1 - \hat{p}_2)$ is approximately normal.

There are no Minitab commands to construct a confidence interval or do a hypothesis test about the difference between two proportions. You can use the LET command to evaluate the formulas. The confidence interval for $(p_1 - p_2)$ is

$$(\hat{p}_1 - \hat{p}_2) \pm z_{\alpha/2}\sqrt{p_1 q_1 / n_1 + p_2 q_2 / n_2}$$

The null and alternative hypotheses for a two-tailed test are

H_0: $p_1 - p_2 = 0$
H_a: $p_1 - p_2 \neq 0$

and the test statistic is

$$z = \frac{(\hat{p}_1 - \hat{p}_2)}{\sqrt{\hat{p}\hat{q}(1/n_1 + 1/n_2)}}$$

where $\hat{p} = (x_1 + x_2)/(n_1 + n_2)$. Use CDF to determine the p-value for the hypothesis test.

■ **Example 3** Consider the comparison of the percentage of smokers in 1985 and 1990 described in Example 9.9 on page 447 of the text. In a random sample of 1,500 adults in 1985, the American Cancer Society found 576 smokers and in a random sample of 2,000 adults in 1990, 652 were smokers. Is there sufficient evidence that the proportion of smokers decreased over the five years? Test using $\alpha = .05$.

Solution The null and alternative hypotheses are

H_0: $p_1 - p_2 = 0$
H_a: $p_1 - p_2 > 0$

where p_1 and p_2 are the proportion of smokers in 1985 and 1990.

```
MTB > LET K1 = 576/1500     # 1985 PROPORTION OF SMOKERS
MTB > LET K2 = 652/2000     # 1990 PROPORTION OF SMOKERS
MTB > LET K3=(576+652)/(1500+2000)     # AVERAGE PROPORTION
MTB > LET K4=(K1-K2)/SQRT(K3*(1-K3)*(1/1500+1/2000))   # Z TEST STATISTIC
MTB > PRINT K4
K4        3.55811
MTB > # THE P-VALUE
MTB > CDF K4
     3.5581     0.9998
MTB > LET K5=1-.9998
K5        0.0002
MTB > STOP
```

The observed significance level, or p-value, of .0002 is less than $\alpha = .05$. The samples provide sufficient evidence that the proportion of smokers has decreased from 1985 to 1990. ■

COMPARING TWO POPULATION VARIANCES: INDEPENDENT RANDOM SAMPLES

The small sample t test assumes that the population variances are equal. The statistical procedure to test equal variances is an F test based on the ratio, σ_1^2/σ_2^2. The test statistic is $F = s_1^2/s_2^2$. If the two samples are randomly and independently selected from normal population distributions with equal variances, the sampling distribution of s_1^2/s_2^2 is the F distribution with respective numerator and

174   Chapter 9

denominator degrees of freedom, $v_1 = n_1 - 1$ and $v_2 = n_2 - 2$. Minitab does not have a command to do this F test. Use LET to calculate the F test statistic and CDF to obtain the p-value. The p-value is the upper tail area of the F distribution for a one-tailed test, and twice that for a two-tailed test.

■ **Example 4** Consider the rental data that are given in Example 1 of this chapter. A large international corporation rents several apartments for relocating employees in Phoenix, Arizona. The following random samples of apartment rentals were independently selected from the city and suburbs.

City			Suburb		
$515	$995	$345	$575	$700	$850
525	600	575	640	650	495
475	550	725	595	650	750
850	465	435	725	600	1075
750	465	625	975	625	650
495	750	500	700	685	100
535	575	549	1175	650	545
515	515	495	850	825	750
525	525	550	50	575	550
475	1150	650	1100	585	1000

Use the F test to check whether the variances of the monthly rents of apartments within the city and the surrounding suburbs are equal. Calculate and interpret the p-value.

Solution The rental data are saved in the file RENTS. The numerator and denominator degrees of freedom both equal 29. The hypothesis test is

$$H_0: \sigma_1^2 = \sigma_2^2$$
$$H_a: \sigma_1^2 \neq \sigma_2^2$$

```
MTB > RETRIEVE 'RENTS'
  WORKSHEET SAVED   1/10/1994

Worksheet retrieved from file: RENTS.MTW
MTB > INFORMATION

COLUMN      NAME        COUNT
C1          CITY          30
C2          SUBURB        30

CONSTANTS USED: NONE

MTB > STDEV 'CITY' K1
   ST.DEV. =        168.71
MTB > STDEV 'SUBURB' K2
   ST.DEV. =        189.21
```

```
MTB > # CALCULATE THE F TEST STATISTIC
MTB > LET K3 = K2**2/K1**2
MTB > PRINT K1 K2 K3
K1        168.705
K2        189.211
K3        1.25787
MTB > CDF K3;           # CALCULATE THE P-VALUE
SUBC> F WITH 29 AND 29 DF.
      1.2579     0.7297
MTB > LET K4 = 1 - .7297
MTB > PRINT K4
K4        0.270300
MTB > STOP
```

The p-value is the probability of obtaining a value of F as large as 1.25787, assuming H_0 is true. For this two-tailed test, the p-value = 2(.2703) = .5406. This is greater than an α level as high as .10. We do not have sufficient evidence in the samples to conclude the variances of city and suburb rents differ. We can assume the population variances are equal for the two sample t test. ∎

EXERCISES

1. The Minnesota Real Estate Research Center compiles information on homes sold in several areas of Minnesota. The following table gives the 1989 selling prices of twenty homes randomly selected from the homes sold in St. Cloud and Rochester.

St. Cloud		Rochester	
$85,000	$104,400	$103,925	$129,900
46,000	90,600	66,000	47,800
78,900	53,500	90,900	96,000
123,000	119,500	53,000	92,330
116,000	54,000	125,500	54,900
46,600	64,500	61,500	144,000
99,875	71,900	64,000	89,000
64,000	69,900	80,750	85,900
52,000	111,900	74,500	135,000
52,500	64,500	129,195	58,000

    a. Use DESCRIBE and DOTPLOT to compare the samples.
    b. Construct a 95% confidence interval for $(\mu_1 - \mu_2)$. Interpret.
    c. Do the data provide sufficient evidence to conclude that the mean sales price of homes in St. Cloud is less than the mean sales price in Rochester? Test using $\alpha = .05$.

2. Consider the 1988 top twenty-five newspaper advertisers reported in the *Leading National Advertisers/Media Records*. The advertising dollars (in thousands) are listed for 1987 and 1988.

Advertisers	1987	1988
May Dept. Stores	$230,174	$224,282
Macy Acquiring Corp.	176,033	182,212
Sears, Roebuck & Co.	175,052	160,969
Campeau Corp.	134,833	124,488
Dayton-Hudson Corp.	81,062	90,630
General Motors Corp.	83,841	86,835
Texas Air Corp.	66,647	82,441
Philip Morris Cos.	46,048	76,711
Carter Hawley Hale Stores	52,353	75,554
J.C.Penney Co.	90,721	74,568
K Mart Corp.	82,033	73,073
Montgomery Ward & Co.	67,176	63,500
American Stores Co.	65,659	61,803
Time Warner	36,748	42,037
Woodward & Lothrop	36,625	38,856
RJR Nabisco	12,934	37,704
B.A.T. Industries PLC	61,875	35,915
Tandy Corp	38,900	32,911
AT&T	32,610	32,690
Dillard Department Stores	29,057	32,073
American Express Co.	29,286	30,251
Circuit City Stores	25,124	29,654
General Electric Co.	33,121	29,357
Safeway Stores	34,403	28,738
AMR Corp.	36,654	28,671

    a. Numerically and graphically summarize the 1987 and 1988 newspaper advertising dollars.
    b. Construct a 95% confidence interval for the mean difference between advertising expenditures for 1987 and 1988. Explain why a paired difference procedure is required.
    c. Test whether the samples provide sufficient evidence to conclude that the mean advertising expenditure for the 25 companies has increased from 1987 to 1988. Use $\alpha = .05$.
    d. What assumptions are necessary to apply a paired difference analysis to the data?

3. The Census Bureau Foreign Trade Division compiles export and import data of the United States. The following is a list of some exports and imports (in billion dollars) for January through August of 1988 and 1989.

	U.S. Exports		U.S. Imports	
	1989	1988	1989	1988
Office Computers	$14.97	$14.92	$16.08	$14.42
Electrical machinery	15.70	14.03	21.25	19.67
Petroleum & Products	3.10	2.52	32.32	26.56
Medicinal chemicals	2.42	2.54	1.41	1.24
Clothes & footwear	1.61	1.16	21.73	19.77
Furniture & parts	.64	.53	3.25	3.14
Paper & paperboard	2.82	2.53	5.73	5.58
Power machinery	9.34	8.21	9.54	8.21

Scientific instruments	7.20	5.76	3.82	3.36
Telecommunications	4.87	4.20	15.07	13.91
Toys, games, sports	.96	.81	4.91	4.05
Cars-Canada	4.08	4.10	8.30	8.83
Cars-Japan	.19	.12	12.95	12.29
Auto parts	7.50	8.04	10.30	9.57
Industrial machinery	8.67	6.84	9.93	8.59
Iron & steel	2.15	1.09	6.30	6.87
Cotton	1.47	1.48	-	-
Soybeans	2.54	3.18	.02	-
Rice	.61	.49	.04	.04
Lumber, wood, cork	3.32	2.93	2.42	2.27

a. Test whether the data provide sufficient evidence to conclude more dollars worth of commodities and merchandise were imported to the United States than were exported during the eight months of 1989. Use $\alpha = .05$.

b. Test whether the data provide sufficient evidence to conclude that more dollars worth of commodities and merchandise were imported to the United States than were exported during the eight months of 1988. Use $\alpha = .05$.

c. Construct a 90% confidence interval for the mean difference between 1988 and 1989 exports. Interpret.

d. Construct a 90% confidence interval for the mean difference between 1988 and 1989 imports. Interpret.

4. The price earnings ratio or PE ratio is the price of a share of stock divided by the earnings per share. The PE ratio reflects the price an investor is willing to pay for a dollar of earnings. The following table gives the reported PE ratios as of February 1, 1990, of two random samples of stocks in the financial industry and computers and technology industry.

Financial	PE	Computers and Technology	PE
Alex&Alex Svc	17.9	Mtn Med Equip	17.3
Aetna Life	8.0	Apple Cptr	10.1
Am Express	11.2	TSI Inc.	13.3
Wells Fargo	6.3	IBM	15.2
Norwest Cp	8.1	Hewlett-Pack	12.7
Am Intl Grp	11.8	Medtronic Inc	17.2
Old Rep Intl	12.5	NCR	12.6
Teledyne Inc	15.1	Cray Research	14.8
Marsh&McLn	18.6	Digital Equip	10.9
Am Gen Corp	8.8	Research Inc	10.0

a. Graphically and numerically compare the two samples.

b. Construct a 95% confidence interval for the difference between mean price earnings ratios for the two industries. Interpret. What assumptions are necessary for this analysis?

c. Test whether the data provides sufficient evidence that the mean price earnings ratio of the financial industry is less than that of the computers and technology industry. Use $\alpha = .05$.

5. Product quality is thought to be related to employee turnover rates. Consider the following data on annual turnover rates in U.S. and Japanese air conditioner manufacturing plants given in Exercise 9.29 on page 426 of the text.

U.S. Plants	Japanese Plants
7.11%	3.52%
6.06%	2.02%
8.00%	4.91%
6.87%	3.22%
4.77%	1.92%

   a. Use DOTPLOT and DESCRIBE to summarize the samples.
   b Test whether the data provide sufficient evidence that the mean turnover rate for U.S. plants exceeds the mean turnover rate for Japanese plants. Use $\alpha = .05$. Interpret the p-value.
   c. What assumptions must be satisfied for the test in part b? Use Minitab to test whether the population variances are equal. Use $\alpha = .05$.

6. An EPA inspector has collected six water specimens from the discharge of a plant purification system and six water specimens from the river, upstream from the plant. The following bacteria counts for each specimen are determined and reported in Exercise 9.32 on page 427 of the text.

Plant Discharge	Upstream
30.1	29.7
36.2	30.3
33.4	26.4
28.2	27.3
29.8	31.7
34.9	32.3

   a. Use DOTPLOT and DESCRIBE to summarize the samples. Do you think that bacteria counts are approximately normal distributed?
   b Test whether the samples provide sufficient evidence that the mean bacteria count for the plant discharge exceeds that for the upstream location. Use $\alpha = .05$. Interpret the p-value.
   c. What assumptions must be satisfied for the test in part b? Discuss.

7. Refer to Exercise 9.47 on page 442 of the text. To compare shocks produced by a manufacturer and a competitor, shocks of each company were put on the rear wheels of each of six cars. After 20,000 miles, the strengths of the shocks were recorded. The results are given in the table.

Car	Manufacturer	Competitor
1	8.8	8.4
2	10.5	10.1
3	12.5	12.0
4	9.7	9.3
5	9.6	9.0
6	13.2	13.0

   a. Graphically compare the strengths of shocks for the two companies.

b. Construct a 95% confidence interval for μ_D. Interpret.
c. Do the data provide sufficient evidence that there is a difference between the mean strength of the two types of shocks after 20,000 miles? Use $\alpha = .05$.

8. Exercise 9.59 on page 449 of the text describes a study to compare two surgical procedures to treat a certain type of cancer. Random samples of 100 patients of each type of surgical procedure were obtained and the numbers of patients with no cancer reoccurrence after one year were recorded. For one procedure, 78 patients had no reoccurrence and for another procedure, 87 had no reoccurrence. Test whether there is sufficient evidence of a difference in success rates of the two surgical procedures. Use $\alpha = .05$. Calculate and interpret the p-value.

9. Exercise 9.91 on page 467 of the text describes a study comparing quality performance in terms of variance in inspection errors of novice and experienced inspectors. A sample of 12 novice and a sample of 12 experienced inspectors evaluated the same 200 finished products. The following table lists the number of inspection errors made by each inspector.

Novice		Experienced	
30	45	31	19
35	31	15	18
26	33	25	24
40	29	19	10
36	21	28	20
20	48	17	21

a. Graphically and numerically describe the two samples.
b. Do the data provide sufficient evidence that the variance in inspection errors was lower for experienced inspectors than for novice inspectors. Calculate and interpret the p-value.

10. A class data set containing information on 200 students enrolled in a statistics course is given in Appendix B. Two samples of business and nonbusiness majors were randomly selected from the data set. The grade point averages of the students are given as follow:

Business Major		Nonbusiness Major	
3.440	2.650	2.790	3.100
3.941	3.360	3.400	2.890
2.730	3.680	4.000	2.560
3.625	2.760	2.666	3.000
3.330	3.100	3.300	3.900
3.460	3.890	3.750	2.680
2.700	2.500	2.000	2.900
2.690	3.010	3.428	3.650
3.450	3.200	2.750	2.850
3.125	2.800	3.800	2.768

a. Compare the grade point averages for business and nonbusiness majors using numerical descriptive measures.
b. Graphically compare the two samples.
c. Construct a 99% confidence interval for the difference between mean grade point averages.

Interpret.
d. Test whether the samples provide sufficient evidence of a difference between mean grade point averages for the two majors. Use $\alpha = .05$.

CHAPTER 10

ANALYSIS OF VARIANCE: COMPARING MORE THAN TWO MEANS

The previous chapter presented methods for making inferences about the difference between two population means. This chapter presents methods to compare more than two populations means. The appropriate method to use in a given analysis depends on the design of the experiment. We describe several important experimental designs, and show how to use Minitab for the analysis of variance.

NEW COMMANDS

ANOVA   AOVONEWAY   LPLOT   ONEWAY   TWOWAY

THE COMPLETELY RANDOMIZED DESIGN

The completely randomized design is one in which independent samples are randomly selected from each treatment population. This design is perhaps the simplest experimental design for comparing more than two population means. For example, if a financial analyst is interested in the annual rate of return on investment in stock, bond, and foreign mutual funds, the analyst would independently select a random sample from each type of fund. Or if a data processing manager wants to compare four financial computer software programs, the manager would independently select a random sample of financial analysts for each software program.

An objective of a completely randomized design is to compare treatment means. The variable of interest is the dependent or response variable, and the independent variables are called factors. A treatment is a factor level in an experiment with one factor, or a combination of levels if two or more factors are studied simultaneously. Treatments are applied to objects called experimental units.

The analysis of variance is used to determine whether a factor has a significant effect on the response variable. If the factor is significant, there is a difference in treatment mean responses. Two Minitab commands to do an analysis of variance for balanced and unbalanced completely randomized designs with a single factor are AOVONEWAY and ONEWAY. A balanced design has an equal number of observations per treatment and an unbalanced design has an unequal number. Use AOVONEWAY if the data is unstacked; that is, if the sample data are in different columns. Use ONEWAY if the data are organized as stacked data, or if you want multiple comparisons, such as the Tukey and Fisher procedures. Some descriptive statistics and graphical analyses, for example, those obtained with DESCRIBE and DOTPLOT, are useful in analyzing observational and experimental data and in verifying the assumptions.

Chapter 10

> AOVONEWAY ON THE DATA IN C,...,C
>
> This command does a single factor analysis of variance for unstacked data. The first column contains the observations for the first treatment, the second column for the second treatment, and so on. The printout includes the analysis of variance table; the number of observations, mean, standard deviation, and 95% confidence interval for each treatment mean; and the pooled standard deviation.
>
> ONEWAY, DATA IN C, LEVELS IN C (RESIDUALS IN C (FITS IN C))
>   TUKEY (FAMILY ERROR RATE K)
>   FISHER K (INDIVIDUAL ERROR RATE K)
>
> This command does a single factor analysis of variance for stacked data. Enter the observations in one column, and codes, usually 1, 2, ..., identifying the treatments in a second column. The output is similar to that of AOVONEWAY. The option stores the residuals and fitted values, or level means, in columns.
>
> The subcommands provide confidence intervals for all pairwise differences between treatment level means. TUKEY requires a family error rate, and calculates the individual rate. FISHER requires an individual error rate, and calculates a family rate. If the corresponding family and individual rates are identical, the TUKEY and FISHER results are similar.

■ **Example 1** Refer to Example 10.3 on page 493 of the text. The United States Golf Association (USGA) compares distances traveled by four different brands of golf balls when struck with a driver. A completely randomized design resulted in the following table of distances.

Brand A	Brand B	Brand C	Brand D
251.2	263.2	269.7	251.6
245.1	262.9	263.2	248.6
248.0	265.0	277.5	249.4
251.1	254.5	267.4	242.0
265.5	264.3	270.5	246.5
250.0	257.0	265.5	251.3
253.9	262.8	270.7	262.8
244.6	264.4	272.9	249.0
254.6	260.6	275.6	247.1
248.8	255.9	266.5	245.9

a. Numerically and graphically analyze the distances traveled by the four brands.
b. Use AOVONEWAY to obtain the analysis of variance table. Test whether the mean distances differ for the four brands. Use $\alpha = .10$. Interpret the p-value.
c. Interpret the 95% confidence intervals given on the printout.

Solution To use AOVONEWAY, we enter the distances for the four brands in different columns.

a. The SAME subcommand is used with DOTPLOT so that the distances of the four groups are plotted on the same scale.

```
MTB > NAME  C1 'BRAND A'  C2 'BRAND B'  C3 'BRAND C'  C4 'BRAND D'
MTB > SET 'BRAND A'
DATA> 251.2   245.1   248.0   251.1   265.5   250.0   253.9   244.6   254.6   248.8
DATA> END
MTB > SET 'BRAND B'
DATA> 263.2   262.9   265.0   254.5   264.3   257.0   262.8   264.4   260.6   255.9
DATA> END
MTB > SET 'BRAND C'
DATA> 269.7   263.2   277.5   267.4   270.5   265.5   270.7   272.9   275.6   266.5
DATA> END
MTB > SET 'BRAND D'
DATA> 251.6   248.6   249.4   242.0   246.5   251.3   262.8   249.0   247.1   245.9
DATA> END
MTB > SAVE 'GOLF-CRD'

Worksheet saved into file: GOLF-CRD.MTW
MTB > DESCRIBE 'BRAND A'-'BRAND D'

                N       MEAN     MEDIAN     TRMEAN      STDEV     SEMEAN
BRAND A        10      251.28     250.55     250.34       5.98       1.89
BRAND B        10      261.06     262.85     261.39       3.87       1.22
BRAND C        10      269.95     270.10     269.85       4.50       1.42
BRAND D        10      249.42     248.80     248.67       5.47       1.73

              MIN       MAX         Q1         Q3
BRAND A    244.60    265.50     247.27     254.07
BRAND B    254.50    265.00     256.73     264.32
BRAND C    263.20    277.50     266.25     273.58
BRAND D    242.00    262.80     246.35     251.37

MTB > DOTPLOT 'BRAND A'-'BRAND D';
SUBC> SAME.
```

```
                  . .      ... .     . .               .
        ------+---------+---------+---------+---------+---------BRAND A

                              . ..       . .: :.
        ------+---------+---------+---------+---------+---------BRAND B

                                        .   . ..  ...    .    . .
        ------+---------+---------+---------+---------+---------BRAND C

              .   ... .: :             .
        ------+---------+---------+---------+---------+---------BRAND D
            245.0     252.0     259.0     266.0     273.0     280.0
```

The descriptive statistics and dot plots indicate that brand C seems to travel the greatest distance. Brands A and D are about equal and travel the shortest distances. However, there is some overlap among the distances traveled by the brands.

b. To test whether the mean distances for the four brands differ, the null and alternative hypotheses are:

$H_0: \mu_1 = \mu_2 = \mu_3 = \mu_4$
$H_a:$ At least two of the mean distances differ

```
MTB > AOVONEWAY ON 'BRAND A'-'BRAND D'

ANALYSIS OF VARIANCE
SOURCE     DF        SS        MS        F         p
FACTOR      3    2709.2     903.1    35.81     0.000
ERROR      36     907.9      25.2
TOTAL      39    3617.1
                                   INDIVIDUAL 95 PCT CI'S FOR MEAN
                                   BASED ON POOLED STDEV
LEVEL       N      MEAN     STDEV   ---+---------+---------+---------+---
BRAND A    10    251.28      5.98   (---*---)
BRAND B    10    261.06      3.87                   (---*---)
BRAND C    10    269.95      4.50                                (---*---)
BRAND D    10    249.42      5.47   (---*---)
                                   ---+---------+---------+---------+---
POOLED STDEV =    5.02            248.0     256.0     264.0     272.0

MTB > STOP
```

The test statistic, $F = 35.81$ has a p-value of 0. The p-value is the probability of obtaining a test statistic value as large as $F = 35.81$, assuming H_0 is true. Since the p-value is less than $\alpha = .10$, the null hypothesis is rejected. We have sufficient evidence to conclude that the true mean distances differ for at least two of the four brands.

c. The 95% confidence intervals on the printout are determined using the formula,

$$\bar{x} \pm t_{.025}(\text{Pooled Stdev})/\sqrt{n}$$

The confidence intervals give an indication of which means differ. If two intervals do not overlap, it is possible that the corresponding mean distances differ significantly. The confidence intervals for brand B and C do not overlap with any other confidence intervals. The mean distance traveled for brand C seems to exceed the mean for brand B; brand B seems to exceed brands A and D. A procedure for comparing treatment means is illustrated in the next section. ■

Multiple Comparisons of Means

If the conclusion of the F test is a significant difference in treatment means, we want to know which pairs of treatment means differ. There are several procedures for multiple comparisons, some of which are can be done with ONEWAY subcommands. The following example illustrates the TUKEY and FISHER procedures. TUKEY requires a family error rate α, and FISHER, the corresponding individual error rate α/c, where c equals the number of pairs of means. The TUKEY and FISHER results are comparable for corresponding family and individual rates. Other subcommands with ONEWAY are DUNNETT and MCB. Refer to the HELP facility for specific information.

■ **Example 2** In Example 1, we found that the mean distances differ for at least two golf brands. Use $\alpha = .10$ and the TUKEY and FISHER multiple comparison procedures to determine which pairs of means differ. Obtain the residuals and fitted values.

Solution ONEWAY requires stacked data. The distances are put in one column, and codes for the four golf brands are put in a second column. For TUKEY, we enter the family error rate $\alpha = .10$, and for FISHER, the corresponding individual error rate, $\alpha/c = .10/6 = .0167$, for 6 pairs of means.

```
MTB > RETRIEVE 'GOLF-CRD'
  WORKSHEET SAVED   1/10/1994

Worksheet retrieved from file: GOLF-CRD.MTW
MTB > INFO

Column    Name         Count
C1        BRAND A      10
C2        BRAND B      10
C3        BRAND C      10
C4        BRAND D      10

MTB > NAME C5 'ALL DIST' C6 'BRAND' C7 'RESIDS' C8 'FITS'
MTB > STACK C1-C4 IN 'ALL DIST';
SUBC> SUBSCRIPTS IN 'BRAND'.
MTB > ONEWAY 'ALL DIST' TREATMENT 'BRAND' STORE 'RESIDS' 'FITS';
SUBC> TUKEY .10;
SUBC> FISHER .0167.

ANALYSIS OF VARIANCE ON ALL DIST
SOURCE       DF         SS         MS        F         p
BRAND         3     2709.2      903.1    35.81     0.000
ERROR        36      907.9       25.2
TOTAL        39     3617.1
                                       INDIVIDUAL 95% CI'S FOR MEAN
                                       BASED ON POOLED STDEV
LEVEL     N       MEAN      STDEV    ---+---------+---------+---------+---
    1    10     251.28       5.98    (---*---)
    2    10     261.06       3.87                   (---*---)
    3    10     269.95       4.50                                (---*---)
    4    10     249.42       5.47    (---*---)
                                     ---+---------+---------+---------+---
POOLED STDEV =    5.02             248.0     256.0     264.0     272.0
```

Tukey's pairwise comparisons

    Family error rate = 0.100
Individual error rate = 0.0229

Critical value = 3.36
Intervals for (column level mean) - (row level mean)

```
              1           2           3
    2    -15.116
          -4.444

    3    -24.006     -14.226
         -13.334      -3.554


    4     -3.476       6.304      15.194
           7.196      16.976      25.866
```

Fisher's pairwise comparisons

    Family error rate = 0.0753
Individual error rate = 0.0167

Critical value = 2.510

Intervals for (column level mean) - (row level mean)

```
              1           2           3
    2    -15.417
          -4.143

    3    -24.307     -14.527
         -13.033      -3.253

    4     -3.777       6.003      14.893
           7.497      17.277      26.167
```

PRINT 'RESIDS' 'FITS'

ROW	ALL DIST	BRAND	RESIDS	FITS
1	251.2	1	-0.0800	251.28
2	245.1	1	-6.1800	251.28
3	248.0	1	-3.2800	251.28
4	251.1	1	-0.1800	251.28
5	265.5	1	14.2200	251.28
6	250.0	1	-1.2800	251.28
7	253.9	1	2.6200	251.28
8	244.6	1	-6.6800	251.28
9	254.6	1	3.3200	251.28
10	248.8	1	-2.4800	251.28
11	263.2	2	2.1400	261.06
12	262.9	2	1.8400	261.06
13	265.0	2	3.9400	261.06
14	254.5	2	-6.5600	261.06
15	264.3	2	3.2400	261.06
16	257.0	2	-4.0600	261.06
17	262.8	2	1.7400	261.06
18	264.4	2	3.3400	261.06
19	260.6	2	-0.4600	261.06
20	255.9	2	-5.1600	261.06

```
21      269.7      3    -0.2500    269.95
22      263.2      3    -6.7500    269.95
23      277.5      3     7.5500    269.95
24      267.4      3    -2.5500    269.95
25      270.5      3     0.5500    269.95
26      265.5      3    -4.4500    269.95
27      270.7      3     0.7500    269.95
28      272.9      3     2.9500    269.95
29      275.6      3     5.6500    269.95
30      266.5      3    -3.4500    269.95
31      251.6      4     2.1800    249.42
32      248.6      4    -0.8200    249.42
33      249.4      4    -0.0200    249.42
34      242.0      4    -7.4200    249.42
35      246.5      4    -2.9200    249.42
36      251.3      4     1.8800    249.42
37      262.8      4    13.3800    249.42
38      249.0      4    -0.4200    249.42
39      247.1      4    -2.3200    249.42
40      245.9      4    -3.5200    249.42

MTB > STOP
```

The column and row numbers 1 through 4 in the TUKEY and FISHER confidence interval tables correspond to brands A through D. For example, the TUKEY interval for the difference between mean distances traveled by brands A and B is (-15.116, -4.444). With both procedures, the interval for the difference between mean distances traveled by brands A and D contains 0, and does not indicate a difference in these means. Since the other intervals do not contain 0, we can conclude that there are differences between mean distances traveled by all other pairs of brands. The fitted values are the brand means and the residuals are the differences between the distances and the means. ■

RANDOMIZED BLOCK DESIGNS

In a randomized block design, the experimental units are grouped into blocks, and treatments are randomly assigned within the blocks. The objective of a blocked experiment is to increase the precision of the experimental results by reducing the sampling variability. The variable selected for blocking should be one which accounts for some variability among the responses; that is, one which is correlated with the response or in some way affects the response. A blocking variable can be one associated with the experiment itself, such as the subject, time, or the measuring instrument. Or a blocking variable can be a characteristic of the experimental unit, such as the age and education of an individual, or the population size of a geographical area.

As an example, consider a study of three different advertising campaigns on sales volume. Suppose the size of the city selected for a campaign is correlated with sales volume. If you had 12 cities in our study, you could group the cities in four blocks according to city size, and randomly assign advertising campaigns to the three cities within each block.

The TWOWAY and ANOVA commands construct the analysis of variance table for the randomized block design. Both commands require stacked data.

> TWOWAY, DATA IN C, FACTOR LEVELS IN C, C (RESIDS IN C (FITS IN C))
>   MEANS FOR C (C)
>   ADDITIVE
>
> The response data are stacked in the first column and codes for the factor levels in the other columns. The factor levels are usually coded 1, 2, and so on; however, any integer between -10,000 and +10,000 may be used for the code.
>
> The output of TWOWAY is the analysis of variance table. The MEANS subcommand gives the confidence intervals for the factor level means. The ADDITIVE subcommand is used for an analysis of variance if there are no interactions.

The ANOVA command does an analysis of variance for single factor balanced and unbalanced designs, and for more complex balanced designs. Factors may be fixed or random for ANOVA. A factor is considered fixed if the conclusions pertain only to those factor levels included in the study. A factor is random if the levels constitute a sample from a population, and interest is in the population. The single factor analysis of variance is the same for a fixed or random factor. In a multifactor analysis of variance, the F test statistic differs for a random factor.

> ANOVA C = C,...,C
>   MEANS C,...,C
>   RANDOM C,...,C
>   RESIDUALS C,...,C
>   FITS C,...,C
>
> This command does an analysis of variance for balanced and unbalanced single factor designs and balanced multifactor designs. All the observations are stacked in the first column, and codes identifying the factor levels are in separate columns. Usually the levels are coded 1, 2, and so on; however, any integer between -9999 and +9999 may be used.
>
> Use an asterisk * to denote an interaction term, and a vertical line | or exclamation point to specify all interactions. If the column names for variables have no special symbols, the single quotes around the names may be omitted. Extra text may appear on the command line only after the # comment symbol.
>
> The output includes a factor table and an analysis of variance table. The factor table lists the factors and codes for each factor. The analysis of variance table includes the F test statistics and p-values. Use the MEANS subcommand to obtain a summary table of factor means, RANDOM to specify random factors, RESIDUALS to store residuals, and FITS to store fitted values. The fitted values are cell means if you specify a full model.

The LPLOT command adds information about a third variable to a scattergram. Letters are used to plot different observations of the third variable. The observations may be levels of a quantitative or a qualitative variable.

LPLOT C VS C, USING LETTERS AS CODED IN C
  XINCREMENT = K
  XSTART AT K (END AT K)
  YINCREMENT = K
  YSTART AT K (END AT K)
  TITLE = 'TEXT'
  FOOTNOTE = 'TEXT'
  XLABEL = 'TEXT'
  YLABEL = 'TEXT'

LPLOT puts the first column on the vertical axis, and the second on the horizontal axis. The plotting letters are determined by the numerical code given in the third column, and the following correspondence:

$$...-2:X,\ -1:Y,\ 0:Z,\ 1:A,\ 2:B,\ 3:C,...$$

If several points fall at the same location, a count is given on the plot. The subcommands are the same as those defined with MPLOT.

■ **Example 3** Refer to Example 10.6, page 515, of the text. A randomized block design is employed with a random sample of ten golfers to study four brands of golf balls. Each golfer used a driver to hit one ball of each brand, in a random sequence. The distances traveled by the golf balls are given in the accompanying table.

Golfer	Brand A	Brand B	Brand C	Brand D
1	202.4	203.2	223.7	203.6
2	242.0	248.7	259.8	240.7
3	220.4	227.3	240.0	207.4
4	230.0	243.1	247.7	226.9
5	191.6	211.4	218.7	200.1
6	247.7	253.0	268.1	244.0
7	214.8	214.8	233.9	195.8
8	245.4	243.6	257.8	227.9
9	224.0	231.5	238.2	215.7
10	252.2	255.2	265.4	245.2

a. Construct a labeled plot of distance versus golfer using letters for the brands. Does golfer seem to be a good blocking variable? Do some brands seem to have consistently greater or smaller distances?

b. Use the ANOVA command to obtain the analysis of variance table. Test whether golfer is a

190 *Chapter 10*

significant blocking variable. Use $\alpha = .05$.

c. Test whether the mean distances differ significantly for the four brands. Use $\alpha = .05$.

Solution

a. We stack the distances in one column, and use codes 1 to 4 for brands, and codes 1 to 10 for golfers. TABLE organizes the distance data.

```
MTB > NAME C1 'DISTANCE'   C2 'GOLFER' C3 'BRAND'
MTB > SET 'DISTANCE'
DATA> 202.4   242.0   220.4   230.0   191.6   247.7   214.8   245.4   224.0   252.2
DATA> 203.2   248.7   227.3   243.1   211.4   253.0   214.8   243.6   231.5   255.2
DATA> 223.7   259.8   240.0   247.7   218.7   268.1   233.9   257.8   238.2   265.4
DATA> 203.6   240.7   207.4   226.9   200.1   244.0   195.8   227.9   215.7   245.2
DATA> END
MTB > SET 'GOLFER'
DATA> 4(1:10)
DATA> END
MTB > SET 'BRAND'
DATA> (1:4)10
DATA> END
MTB > SAVE 'GOLF-RBD'

Worksheet saved into file: GOLF-RBD.MTW
MTB > TABLE 'GOLFER' BY 'BRAND';
SUBC> DATA FOR 'DISTANCE'.

     ROWS: GOLFER    COLUMNS: BRAND

                1        2        3        4

      1     202.40   203.20   223.70   203.60
      2     242.00   248.70   259.80   240.70
      3     220.40   227.30   240.00   207.40
      4     230.00   243.10   247.70   226.90
      5     191.60   211.40   218.70   200.10
      6     247.70   253.00   268.10   244.00
      7     214.80   214.80   233.90   195.80
      8     245.40   243.60   257.80   227.90
      9     224.00   231.50   238.20   215.70
     10     252.20   255.20   265.40   245.20

     CELL CONTENTS --
           DISTANCE:DATA
```

```
MTB > LPLOT 'DISTANCE' VS 'GOLFER' USING LETTERS FOR 'BRAND';
SUBC> TITLE 'GOLF BALL DISTANCES';
SUBC> FOOTNOTE 'Letters represent Brands A, B, C, and D'.

                            GOLF BALL DISTANCES
            -
            -                               C
 DISTANCE-                                                          C
            -                 C                       C
            -                                 B                     B
       250+                 B           C     A                     A
            -                           B     D              2      D
            -         2     C
            -                                         C
            -                 A                       D       B
       225+         C       B D                                A
            -               A         C
            -                                                 D
            -                         B                2
            -     2       D
       200+       A                   D
            -                                 D
            -                 A
          --+---------+---------+---------+---------+---------+----GOLFER
           0.0       2.0       4.0       6.0       8.0      10.0
                Letters represent Brands A, B, C, and D
```

The labeled plot indicates considerable variability between golfers in the distance traveled. Golfer appears to be a good blocking variable. There is some consistency in the order of distances traveled by the brands. With most of the golfers, the order in terms of greatest to least distance is C, B, A, D.

b. The ANOVA command provides the analysis of variance table.

```
MTB > ANOVA DISTANCE = BRAND GOLFER;
SUBC> MEANS BRAND.

Factor     Type Levels Values
BRAND      fixed    4    1    2    3    4
GOLFER     fixed   10    1    2    3    4    5    6    7    8    9   10
Analysis of Variance for DISTANCE

Source     DF         SS        MS        F       P
BRAND       3     3298.7    1099.6    54.31   0.000
GOLFER      9    12073.9    1341.5    66.26   0.000
Error      27      546.6      20.2
Total      39    15919.2

        MEANS

   BRAND     N   DISTANCE
       1    10     227.05
       2    10     233.18
       3    10     245.33
       4    10     220.73

MTB > STOP
```

The hypothesis test for block effects is:

H_0: The golfer mean distances are equal
H_a: At least two golfer mean distances differ

From the ANOVA printout, the test statistic is $F = 66.26$. The p-value of 0 is less than $\alpha = .05$. Golfer is a useful blocking variable.

c. To test whether the mean distances differ significantly for the four brands, we test

H_0: $\mu_1 = \mu_2 = \mu_3 = \mu_4$
H_a: At least two brand mean distances differ

The test statistic is $F = 54.31$, and the p-value $= 0$ is less than $\alpha = .05$. The mean distances differ for at least two brands.

■

Multiple Comparisons of Means for Randomized Block Experiments

There are no subcommands with TWOWAY and ANOVA for multiple comparison procedures. You can, however, sort the means and compare differences between means with the half-width for the Bonferroni multiple comparison procedure. To compare c pairs of treatment means, the $(1 - \alpha/c)100\%$ confidence interval for each pair is given by

$$(\bar{y}_i - \bar{y}_j) \pm t_{\alpha/2c} s \sqrt{1/n_i + 1/n_j}$$

where $s = \sqrt{MSE}$. To obtain an overall confidence level of at least $(1 - \alpha)$ for c pairwise comparisons, the confidence coefficient for each pair is $(1 - \alpha/c)$. For the randomized block design, $n_i = n_j = b$.

Use INVCDF to obtain the critical value of t. For example, suppose we want $c = 8$ confidence intervals with an overall confidence of at least 95%. The commands to calculate the value of t for 20 degrees of freedom are

```
MTB > LET K1 = 1 - .05/(2*8)
MTB > INVCDF K1;
SUBC> T WITH 20 DF.
```

Minitab prints the cumulative probability and the critical t value.

■ **Example 4** In Example 3, we found that the mean distances differ for at least two golf brands. Use $\alpha = .05$ and the Bonferroni multiple comparison procedure to determine which pairs of means differ.

Solution We enter the means from the ANOVA output of Example 3, and use LET to calculate the half-width for the Bonferroni confidence interval. If the difference between any two means exceeds the half-width, we conclude that the means differ.

```
MTB > NAME C1 'MEANS'
MTB > SET 'MEANS'
DATA> 227.05   233.18   245.33   220.73
DATA> END
MTB > SORT 'MEANS' PUT IN 'MEANS'
MTB > PRINT 'MEANS'

MEANS
   220.73      227.05      233.18      245.33

MTB > LET K1=1 - .05/(2*6)     # CALCULATE THE T VALUE
MTB > INVCDF K1;
SUBC> T WITH 27 DF.
     0.9958      2.8469
MTB > LET K2=2.8469*SQRT(20.2)*SQRT(1/10+1/10)    # THE HALF-WIDTH
MTB > PRINT K2
K2           5.72220
MTB > STOP
```

All pairs of means differ by more than 5.7 yards. We have sufficient evidence that all brand means differ significantly. ∎

FACTORIAL EXPERIMENTS

Often times, a response variable is affected by more than one factor. For example, the sales of a product is affected by the price of the product and the type of advertising campaign. Job satisfaction is affected by work environment and perceived job security. A factorial experiment studies the simultaneous effects of two or more factors on a response variable. In this section, we describe complete factorial experiments with two factors, referred to as factors A and B. A treatment is a combination of a factor level of A and a factor level of B. If factor A has a levels, and factor B has b levels, a complete factorial experiment consists of all ab possible treatments.

For example, the rate of return on investment of a mutual fund may be affected by the type of fund and the fee structure. Suppose a financial analyst is interested in three types of funds, namely stock, bond, and foreign, and two types of fee structures, load and no load. Load refers to the sales commissions and other distribution expenses. Since the type of fund has three levels and fee structure has two levels, there are six treatments. These are load stock fund, no load stock fund, load bond fund, no load bond fund, load foreign fund, and no load foreign fund. The response variable is the rate of return over a fixed period of time.

Consider a supermarket chain that is interested in studying the effects of price and the type of display on the weekly sales of a product. The levels of price are regular price, a 10% reduction, and a 25% reduction. The types of display are current, enlarged, and end-of-aisle displays. A complete factorial experiment investigates the nine possible treatments, randomly assigned to nine supermarkets. The response variable is weekly sales.

If the treatments are randomly assigned to experimental units, the design is a completely randomized design. The analysis of variance for a two factor experiment is an extension of the analysis of

194 *Chapter 10*

variance for a single factor experiment. The analysis evaluates the effects of factor A and B, called main effects, and the joint effects of the two factors, called interactions. Two factors interact if the effect of one factor on the response depends on the level of the other factor.

Use TWOWAY or ANOVA to construct the analysis of variance table for the factorial design. If you have a random factor, the ANOVA command, described earlier in this chapter, must be used. Both commands require stacked data. Use TABLE to organize the data in cells according to factor levels, and to calculate treatment means, factor level means, and the overall mean. Use LPLOT to graphically summarize the data and to check whether interactions exist.

■ **Example 5** Refer to Example 10.8 on page 533 of the text. The USGA is interested in studying four brands of golf balls and two different clubs with respect to distances traveled by brands of golf balls. The results are shown in the following table.

	Brand A	Brand B	Brand C	Brand D
Driver	226.4	238.3	240.5	219.8
	232.6	231.7	246.9	228.7
	234.0	227.7	240.3	232.9
	220.7	237.2	244.7	237.6
Five-iron	163.8	184.4	179.0	157.8
	179.4	180.6	168.0	161.8
	168.6	179.5	165.2	162.1
	173.4	186.2	156.5	160.3

a. Use the TABLE command to print a table of the data and the treatment means.
b. Graphically analyze the data.
c. Analyze the factorial experiment. Use $\alpha = .05$ for each F test.

Solution The two factors of interest to the USGA are brand and type of golf club. Since there are two types of clubs and four brands of golf balls, the experiment consists of 8 treatment combinations and is called a 2 X 4 factorial experiment.

a. The DATA and MEANS subcommands are used with TABLE to produce a table of the data, the cell means, and the means for each type of club and ball brand.

```
MTB > NAME C1 'CLUB' C2 'BRAND' C3 'DISTANCE'
MTB > SET 'CLUB'    #  DRIVER IS CODED 1, FIVE-IRON IS CODED 2
DATA> 4(1:2)4
DATA> END
MTB > SET 'BRAND'      # BRAND A - BRAND D ARE CODED 1 - 4
DATA> (1:4)8
DATA> END
MTB > SET 'DISTANCE'
DATA> 226.4    232.6    234.0    220.7    163.8    179.4    168.6    173.4
DATA> 238.3    231.7    227.7    237.2    184.4    180.6    179.5    186.2
DATA> 240.5    246.9    240.3    244.7    179.0    168.0    165.2    156.5
DATA> 219.8    228.7    232.9    237.6    157.8    161.8    162.1    160.3
DATA> END
```

Analysis of Variance: Comparing More Than Two Means

```
MTB > TABLE 'CLUB' BY 'BRAND';
SUBC> DATA 'DISTANCE';
SUBC> MEAN 'DISTANCE'.

 ROWS: CLUB      COLUMNS: BRAND

              1         2         3         4       ALL
    1     226.40    238.30    240.50    219.80      --
          232.60    231.70    246.90    228.70
          234.00    227.70    240.30    232.90
          220.70    237.20    244.70    237.60
          228.43    233.73    243.10    229.75    233.75

    2     163.80    184.40    179.00    157.80      --
          179.40    180.60    168.00    161.80
          168.60    179.50    165.20    162.10
          173.40    186.20    156.50    160.30
          171.30    182.68    167.18    160.50    170.41
 ALL        --        --        --        --        --
          199.86    208.20    205.14    195.12    202.08

     CELL CONTENTS --

                 DISTANCE:DATA
                          MEAN
```

The table gives the means of the treatment combinations, the mean levels of club and brand, and the overall mean. When a driver (coded 1) is used, Brand C (coded 3) has the highest average distance, 243.10 yards. When a five-iron (coded 2) is used, brand B (coded 2) has the highest average distance, 182.68 yards.

b. We use a labeled plot to compare the distances when the different golf balls are hit by a driver and a five-iron.

```
MTB > LPLOT 'DISTANCE' BY 'BRAND' USING LETTERS FOR 'CLUB';
SUBC> TITLE 'GOLF BALL DISTANCES';
SUBC> FOOTNOTE 'Driver is labeled A and Five-iron B'.
```

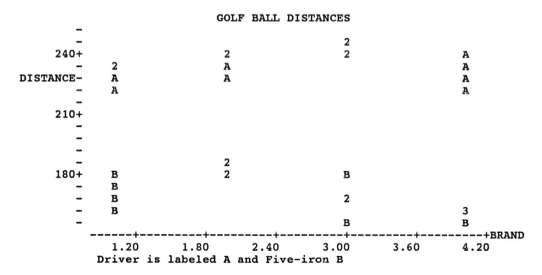

The plot shows that the driver tended to hit brand C the greatest distance while the five-iron tended to hit brand B the greatest distance. The variation in distance within the eight treatments is fairly constant.

We can also use LPLOT to plot the means for the eight treatment combinations. We enter the means from the TABLE printout and the corresponding codes for the type of club and brand.

```
MTB > NAME C4 'MEANS' C5 'CLUBCODE' C6 'BRANCODE'
MTB > READ 'MEANS' 'CLUBCODE' 'BRANCODE'
DATA> 228.43  1  1
DATA> 233.73  1  2
DATA> 243.10  1  3
DATA> 229.75  1  4
DATA> 171.30  2  1
DATA> 182.68  2  2
DATA> 167.18  2  3
DATA> 160.50  2  4
DATA> END
      8 ROWS READ
MTB > LPLOT 'MEANS' VS 'BRANCODE' USING LETTERS FOR 'CLUBCODE';
SUBC> TITLE 'MEAN DISTANCES FOUR BRANDS OF GOLF BALLS TRAVELED';
SUBC> FOOTNOTE 'Driver is labeled A and Five-iron B'.
```

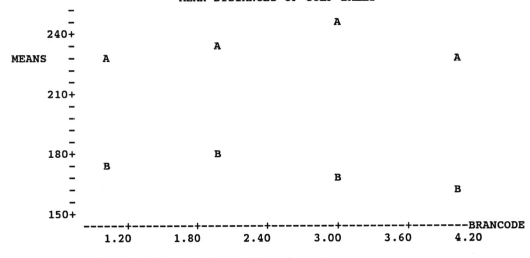

Notice that Brand B has the greatest mean distance for the five-iron, but brand C has the greatest mean distance for the driver. This indicates that the mean distance traveled depends on the combination of brand and club; that is, there is some interaction between brand and club.

c. We illustrate both TWOWAY and ANOVA for the analysis of variance.

```
MTB > TWOWAY ON 'DISTANCE' USING 'BRAND' AND 'CLUB'

ANALYSIS OF VARIANCE   DISTANCE

SOURCE          DF        SS         MS
BRAND            3       800.7      266.9
CLUB             1     32093.1    32093.1
INTERACTION      3       766.0      255.3
ERROR           24       822.2       34.3
TOTAL           31     34482.0

MTB > ANOVA DISTANCE = BRAND|CLUB;
SUBC> MEANS BRAND, CLUB, BRAND*CLUB.

Factor      Type  Levels  Values
BRAND       fixed     4      1     2     3     4
CLUB        fixed     2      1     2

Analysis of Variance for DISTANCE

Source         DF        SS         MS          F       P
BRAND           3       800.7      266.9      7.79   0.001
CLUB            1     32093.1    32093.1    936.75   0.000
BRAND*CLUB      3       766.0      255.3      7.45   0.001
Error          24       822.2       34.3
Total          31     34482.0

       MEANS

BRAND    N    DISTANCE
  1      8     199.86
  2      8     208.20
  3      8     205.14
  4      8     195.12

CLUB     N    DISTANCE
  1     16     233.75
  2     16     170.41

BRAND  CLUB    N   DISTANCE
  1     1      4    228.43
  1     2      4    171.30
  2     1      4    233.73
  2     2      4    182.68
  3     1      4    243.10
  3     2      4    167.18
  4     1      4    229.75
  4     2      4    160.50
```

The TWOWAY output includes the analysis of variance table. ANOVA adds the F test statistics and p-values for all exact F tests. The MEANS subcommand gives factor levels means and treatment means.

To test whether there are significant treatment effects, the null and alternative hypotheses are:

H_0: There is no difference among the ab treatment means
H_a: At least two treatment means differ

The test statistic is not given on the Minitab printout, but can be calculated by

$$F = \frac{(SSA + SSB + SSAB)/7}{MSE}$$

```
MTB > LET K1 = (800.7+32093.1+766.0)/7/(34.3)    # F TEST STATISTIC
MTB > PRINT K1
K1        140.191
MTB > CDF K1 PUT IN K2;
SUBC> F, 7 AND 24 DF.
    140.1908      1.0000
MTB > LET K3 = 1 - K2      # THE P-VALUE
MTB > PRINT K3
K3         0.0000
MTB > STOP
```

Since the *p*-value is 0, at least two treatment means differ.

The test for differences in the mean levels of brand and club are relevant only if the factors do not interact. To test whether the interactions are significant, the null and alternative hypotheses are:

H_o: There is no interaction between brand and club
H_a: Brand and club interact

The test statistic from the analysis of variance table is $F = 7.45$ and the *p*-value $= .001$. Since the *p*-value is less than $\alpha = .05$, there is significant interaction between brand and club. The tests to determine whether there are differences in the means for brands or clubs are irrelevant. ∎

■ **Example 6** In the previous example, brand and club were found to interact. Use the Bonferroni multiple comparison procedure to determine which pairs of treatment means differ. Use $\alpha = .10$.

Solution The analysis of variance table and treatment means that were given in Example 5 are repeated below.

```
Analysis of Variance for DISTANCE

Source        DF         SS         MS          F        P
BRAND          3        800.7      266.9       7.79    0.001
CLUB           1      32093.1    32093.1     936.75    0.000
BRAND*CLUB     3        766.0      255.3       7.45    0.001
Error         24        822.2       34.3
Total         31      34482.0

BRAND  CLUB   N   DISTANCE
  1     1     4    228.43
  1     2     4    171.30
  2     1     4    233.73
  2     2     4    182.68
  3     1     4    243.10
  3     2     4    167.18
  4     1     4    229.75
  4     2     4    160.50
```

Analysis of Variance: Comparing More Than Two Means

We assume club distances differ, and compare pairs of brands within each club. All pairs of treatment means that differ by more than a constant half-width are significantly different. For each club, there are 4 treatments and $4(3)/2 = 6$ pairwise comparisons; thus, the number of pairwise comparisons is $c = 2(6) = 12$. We enter and sort the four treatment means within each club in descending order.

```
MTB > NAME C1 'DRIVER' C2 'FIVEIRON' C3 'ORDER1' C4 'ORDER2'
MTB > SET 'DRIVER'
DATA> 228.43 233.73 243.10 229.75
DATA> END
MTB > SET 'FIVEIRON'
DATA> 171.30 182.68 167.18 160.50
DATA> END
MTB > SET 'ORDER1'
DATA> 1:4
DATA> END
MTB > SET 'ORDER2'
DATA> 1:4
DATA> END
MTB > SORT 'DRIVER' CARRY 'ORDER1' PUT IN 'DRIVER' 'ORDER1';
SUBC> DESCENDING 'DRIVER'.
MTB > SORT 'FIVEIRON' CARRY 'ORDER2' PUT IN 'FIVEIRON' 'ORDER2';
SUBC> DESCENDING 'FIVEIRON'.

MTB > PRINT 'ORDER1' 'DRIVER' 'ORDER2' 'FIVEIRON'

  ROW    ORDER1    DRIVER    ORDER2    FIVEIRON

    1        3     243.10       2       182.68
    2        2     233.73       1       171.30
    3        4     229.75       3       167.18
    4        1     228.43       4       160.50

MTB > LET K1 = 1 - .10/(2*12)    # FOR THE T VALUE
MTB > INVCDF K1;
SUBC> T WITH 24 DF FOR ERROR
       0.9958     2.8751
MTB > LET K2 = 2.8751*SQRT(34.3)*SQRT(1/4 + 1/4)    # THE HALF-WIDTH
MTB > PRINT K2
K2        11.9065
MTB > STOP
```

All pairs of means within each club that differ by more than 11.9 yards are significantly different. For the driver, the mean for brand C is significantly greater than the means for brands D and A. For the five-iron, the mean distance for brand B is significantly greater than the means for brands C and D.

■

EXERCISES

1. The price earnings or PE ratio is the price of a share of stock divided by the earnings per share. The ratio reflects the price an investor is willing to invest for a dollar of earnings. To test whether the mean PE ratios differ for the insurance, medical technology, computer software, and chemical production industries, random samples are selected from each industry. The PE ratios are from the Standard and Poor's Stock Guide, December 20, 1989.

Insurance	Medical	Computer	Chemical
7	23	12	12
11	19	14	14
12	34	13	5
18	20	19	4
9	19	24	11
13	28	12	12
10		11	15

   a. Construct and interpret dot plots of the PE ratios for the four industries.
   b. Obtain the analysis of variance table. Test whether the mean PE ratios differ for the four industries. Use $\alpha = .05$. Interpret the p-value.
   c. Interpret the 95% confidence intervals given on the printout.
   d. If there is a difference in mean PE ratios, use either the Tukey or Fisher procedure to determine which means differ significantly. Use $\alpha = .05$.

2. A large mail order company recently purchased a new integrated computer software package, and is considering three different training methods for its employees. Fifteen employees with comparable computer experience were selected. Five were randomly assigned to each training method. After completing the training, each employee was given an identical computer task. The times to complete the task were recorded as follow.

Method 1	Method 2	Method 3
26	19	30
14	29	38
26	31	30
21	11	19
33	13	35

   a. Construct and interpret dot plots of the times to complete the task for each method.
   b. Test whether there are significant differences between the mean task times for the three training methods. Use $\alpha = .05$.
   c. If there is a difference in mean task times, use the Tukey or Fisher procedure to determine which means differ. Use $\alpha = .05$.

3. A mutual fund is a financial instrument that pools the investments of many people, and invests in stocks and bonds. Many types of mutual funds are available. These include stocks, bonds, small company stocks, and gold stocks. A sound mutual fund performs well in both strong and weak markets. Because of business cycles, the ten year performance record of a fund provides

a good indication of the performance of a fund. Suppose an investor wants to compare the rates of return for the stock, stock and bond, and foreign funds. Random samples of the three types of funds were selected, and the following average annual returns for the past ten years were reported in Forbes, September 4, 1989.

Stock		Stock and Bond	Foreign	
14.8	16.8	16.2	11.5	23.8
16.2	20.1	12.0	0.7	16.0
14.5	12.2	14.4	4.7	14.7
14.7	17.0	12.6	17.0	17.3
8.5	16.2	14.3	20.5	16.0
14.8	11.0	15.7	12.4	3.5
12.0	8.4	9.7	12.5	
5.0	11.3	13.1		
15.3	16.3	13.7		
12.5				

a. Construct and interpret dot plots of the annual rates of return for the three funds.
b. Obtain the analysis of variance table. Test whether the mean annual rates differ for the three funds. Use $\alpha = .05$. Interpret the p-value.
c. Interpret the 95% confidence intervals given on the printout.
d. If there is a difference in mean annual rates, use a multiple comparison procedure to determine which means differ significantly. Use $\alpha = .05$.

4. Refer to Exercise 10.23 on page 508 of the text. A company that employs a large number of salespeople wants to determine the differences in sales for three types of compensation, namely, commission, fixed salary, and reduced fixed salary and commission. The amount of sales is given for a sample of salespeople for each type of compensation.

Commission	Fixed Salary	Salary and Commission
$425	$420	$430
507	448	492
450	437	470
483	432	501
466	444	
492		

a. Construct and interpret dot plots of the sales for the three types of compensation.
b. Obtain the analysis of variance table. Test whether the mean sales differ for the three types of compensation. Use $\alpha = .05$. Interpret the p-value.
c. Interpret the 95% confidence intervals given on the printout.
d. If there is a difference in mean sales, use a multiple comparison procedure to determine which means differ. Use $\alpha = .05$.

5. *The University Chronicle*, the St. Cloud State University newspaper, conducted a survey of prices at four local grocery stores during the week of October 29, 1989. The objective was to compare the four stores in terms of grocery prices. The following prices of eight items commonly purchased by college students were recorded for each store.

	Grocery Stores			
	A	B	C	D
Ground beef/lb	1.78	1.69	1.76	1.59
Mac. & Cheese	0.49	0.56	0.47	0.69
Heinz Ketchup	0.99	1.19	0.99	1.85
Gallon 1% Milk	2.33	2.33	2.48	2.33
Fruit loops	2.33	2.33	2.33	2.79
Campbell's Soup	0.39	0.43	0.39	0.55
Northern t.p.	0.99	1.19	0.99	1.49
Ragu Sauce	1.86	2.00	1.86	2.25

a. Construct an LPLOT of the price versus store using letters for the product. Does product seem to be a good blocking variable? Do some stores have consistently higher or lower prices?
b. Obtain the analysis of variance table. Test whether product is a significant blocking variable. Use $\alpha = .10$.
c. Test whether the mean prices differ significantly for the four stores. Use $\alpha = .10$.
d. If there are significant store effects, use the Bonferroni procedure to determine which means for the four stores differ. Use $\alpha = .05$.

6. Exercise 10.35 on page 526 of the text describes a study which compares job estimates made by three cost estimators to check on consistency. The following estimates in thousands of dollars were made by the three estimators for several jobs.

Job	Estimator A	Estimator B	Estimator C
1	$ 27.3	$ 26.5	$ 28.2
2	66.7	67.3	65.9
3	104.8	102.1	100.8
4	87.6	85.6	86.5
5	54.5	55.6	55.9
6	58.7	59.2	60.1

a. Construct an LPLOT of the estimates versus job using letters for the three estimators. Does job seem to be a good blocking variable? Do some jobs have consistently higher or lower estimates?
b. Obtain the analysis of variance table. Test whether job is a significant blocking variable. Use $\alpha = .05$.
c. Test whether the mean estimates for at least two of the estimators differ. Use $\alpha = .05$.
d. If necessary, use the Bonferroni procedure to compare all estimate means. Use $\alpha = .05$.

7. A foods company recently produced a new snack food and wants to determine whether sales are affected by age and type of television advertising. Three age groups and two types of television advertising to promote the product have been identified for the study. One type included the endorsement of a professional football player; the other type emphasized the nutritional value of the snack food. Twenty consumers from each age group were randomly selected for the study, and each type of TV ad was randomly assigned to ten consumers in each group. After viewing the ad, each person was asked to complete a questionnaire concerning interest in the product. The composite scores, based on a 100-point scale, are given in the following table. A higher

score indicates more interest in the product.

| | Endorsement Ad | | | Nutritional Ad | |
Age 1	Age 2	Age 3	Age 1	Age 2	Age 3
26	13	16	29	14	29
24	15	16	20	15	34
11	24	8	21	9	26
21	23	18	10	25	26
28	16	32	14	28	48
32	7	22	2	11	26
35	41	28	8	12	20
21	26	14	3	19	10
37	24	7	20	37	14
36	23	15	19	24	31

   a. Use the TABLE command to calculate the treatment means, factor level means, and overall mean. Summarize.
   b. Construct a labeled plot of the treatment means. Interpret.
   c. Test for factor interactions. If there is not significant interaction, test for differences in factor level means. Use $\alpha = .10$.
   d. If warranted, use the Bonferroni technique to compare relevant means. Use $\alpha = .10$.

8. A order company recently purchased a new integrated computer software package. Suppose the company wants to study the effects of the amount of employee computer experience and the type of training method on task times. The company identified fifteen employees with little or no experience and fifteen with considerable experience. Each type of training method was randomly assigned to five employees in each experience category. After completing the training, the employees were given the same computer task. The times to complete the task were recorded as follow.

| Considerable Experience | | | Little/No Experience | | |
Employee	Time	Training	Employee	Time	Training
1	19	2	16	18	2
2	27	1	17	27	1
3	18	1	18	20	1
4	12	2	19	22	3
5	26	3	20	26	3
6	20	3	21	30	3
7	17	2	22	23	3
8	25	1	23	21	2
9	23	1	24	23	2
10	27	3	25	30	2
11	8	2	26	24	1
12	23	2	27	24	2
13	22	3	28	28	1
14	17	1	29	35	3
15	30	3	30	22	1

204   *Chapter 10*

   a. Use TABLE to construct a table of the data grouped by experience and method of training, and a table of treatment means, factor level means, and the overall mean. Summarize.
   b. Construct an LPLOT of the treatment means. Interpret.
   c. Construct an analysis of variance table. Test whether there are significant treatment effects. Use $\alpha = .05$.
   d. Test whether there are significant interactions between the two factors. If there are no significant interactions, test whether there are significant factor effects. Use $\alpha = .05$.
   e. If there are significant method effects, use the Bonferroni procedure to determine which means for the methods differ. Use $\alpha = .05$.

9. An investor is interested in studying the effects of the type of mutual fund and fee structure on fund performance. The fund types of interest are stock, bond, and foreign. Fee structures of mutual funds are of two types: load or no load. Load refers to the sales commissions and other distribution expenses charged to the investor. The load is usually between 4% and 8% of the invested amount. No load refers to no fees at the time of purchase. For the study, random samples of five funds of each type, load and no load, were observed. The average annual returns are given in the table.

	Stock	Stock/Bond	Foreign
Load Funds	16.2	16.2	11.5
	14.7	12.0	17.0
	5.0	12.6	12.4
	15.3	14.3	23.8
	12.5	13.1	16.0
No load Funds	14.8	14.4	0.7
	14.5	15.7	20.5
	8.5	9.7	12.5
	14.8	16.4	16.0
	12.0	15.1	14.2

   a. Use TABLE to construct a table of the data grouped by type of fund and fee structure, and a table of treatment means, factor level means, and the overall mean. Summarize.
   b. Construct an LPLOT of the treatment means. Interpret.
   c. Construct the analysis of variance table. Test whether there are significant treatment effects. Use $\alpha = .10$.
   d. Test whether there are significant interactions between the two factors. If there are no significant interactions, test whether there are significant main effects. Use $\alpha = .10$.
   e. If warranted, use the Bonferroni technique to compare the relevant pairs of treatment means. Use $\alpha = .10$.

10. Consider Exercise 10.43 on page 543 of the text. The accompanying table gives the results of a factorial experiment to evaluate the effects of price level and display space on unit sales. The price levels were regular, reduced price, and cost to supermarket price. Display spaces were normal, normal plus end-of-the-aisle, and twice the normal space.

	Regular Price	Reduced Price	Cost Price
Normal Display	989	1,211	1,577
	1,025	1,215	1,559
	1,030	1,182	1,598
Normal Plus	1,191	1,860	2,492
	1,233	1,910	2,527
	1,221	1,926	2,511
Twice Normal	1,226	1,516	1,801
	1,202	1,501	1,833
	1,180	1,498	1,852

a. Construct a labeled plot of the data. Describe the plot.
b. Obtain the analysis of variance table. Do the data indicate that the treatment means differ among the nine treatments? Use $\alpha = .10$.
c. If warranted, test whether there is significant interaction between price and display. Use $\alpha = .10$.
d. If warranted, test whether there are differences in price and display. Use $\alpha = .10$.
e. If warranted, use the Bonferroni technique to compare the relevant pairs of treatment means. Use $\alpha = .10$.

CHAPTER 11
NONPARAMETRIC STATISTICS

Most statistical procedures presented in earlier chapters are parametric procedures which assume that the sampled populations have approximately normal distributions. Alternative nonparametric procedures require less restrictive assumptions. This chapter presents some nonparametric procedures to analyze one or more sets of data. Generally, nonparametric procedures compare the distributions of the sampled populations by using the relative ranks of the observations instead of the actual data.

NEW COMMANDS

| CORRELATION | FRIEDMAN | KRUSKAL-WALLIS | MANN-WHITNEY |
| STEST | WTEST | | |

THE SIGN TEST FOR SINGLE POPULATION INFERENCES

The parametric test based on the t test statistic for testing a hypothesis about a population mean assumes that the sampled population is approximately normal. If this condition is not satisfied, a nonparametric procedure, such as the sign test, can be used.

The sign test, or test of location, is a nonparametric procedure designed to test a hypothesis about the median M of a continuous population. The null and alternative hypotheses for a two-tailed test are

H_0: $M = M_0$
H_a: $M \neq M_0$

The test statistic is determined by counting the number of sample observations less than M_0 and the number greater than M_0. The larger of these is the test statistic S. Under H_0, S has a binomial distribution with parameters n, and $p = .5$. The Minitab command for the sign test is STEST.

208 Chapter 11

> STEST (OF MEDIAN = K) ON DATA IN C,...,C
> ALTERNATIVE = K
>
> This command does a sign test on the median of the population for each specified column. The output includes the number of observations below, equal to, and above the hypothesized median, the *p*-value, and the sample median. If the median is not specified, K = 0 is used.
>
> Use the ALTERNATIVE subcommand for a one-tailed test.
>
> ALTERNATIVE = +1 to test H_a: $M > K$
> ALTERNATIVE = -1 to test H_a: $M < K$

■ **Example 1** Consider the laboratory testing for substance abuse described on page 564 of the text. Eight independent measurements for an individual are given in the table.

Substance Abuse Test
.78   .51   3.79   .23   .77   .98   .96   .89

The laboratory reports a normal result if the median level is less than 1.00. Do the data provide sufficient evidence to indicate that the median measurement for this individual is less than 1.00? Use $\alpha = .05$.

Solution The hypotheses to determine whether the median abuse measurement is less than 1.00 are

H_0: $M = 1.00$
H_a: $M < 1.00$

```
MTB > NAME C1 'ABUSE'
MTB > SET 'ABUSE'
DATA> .78 .51 3.79 .23 .77 .98 .96 .89
DATA> END
MTB > DOTPLOT FOR 'ABUSE'
```

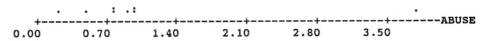

The dot plot shows that most substance abuse measurements are low. The unusually high measurement suggests that the underlying distribution of measurements may not be normal. The nonparametric sign test is the recommended procedure.

```
MTB > STEST OF MEDIAN = 1.00 ON 'ABUSE';
SUBC> ALTERNATIVE = -1.
```

```
SIGN TEST OF MEDIAN = 1.000 VERSUS   L.T.   1.000

              N    BELOW   EQUAL   ABOVE   P-VALUE      MEDIAN
ABUSE         8      7       0       1      0.0352      0.8350
MTB > STOP
```

Seven of the eight measurements are below 1.00; the *p*-value of .0352 is less than the α level of 0.05. There is sufficient evidence in the sample to conclude that the median substance abuse measurement for this individual is less than 1.00.

∎

WILCOXON RANK SUM TEST FOR COMPARING TWO POPULATIONS: INDEPENDENT SAMPLES

The Wilcoxon rank sum test is a nonparametric counterpart of the z or t test for the difference between two population means based on independent random samples. The test is used to determine whether one population distribution is shifted to the right or left of another distribution. The equivalent of the Wilcoxon rank sum test is the Minitab MANN-WHITNEY test.

To use the rank sum test, rank all the observations from both samples, and then sum the ranks for each sample. If the two distributions for equal sample sizes are identical, the rank sums should be about the same. If the rank sum of one sample is significantly different then the rank sum of a another sample, the data suggest that one population distribution is shifted to the right or left of another. The hypotheses to test whether one population is shifted to the right or left of another population are

H_0: The population distributions are identical
H_a: One of the population distributions is shifted to the right or to the left of the other population

---

MANN-WHITNEY (PERCENT CONFIDENCE K) C,C
ALTERNATIVE K

This command performs a two-sample rank test for the differences between two population medians. The output includes the point estimate, confidence interval for the difference between population medians, and the hypothesis test. The PERCENT CONFIDENCE option specifies the confidence coefficient. The default coefficient is 95%.

Use the ALTERNATIVE subcommand for one-sided tests. ALTERNATIVE = +1 (-1) tests whether the distribution in the first listed column is shifted to the right (left) of the distribution in the second listed column.

Chapter 11

■ **Example 2** In Example 11.2 on page 572 of the text, a psychologist wants to compare reaction times for adult males under the influence of drugs A and B. Seven subjects are randomly assigned to each of two groups to receive either drug A or B. The reaction times, with the exception of one subject, are given in the accompanying table.

Drug A	Drug B
1.96	2.11
2.24	2.43
1.71	2.07
2.41	2.71
1.62	2.50
1.93	2.84
	2.88

Is there sufficient evidence in the data to indicate that the probability distributions of reaction times differ for the two drugs? Use the Mann-Whitney test with $\alpha = .05$.

Solution We use DOTPLOT with the SAME subcommand to graphically compare the drug reaction times.

```
MTB > C1 'DRUG A' C2 'DRUG B'
MTB > SET 'DRUG A'
DATA> 1.96 2.24 1.71 2.41 1.62 1.93
DATA> END
MTB > SET 'DRUG B'
DATA> 2.11 2.43 2.07 2.71 2.50 2.84 2.88
DATA> END
MTB > DOTPLOT OF 'DRUG A' AND 'DRUG B';
SUBC> SAME.
```

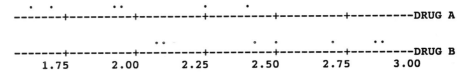

The reaction times for drug A tend to be lower than reaction times for drug B. However there is some overlap between the two groups. The hypotheses to determine whether the distributions differ are

H_o: The two distributions of reaction times are identical
H_a: The distribution of reaction times for drug A is shifted to the right or left of the distribution of reaction times for drug B

```
MTB > MANN-WHITNEY FOR 'DRUG A' AND 'DRUG B'

Mann-Whitney Confidence Interval and Test

DRUG A       N =   6       Median =      1.9450
DRUG B       N =   7       Median =      2.5000
Point estimate for ETA1-ETA2 is     -0.4950
96.2 pct c.i. for ETA1-ETA2 is (-0.9497,-0.1099)
```

```
W = 25.0
Test of ETA1 = ETA2   vs.   ETA1 n.e. ETA2 is significant at 0.0184
MTB > STOP
```

The output gives the sample medians, confidence interval estimates of the difference between population medians, and the hypothesis test. The $W = 25$, denoted T_A on page 573 of the text, is the sum of the ranks of the reaction times for drug A.

Since the observed significance level, p-value $= .0184$, is less than $\alpha = .05$, there is sufficient evidence to conclude that the distribution of reaction times for drug A is shifted to the right or left of the distribution of reaction times for drug B. The test result agrees with the evidence given in the dot plots. Drug B tends to give greater reaction times than drug A.

WILCOXON SIGNED RANK TEST FOR COMPARING TWO POPULATIONS: PAIRED DIFFERENCE EXPERIMENT

The Wilcoxon signed rank test is a nonparametric alternative to the t test for the paired difference experiment that is described in Chapter 9. The Wilcoxon procedure detects shifts in location of probability distributions. To do a signed rank test, the absolute values of the differences between pairs of observations are assigned ranks, and the rank sums of the positive and negative differences are calculated. If the population distributions are identical, the rank sums should be about the same. The nonparametric test is

> H_0: The population distributions are identical
> H_a: One of the population distributions is shifted to the right or to the left of another population

---

WTEST (CENTER = K) DIFFERENCES IN C,...,C
ALTERNATIVE = K

This command does a one sample Wilcoxon signed rank test on each column. The center is assumed to be zero if it is not specified. Use the subcommand, ALTERNATIVE = +1 (-1), to test whether the center is greater (less) than K.

The output includes the number of pairs in the sample, the number of pairs used for the test, the Wilcoxon test statistic, p-value, and estimated median. Any difference equal to the hypothesized center is eliminated.

■ **Example 3**   Refer to Table 11.3 on page 578 of the text. Each of ten judges rates the softness of two paper products on a scale of 1 to 10, higher numbers indicating greater softness. The product softness ratings are shown in the accompanying table.

Judge	Product A	Product B
1	6	4
2	8	5
3	4	5
4	9	8
5	4	1
6	7	9
7	6	2
8	5	3
9	6	7
10	8	2

Do the two products differ significantly with respect to softness? Use the Wilcoxon signed rank test with $\alpha = .05$

Solution We use MPLOT to graphically compare two samples.

```
MTB > NAME C1 'JUDGE' C2 'PROD A' C3 'PROD B'
MTB > READ 'JUDGE' 'PROD A' 'PROD B'
DATA>      1         6         4
DATA>      2         8         5
DATA>      3         4         5
DATA>      4         9         8
DATA>      5         4         1
DATA>      6         7         9
DATA>      7         6         2
DATA>      8         5         3
DATA>      9         6         7
DATA>     10         8         2
DATA> END
     10 ROWS READ
MTB > MPLOT 'PROD A' VS 'JUDGE' AND 'PROD B' VS 'JUDGE';
SUBC> TITLE 'PAPER SOFTNESS RATINGS'.
```

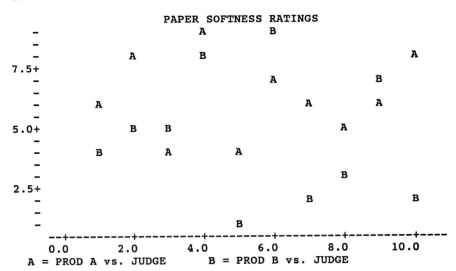

The multiple plot shows that the product A softness rating is greater than the product B softness rating for seven of the ten judges. The plot also shows the ratings depend on the judge; that is, a judge tends to rate both products either high or low.

A matched pairs test is used to compare the softness ratings. The nonparametric test is

H_0: The probability distributions of softness ratings are identical
H_a: The probability distribution of product A softness ratings is shifted to the right or left of that of product B

```
MTB > NAME C4 'A-B'
MTB > LET 'A-B' = 'PROD A' - 'PROD B'
MTB > PRINT C1-C4

ROW    JUDGE    PROD A    PROD B    A-B

 1       1         6         4        2
 2       2         8         5        3
 3       3         4         5       -1
 4       4         9         8        1
 5       5         4         1        3
 6       6         7         9       -2
 7       7         6         2        4
 8       8         5         3        2
 9       9         6         7       -1
10      10         8         2        6

MTB > WTEST ON DIFFERENCES IN 'A-B'

TEST OF MEDIAN = 0.000000 VERSUS MEDIAN N.E. 0.000000

                N FOR    WILCOXON              ESTIMATED
         N      TEST     STATISTIC   P-VALUE    MEDIAN
A-B     10       10        46.0       0.067     2.000
MTB > STOP
```

The Wilcoxon statistic corresponds to $T_+ = 46$ on page 578 of the text. Since the p-value of 0.067 is not less than $\alpha = .05$, there is not sufficient evidence to conclude that the two distributions of softness ratings differ. Although the graph seems to indicate that product A is softer than product B, we do not have enough sample evidence to support this conclusion. If in fact, product A is softer, a larger sample size might result in a statistical difference in product softness. ■

THE KRUSKAL-WALLIS H TEST FOR A COMPLETELY RANDOMIZED DESIGN

The analysis of variance F test described in Chapter 10 assumes that independent random samples are selected from normally distributed populations with equal variances. If these assumptions are not satisfied, the nonparametric Kruskal-Wallis H test can be used to compare the locations of more than two populations.

The Kruskal-Wallis test assumes that independent random samples are selected from continuous distributions. All the observations from k samples are ranked, and rank sums are calculated for each sample. The H test statistic is calculated using the rank sum for each sample. The hypotheses are

H_0: The k probability distributions are identical
H_a: At least two of the k distributions differ in location

Under the null hypothesis, the H statistic has an approximate chi-square distribution, with $(k - 1)$ degrees of freedom. The rejection region is located in the upper tail of the chi-square distribution.

KRUSKAL-WALLIS FOR DATA IN C, LEVELS IN C

This command performs a test of differences in k population locations. Enter the data in the first listed column, and codes for the samples $(1,2,...,k)$ in the second column.

In addition to the H test statistic and H adjusted for ties, the output includes a table of the number of observations, median, average rank, and z value of each sample. The z value measures the difference between the mean rank for each sample and the mean rank for all observations.

■ **Example 4** The following data on the number of available beds in three hospitals for ten randomly selected days are given in Table 11.6 on page 588 of the text. The hospital administrator wants to compare unoccupied bed space.

Hospital 1	Hospital 2	Hospital 3
6	34	13
38	28	35
3	42	19
17	13	4
11	40	29
30	31	0
15	9	7
16	32	33
25	39	18
5	27	24

Do the data provide sufficient evidence to indicate that the three distributions differ in location? Test using $\alpha = .05$.

Solution The data must be stacked in one column and the code in another for the KRUSKAL-WALLIS test. We use the BY subcommand with DOTPLOT to construct dot plots of the data.

```
MTB > NAME C1 'HOSPITAL' C2 'BEDS'
MTB > SET 'HOSPITAL'
DATA> (1:3)10
DATA> END
```

```
MTB > SET 'BEDS'
DATA>  6 38 3 17 11 30 15 16 25 5
DATA>  34 28 42 13 40 31 9 32 39 27
DATA>  13 35 19 4 29 0 7 33 18 24
DATA> END
MTB > DOTPLOT 'BEDS';
SUBC> BY 'HOSPITAL'.

HOSPITAL
1
           . ..          .       ...          .         .            .
         -+---------+---------+---------+---------+---------+------BEDS
HOSPITAL
2
                           .      .              ..    ...            ... .
         -+---------+---------+---------+---------+---------+------BEDS
HOSPITAL
3
           .    .   .      .         . .         .         .  .
         -+---------+---------+---------+---------+---------+------BEDS
         0.0       8.0      16.0      24.0      32.0      40.0
```

The dot plots indicate that hospital 2 tends to have a greater number of unoccupied hospital beds. However, there is considerable overlap in the numbers of beds for the three hospitals.

The hypothesis test is

H_0: The three distributions of the number of hospital beds are identical
H_a: At least two of the three distributions differ in location

```
MTB > KRUSKAL-WALLIS TEST FOR 'BEDS' USING 'HOSPITAL'

LEVEL      NOBS     MEDIAN    AVE. RANK    Z VALUE
  1         10       15.50       12.0       -1.54
  2         10       31.50       21.0        2.44
  3         10       18.50       13.4       -0.90
OVERALL     30                   15.5

H = 6.10   d.f. = 2   p = 0.048
H = 6.10   d.f. = 2   p = 0.048 (adj. for ties)

MTB > STOP
```

The p-value is the probability, assuming the null hypothesis is true, of obtaining an $H > 6.10$. Since the p-value $= .048$ is less than $\alpha = .05$, there is sufficient evidence in the sample to indicate that at least two of the distributions differ.

The z value of 2.44 corresponding to hospital 2 in the table reflects the greater relative difference between the mean rank for that hospital and the mean rank for all observations. This indicates that there tends to be more available beds in hospital 2 than in the other hospitals. ∎

THE FRIEDMAN F_r-TEST FOR A RANDOMIZED BLOCK DESIGN

The nonparametric counterpart of the F test for a randomized block design is the Friedman test. The Friedman F_r-test assumes that the treatments are randomly assigned to experimental units within blocks, and that the probability distributions for the treatments are continuous. The observations within each block are ranked, and rank sums are calculated for each treatment. The Friedman test statistic F_r is based on the rank sum for the treatments. The two-tailed hypothesis test is

H_0: The k probability distributions are identical
H_a: At least two of the k distributions differ in location

Under the null hypothesis, F_r has an approximate chi-square distribution, with $(k - 1)$ degrees of freedom. The rejection region is located in the upper tail of the chi-square distribution. The test statistic is denoted S on the FRIEDMAN output.

FRIEDMAN DATA IN C, TREATMENT IN C, BLOCK IN C

This command does a test of differences in k treatment populations. Enter the data in the first listed column, codes for the treatments $(1,2,...,k)$ in the second column, and codes for the blocks $(1,2,...,b)$ in the third column.

The output includes the test statistic, the test statistic adjusted for ties, and the corresponding degrees of freedom and p-values. The output also gives a table of the number of observations, median, the sum of ranks, and the overall median.

■ **Example 5** Consider the advertising response rate experiment described in Example 11.5 on page 599 of the text. A marketing firm wants to compare the relative effectiveness of direct-mail, newspaper, and magazine advertising. Fifteen clients used each type of advertising over a one year time period. The percentage response rates are shown in the accompanying table.

Company	Direct-Mail	Newspaper	Magazine
1	7.3	15.7	10.1
2	9.4	18.3	8.2
3	4.3	11.2	5.1
4	11.3	19.1	6.5
5	3.3	9.2	8.7
6	4.2	10.5	6.0
7	5.9	8.7	12.3
8	6.2	14.3	11.1
9	4.3	3.1	6.0
10	10.0	18.8	12.1
11	2.2	5.7	6.3
12	6.3	20.2	4.3
13	8.0	14.1	9.1
14	7.4	6.2	18.1
15	3.2	8.9	5.0

Test for a significant difference in the percentage response for the three modes of advertising. Use the Friedman F_r-test with $\alpha = .10$.

Solution   We use LPLOT to graphically compare the response rates for each company. The letters represent the three modes of advertising.

```
MTB > NAME C1 'COMPANY' C2 'TYPE AD' C3 'RESPONSE'
MTB > SET 'COMPANY'
DATA> 3(1:15)
DATA> END
MTB > SET 'TYPE AD'
DATA> (1:3)15
DATA> END
MTB > SET 'RESPONSE'
DATA> 7.3 9.4 4.3 11.3 3.3 4.2 5.9 6.2 4.3 10.0 2.2 6.3  8.0  7.4 3.2
DATA> 15.7 18.3 11.2 19.1 9.2 10.5 8.7 14.3 3.1 18.8 5.7 20.2 14.1 6.2 8.9
DATA> 10.1 8.2 5.1 6.5 8.7 6.0 12.3 11.1 6.0 12.1  6.3 4.3 9.1 18.1 5.0
DATA> END
MTB > LPLOT 'RESPONSE' BY 'COMPANY' USING LETTERS FOR 'TYPE AD';
SUBC> TITLE 'PERCENTAGE RESPONSE TO THREE TYPES OF ADVERTISING';
SUBC> FOOTNOTE 'A = Direct Mail,  B = Newspaper,  C = Magazine'.

              PERCENTAGE RESPONSE TO THREE TYPES OF ADVERTISING
          -                                       B
          -         B                B
  18.0+   B                                              C
          -
RESPONSE- B
          -                      B              B
          -
  12.0+                     C            C
          -     B  A     B      C            C
          - C            B             A           A  B
          -    C         C                      A
          - A                                  A    A
   6.0+         C    C  A  A    C      2  A    B
          -    2              A                     C
          -          A  A            B              A
          -                        A
          -
          ------+---------+---------+---------+---------+---------+COMPANY
              2.5        5.0       7.5      10.0      12.5      15.0
              A = Direct Mail,   B = Newspaper,   C = Magazine
```

Newspaper advertising, coded B on the graph, has the highest response rate for most companies, and direct-mail, coded A, tends to have the lowest response rate.

The null and alternative hypotheses for the Friedman test are

H_0: The response rate distributions are identical
H_a: At least two of the distributions differ in location

```
MTB > FRIEDMAN ON 'RESPONSE', TREATMENT 'TYPE AD', BLOCK 'COMPANY'
Friedman test of RESPONSE by TYPE AD blocked by COMPANY
S = 12.13   d.f. = 2   p = 0.002
                       Est.      Sum of
    TYPE AD      N    Median     RANKS
        1       15    6.300       20.0
        2       15   12.600       39.0
        3       15    8.100       31.0
```

218   Chapter 11

```
Grand median  =     9.000

MTB > STOP
```

The *p*-value is the probability, assuming the null hypothesis is true, of obtaining a test statistic $S \geq 12.13$. The *p*-value of .002 is less than $\alpha = .10$. There is sufficient evidence that the response rate distributions differ for at least two types of advertising.   ∎

SPEARMAN'S RANK CORRELATION COEFFICIENT

The Spearman's rank correlation coefficient r_s, similar to the Pearson product moment correlation, measures the correlation between two sets of ranked data. The Pearson product moment correlation coefficient r is a numerical measure of the strength of the linear relationship between *y* and *x*. The coefficient can have a value between -1 and 1. A value close to -1 or to +1 implies a strong linear relationship in the sample data. A value close to 0 implies no linear relationship. If *r* is positive, there is a direct relationship between *y* and *x*; that is, *y* tends to increase as *x* increases. If *r* is negative, there is an inverse relationship and *y* tends to decrease as *x* increases. CORRELATION calculates the Pearson product moment correlation coefficient.

CORRELATION COEFFICIENT FOR VARIABLES IN C AND C

This command calculates the Pearson product moment correlation between corresponding values stored in two columns.

Use the rank correlation to make an inference about the population correlation if the samples are not drawn from normal populations or if the data are ranked in the first place. To calculate the rank correlation coefficient, use RANK to rank each set of data. Then use CORRELATION to calculate the correlation between the two sets of ranked data.

■ **Example 6**   Refer to the study on the relationship between cigarette smoking during pregnancy and the weight of newborn infants given in Example 11.6 on page 608 of the text. The following table gives the number of cigarettes smoked per day and the birth weight of the baby for each of 15 women smokers.

Woman	Cigarettes	Weight	Woman	Cigarettes	Weight
1	12	7.7	9	20	8.3
2	15	8.1	10	25	5.2
3	35	6.9	11	39	6.4
4	21	8.2	12	25	7.9
5	20	8.6	13	30	8.0
6	17	8.3	14	27	6.1
7	19	9.4	15	29	8.6
8	46	7.8			

Obtain a scattergram of the sample data. Calculate the rank correlation coefficient to measure the relationship between cigarette smoking and the weight of newborn infants.

Solution The scattergram gives information about the relationship between smoking and birth weight.

```
MTB > NAME C1 'WOMAN' C2 'CIGARETS' C3 'BABY WT'
MTB > READ 'WOMAN'-'BABY WT'
DATA>  1 12 7.7
DATA>  2 15 8.1
DATA>  3 35 6.9
DATA>  4 21 8.2
DATA>  5 20 8.6
DATA>  6 17 8.3
DATA>  7 19 9.4
DATA>  8 46 7.8
DATA>  9 20 8.3
DATA> 10 25 5.2
DATA> 11 39 6.4
DATA> 12 25 7.9
DATA> 13 30 8.0
DATA> 14 27 6.1
DATA> 15 29 8.6
DATA> END
     15 ROWS READ
MTB > PLOT 'BABY WT' VS 'CIGARETS';
SUBC> TITLE 'CIGARETTE SMOKING'.
```

```
                         CIGARETTE SMOKING
           -             *
      9.0+
           -                    *           *
BABY WT  -                 *  *
           -              *    *        *
           -         *              *
      7.5+                                                      *
           -
           -                                  *
           -                                        *
      6.0+                         *
           -
           -
           -                 *
           -
             --------+---------+---------+---------+---------+------CIGARETS
                  14.0      21.0      28.0      35.0      42.0
```

The scattergram indicates an inverse relationship between the weight of infants and the number of cigarettes smoked by the women during pregnancy. Weight tends to decrease as smoking increases.

```
MTB > NAME C4 'CIG RANK' C5 'WT RANK'
MTB > RANK 'CIGARETS' PUT IN 'CIG RANK'
MTB > RANK 'BABY WT' PUT IN 'WT RANK'
MTB > PRINT 'WOMAN' 'CIGARETS' 'CIG RANK' 'BABY WT' 'WT RANK'
```

ROW	WOMAN	CIGARETS	CIG RANK	BABY WT	WT RANK
1	1	12	1.0	7.7	5.0
2	2	15	2.0	8.1	9.0

```
 3    3   35  13.0   6.9   4.0
 4    4   21   7.0   8.2  10.0
 5    5   20   5.5   8.6  13.5
 6    6   17   3.0   8.3  11.5
 7    7   19   4.0   9.4  15.0
 8    8   46  15.0   7.8   6.0
 9    9   20   5.5   8.3  11.5
10   10   25   8.5   5.2   1.0
11   11   39  14.0   6.4   3.0
12   12   25   8.5   7.9   7.0
13   13   30  12.0   8.0   8.0
14   14   27  10.0   6.1   2.0
15   15   29  11.0   8.6  13.5
```

MTB > CORRELATION 'CIG RANK' 'WT RANK'

Correlation of CIG RANK and WT RANK = -0.425
MTB > STOP

The rank correlation coefficient, $r_s = -.425$, indicates a moderate inverse relationship. The test to determine whether the rank correlation coefficient is significant is given on page 609 of the text. ∎

EXERCISES

1. The U.S. Department of Labor compiles data on the average state unemployment benefits. The following table gives the 1988 mean unemployment benefits for a random sample of 20 states.

State	Benefit	State	Benefit
Arkansas	$1,557.25	Michigan	$2,862.24
California	1,844.25	Minnesota	2,615.84
Connecticut	1,915.88	Missouri	1,565.76
Delaware	2,113.76	New Jersey	2,653.50
Florida	1,851.08	New York	2,438.09
Idaho	1,705.86	North Carolina	1,071.17
Illinois	2,656.91	Ohio	2,205.94
Kentucky	1,476.99	Pennsylvania	2,434.81
Maine	1,538.05	Vermont	1,608.16
Massachusetts	2,869.55	Wyoming	2,549.71

   a. Construct a dot plot of the data. Interpret.
   b. Do the data provide sufficient evidence to indicate that the median unemployment benefit for all states exceeds $2,000? Use $\alpha = .05$.

2. Exercise 11.6 on page 568 of the text reports that the American Cancer Society funds medical research for finding cancer treatments. One such study is whether a treatment extends the average time a virulent form of cancer can be held in remission. Suppose that seven patients with this cancer have been given this treatment. Assume that the treatment provides a median remission of 4.5 years. The following data are remission times for the seven patients.

Years of Remission

5.3   7.3   3.6   5.2   6.1   4.8   8.4

a. Construct a dot plot of the data. Interpret.
b. Do the data provide sufficient evidence to indicate that the median remission time is increased by the new drug? Use $\alpha = .01$.

3. A survey conducted by the UCLA Graduate School of Management on the status of computer development in business schools is reported in the *Sixth Annual UCLA Survey of Business School Computer Usage*, Jason L. Fraud and Julia A. Britt, September, 1989. The survey queries business schools on hardware, software, and resource commitments. One hundred sixty-three business schools completed the sixth annual UCLA computer survey. The following random sample gives 25 student operating budgets.

Student Operating Budget

41	114	5	*	11
271	43	217	121	*
123	13	605	*	22
34	29	145	27	180
41	875	405	495	400

Do the data provide sufficient evidence to indicate that the median computer operating budget per student exceeds $100? Use $\alpha = .05$.

4. Consider the random samples of August 1989 rental data for three bedroom homes in the city of Phoenix and the surrounding suburbs.

City		Suburb	
$850	$1,000	$ 650	$1,000
585	475	575	475
800	435	1,100	435
695	595	600	595
475	515	750	515
430	405	625	405
575	640	550	640
395		795	

a. Construct and interpret dot plots of the rental data for the city and suburb.
b. Is there sufficient evidence in the data to indicate that the probability distribution of city rents is shifted to the right of the distribution of rents in the suburbs? Use the Wilcoxon rank sum test with $\alpha = .05$.

5. The U.S. Department of Labor compiles data on the mean unemployment insurance taxes. Nationwide, the mean federal and state unemployment insurance tax for 1988 was $224 per employee. The following table gives the mean unemployment insurance taxes for random samples of ten western and eastern states.

222 Chapter 11

Western States	Tax	Eastern States	Tax
California	$217.00	Connecticut	$191.90
Colorado	266.00	Delaware	285.50
Idaho	542.00	Maine	245.00
Montana	295.40	Maryland	161.00
Nevada	237.50	Massachusetts	203.00
New Mexico	250.00	New Hampshire	105.00
Oregon	476.00	New Jersey	308.00
Utah	254.00	New York	252.00
Washington	629.80	Pennsylvania	392.00
Wyoming	392.60	Rhode Island	416.00

a. Construct and interpret dot plots of the mean unemployment insurance taxes withheld from payroll in the sample of states in the West and East.

b. Is there sufficient evidence in the data to indicate that the probability distribution of western taxes is shifted to the right of the distribution of eastern taxes? Use the Wilcoxon rank sum test with $\alpha = .05$.

6. Refer to Exercise 11.20 on page 577 of the text. A clinical psychologist believes that deaf children have greater visual acuity than hearing children. The following data are eye movement rates of ten deaf and ten hearing children.

Deaf Children		Hearing Children	
2.75	1.95	1.15	1.23
3.14	2.17	1.65	2.03
3.23	2.45	1.43	1.64
2.30	1.83	1.83	1.96
2.64	2.23	1.75	1.37

a. Construct and interpret dot plots of the eye movement rates for deaf and hearing children.

b. Is there sufficient evidence in the data to indicate that the probability distribution of eye movement rates for deaf children is shifted to the right of that for hearing children? Use the Wilcoxon rank sum test with $\alpha = .05$.

7. An advertising consulting firm is interested in the differences between advertising dollars spent in printed publications and television. The table gives advertising dollars for 20 companies randomly selected for the study.

Company	Printed	Television
1	$ 4,791	$103,383
2	90,849	27,032
3	15,461	21,240
4	63,321	88,671
5	68,397	5,321
6	98,842	92,597
7	9,425	414,802
8	75,730	258,498
9	37,746	39,208

10	102,640	684,158
11	15,514	92,206
12	44,865	103,050
13	163	84,399
14	23,854	36,223
15	26,593	72,272
16	12,439	20,144
17	16,852	64,930
18	36,427	93
19	67,207	55,270
20	31,506	118,222

a. Use MPLOT to graphically compare advertising dollars for printed and television advertising. Interpret the graph.
b. Is there sufficient evidence in the data to indicate that the probability distribution of television dollars is shifted to the right of the distribution of printed advertising dollars? Use the Wilcoxon signed rank test with $\alpha = .05$.

8. Consider Exercise 11.31 on page 585 of the text. The number of corporate lawyers is reported to be growing rapidly. The following data are the salaries of lawyers with eight years experience for a sample of ten cities in the United States.

City	Corporate	Law Firms
Atlanta	$45,500	$45,500
Chicago	43,000	48,000
Cincinnati	43,500	45,000
Dallas/Fort Worth	49,500	46,500
Los Angeles	47,000	60,000
Milwaukee	37,500	50,000
Minneapolis/St. Paul	47,500	43,500
New York	43,500	54,000
Pittsburgh	42,000	44,000
San Francisco	47,500	59,500

a. Use MPLOT to graphically compare salaries of corporate lawyers and lawyers with law firms. Interpret the graph.
b. Is there sufficient evidence in the data to indicate that the salary probability distribution of corporate lawyers differs from the salary distribution of lawyers working in law firms? Use the Wilcoxon signed rank test with $\alpha = .05$.
c. Under what conditions would a paired difference t test be appropriate? Conduct the t test. Compare the results with that of part b.

9. State revenue information is compiled by the Commerce Clearing House, Inc., a tax and business law research firm in Chicago. The following table gives random samples of revenue data for states from three regions: Western, Midwest, and Eastern. The percent listed is that of state revenue derived from the state income tax.

Western	%	Midwest	%	Eastern	%
Washington	69	Minnesota	49	Pennsylvania	33
California	49	Kansas	42	New Jersey	38
Montana	41	Illinois	37	North Carolina	39
Oregon	69	Missouri	40	New York	60
Utah	44	Kansas	42	Maine	43
Colorado	41	Wisconsin	49	New Hampshire	30

a. Compare the percents for the three regions using dot plots.
b. Do the data provide sufficient evidence to indicate differences in the three regions? Use the Kruskal-Wallis test with $\alpha = .05$.

10. Sixth graders in an elementary school are taught reading by three different methods, programmed instruction, standard memorization techniques, and open classroom approach. Exercise 11.43 on page 594 of the text provides the following increases in reading levels attained by five randomly selected students from each of the three classes.

Programmed	Standard	Open
.9	1.0	1.7
1.5	.8	.5
.7	.9	1.6
1.1	1.2	1.4
.5	1.4	1.0

a. Compare the increases in reading levels for the three methods using dot plots. Interpret.
b. Use the Kruskal-Wallis test to determine whether there are differences in the increases in reading levels for at least two of the methods. Use $\alpha = .05$.

11. The nation's thrift institutions suffered major losses during 1988 and 1989, reported the Office of Thrift Supervision. The 2,903 Savings and Loan Associations reported losses from bad loans and current operations. The following table gives the 1988 and 1989 third quarter profits and losses in millions of dollars for randomly selected states in the west, mid-south, southeast, northeast, and midwest regions of the United States.

	1988	1989
West	$ 8.5	$ 4.7
	287.9	-425.7
	-10.3	-4.0
	12.3	24.0
	-12.5	-153.4
Mid-South	-1706.6	-997.8
	-142.1	-245.7
	-26.5	-17.4
	-33.3	-26.3
	-58.2	-43.8

	1988	1989
Southeast	3.5	-25.3
	-9.2	-14.3
	-180.8	-201.7
	12.7	4.0
	10.0	-16.2
Northeast	-0.1	0.5
	63.2	-223.7
	14.9	-210.5
	46.4	-321.7
	22.6	8.2
Midwest	2.9	-4.3
	27.6	15.9
	2.5	1.2
	53.4	-57.0
	-34.6	-6.8

Use the Friedman F_r-test to determine whether profits or losses differ significantly for the five regions. Use $\alpha = .10$.

12. Refer to Exercise 11.57 on page 604 of the text. Three sealers used to retard the corrosion of metals were tested to determine any differences among them. The sealers were applied to samples of 10 different metal compositions. After being exposed to the same environmental conditions for one month, the amount of corrosion was measured. The data are given in the following table.

Metal	Sealer 1	Sealer 2	Sealer 3
1	4.6	4.2	4.9
2	7.2	6.4	7.0
3	3.4	3.5	3.4
4	6.2	5.3	5.9
5	8.4	6.8	7.8
6	5.6	4.8	5.7
7	3.7	3.7	4.1
8	6.1	6.2	6.4
9	4.9	4.1	4.2
10	5.2	5.0	5.1

Use the Friedman F_r-test to determine whether the probability distributions of the amounts of corrosion differ among the three sealers. Use $\alpha = .05$.

13. Consider the 1988 top twenty-five newspaper advertisers reported in the *Leading National Advertisers/Media Records*. The advertising dollars (in thousands) are listed for 1987 and 1988.

Advertiser	1987	1988
Philip Morris Cos.	$271,178	$270,251
General Motors Corp.	153,926	190,799
RJR Nabisco	105,674	131,463
Ford Motor Co.	125,529	125,532
Chrysler Corp.	100,433	104,527
Proctor & Gamble Co.	79,501	79,279
AT&T	76,270	66,193
Time Warner	53,029	64,989
Nestle SA	56,616	63,477
Grand Metropolitan PLC	54,948	59,545
Unilever NV	58,259	59,449
Franklin Mint	31,733	48,981
Honda Motor Co.	42,701	47,159
Bristol-Myers Squibb	40,154	43,616
Revlon Group	36,562	42,488
Sony Corp.	31,862	41,771
General Electric Co.	37,727	36,747
Schering-Plough Corp.	29,202	36,617
Toyota Motor Corp.	27,105	36,476
American Brands	38,545	35,888
Nissan Motor Co.	19,361	35,826
E.I. du Pont de Nemours & Co.	41,676	34,676
Sara Lee Corp.	24,268	34,107
U.S. Government	44,164	33,436
Sears, Roebuck & Co.	21,608	33,373

Plot the advertising dollars for 1987 and 1988. Calculate and interpret the rank correlation coefficient to measure the relationship between 1987 and 1988 newspaper advertising dollars.

14. In Exercise 11.68 on page 617 of the text, data on references and job performance for eight recently hired employees are given. A company collected the data to study the correlation between the strength of an applicant's references and job performance. Independent evaluations were made on a scale from 1 to 20.

Employee	References	Job Performance
1	18	20
2	14	13
3	19	16
4	13	9
5	16	14
6	11	18
7	20	15
8	9	12

a. Obtain a scattergram of the data. Interpret.
b. Calculate and interpret the rank correlation coefficient to measure the relationship between strength of an applicant's references and job performance.

CHAPTER 12

THE CHI-SQUARE TEST AND THE ANALYSIS OF CONTINGENCY TABLES

This chapter illustrates Minitab commands to analyze one- or two-dimensional classification data. One-dimensional data, which are data classified on a single scale, are analyzed using a hypothesis test about multinomial probabilities. When data are classified in a two-dimensional table, the result is a a contingency table. The analysis is a test of independence.

NEW COMMAND

CHISQUARE

ONE-DIMENSIONAL COUNT DATA: MULTINOMIAL DISTRIBUTION

In a multinomial experiment, a qualitative variable is classified into two or more distinct categories. The objective of the experiment is to make inferences on the probabilities, p_1, p_2, ..., p_k, that observations fall in k categories. Examples include consumer preference studies and opinion polls. The binomial experiment described in Chapter 4 is a special case of the multinomial experiment for $k = 2$ categories.

A test about multinomial probabilities determines whether the observed categorical data fit the hypothesized categorical probabilities. The test statistic X^2 measures the amount of disagreement between the observed and the expected number of responses in each category. The formula for the test statistic is

$$X^2 = \sum_{i=1}^{k} \frac{[n_i - E(n_i)]^2}{E(n_i)}$$

where n_i and $E(n_i)$ are the observed and expected frequencies of category i. The expected frequency is $E(n_i) = np_i$, where n is the number of observations. The null and alternative hypotheses are

H_0: All the multinomial probabilities equal the expected probabilities
H_a: At least one multinomial probability does not equal the expected probability

Under the null hypothesis, the sampling distribution of X^2 is approximately a chi-square distribution, with $(k - 1)$ degrees of freedom. The rejection region is located in the upper tail of the chi-square distribution.

228 *Chapter 12*

Minitab does not have a command for a goodness-of-fit test. However, you can use LET to calculate the expected frequencies and the chi-square test statistic, and CDF with the CHISQUARE subcommand for the *p*-value. Use MPLOT for a multiple plot of the observed and expected frequencies, or observed and expected probabilities. If the null hypothesis is true, there should be little differences between the hypothesized and sample values.

■ **Example 1** Consider the voter preference study described on page 630 of the text. Suppose we wish to compare the percentage of voters favoring each of three political candidates running for the same position. Of 150 voters, 61 preferred candidate 1, 53 preferred candidate 2, and 36 preferred candidate 3.

a. Calculate the expected frequencies assuming no difference in voter preference. Graphically describe the differences between the observed frequencies and expected frequencies.
b. Test whether the data indicate a preference for any of the candidates. Use $\alpha = .05$. Calculate and interpret the *p*-value.

Solution

a. Assuming no difference in voter preference, we expect 1/3, or 50 voters, to prefer each candidate. MPLOT plots the observed and expected frequencies.

```
MTB > NAME C1 'CANDIDAT' C2 'P' C3 'OBSERVED'
MTB > READ 'CANDIDAT' 'P' 'OBSERVED'
DATA> 1   .3333    61
DATA> 2   .3333    53
DATA> 3   .3333    36
DATA> END
      3 ROWS READ
MTB > NAME C4 'EXPECTED'
MTB > LET 'EXPECTED' = 'P'*150   # THE SAMPLE SIZE IS 150
MTB > MPLOT 'EXPECTED' VS 'CANDIDAT' AND 'OBSERVED' VS 'CANDIDAT';
SUBC> TITLE 'VOTER PREFERENCE FOR CANDIDATES';
SUBC> YLABEL 'FREQUENCY';
SUBC> XINCREMENT = 1.
```

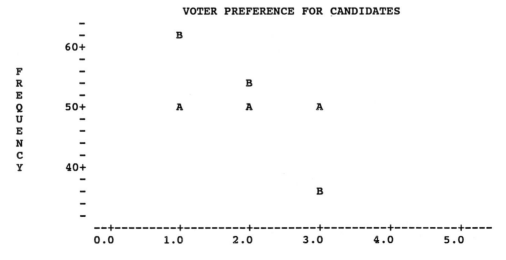

A = EXPECTED vs. CANDIDAT   B = OBSERVED vs. CANDIDAT

The A's on the MPLOT represent the 50 voters that were expected to favor each candidate, assuming there is no preference. We observe more voters who favor candidate 1 and fewer voters who favor candidate 3 than expected.

b. The null and alternative hypotheses to test whether the true proportions differ significantly are

H_0: $p_1 = 1/3$, $p_2 = 1/3$, $p_3 = 1/3$
H_a: At least one proportion differs from 1/3

```
MTB > NAME C5 'OBS-EXP' C6 'VALUE' C7 'CHISQ' C8 'P-VALUE'
MTB > LET 'OBS-EXP' = 'OBSERVED'-'EXPECTED'
MTB > LET 'VALUE' = 'OBS-EXP'**2/'EXPECTED'
MTB > LET 'CHISQ' = SUM('VALUE')
MTB > CDF 'CHISQ', PUT IN K1;
SUBC> CHISQUARE WITH 2 DF.
MTB > LET 'P-VALUE' = 1 - K1
MTB > PRINT 'CANDIDAT' 'OBSERVED'-'P-VALUE'
```

ROW	CANDIDAT	OBSERVED	EXPECTED	OBS-EXP	VALUE	CHISQ	P-VALUE
1	1	61	49.995	11.005	2.42244	6.52065	0.0383757
2	2	53	49.995	3.005	0.18062		
3	3	36	49.995	-13.995	3.91759		

MTB > STOP

Since the rejection region is the upper tail of the chi-square distribution, the *p*-value is the probability, assuming the null hypothesis is true, that $\chi^2 > 6.52$. The *p*-value = .038 is less than $\alpha = .05$. There is significant evidence to indicate voter preference for one or more of the candidates. ■

CONTINGENCY TABLES

A contingency table contains data classified according to two categorical variables. An objective of a contingency table analysis is to test whether the variables are not independent. The general form of the hypothesis test is

H_0: The two classifications are independent
H_a: The two classifications are not independent

The formula for the test statistic for a contingency table of *r* rows and *c* columns is

$$X^2 = \sum_{i=1}^{r} \sum_{j=1}^{c} \frac{[n_{ij} - \hat{E}(n_{ij})]^2}{\hat{E}(n_{ij})}$$

where n_{ij} and $\hat{E}(n_{ij}) = r_i c_j/n$ are the observed and expected frequencies of the cell in row *i* and column *j*. Under the null hypothesis, the sampling distribution of X^2 is approximately a chi-square distribution, with $(r-1)(c-1)$ degrees of freedom. The rejection region is located in the upper tail of the chi-square distribution.

CHISQUARE calculates the chi-square statistic for a contingency table. Use LPLOT for a multiple plot of the sample data in a contingency table. The plot graphically describes the relationship between the observed responses of two categorical variables.

> CHISQUARE ANALYSIS OF THE TABLE IN C,...,C
>
> This command does a chi-square test of independence between the rows and columns of a contingency table. The output includes a table of observed and expected frequencies, the calculations for the chi-square test statistic, and the degrees of freedom.

■ **Example 2** Refer to the contingency table given in Table 12.3 on page 639 of the text. A random sample of 1,000 recently purchased American-made cars were classified according to size and manufacturer. The results are given in the accompanying table.

	Manufacturer			
	A	B	C	D
Small	157	65	181	10
Intermediate	126	82	142	46
Large	58	45	60	28

a. For each manufacturer, calculate the percentages of the total car sales in each size category. Graphically compare the four percentage distributions.
b. Test whether the two classifications are not independent, using $\alpha = .05$. Calculate and interpret the p-value.

Solution

a. LPLOT gives a labeled plot of the percentages of cars for each manufacturer in each size category.

```
MTB > NAME C1 'A' C2 'B' C3 'C' C4 'D'
MTB > READ 'A'-'D'
DATA> 157 65 181 10
DATA> 126 82 142 46
DATA> 58 45 60 28
DATA> END
     3 ROWS READ
MTB > NAME C5 'A %' C6 'B %' C7 'C %' C8 'D %'
MTB > LET 'A %' = 100*'A'/SUM('A')
MTB > LET 'B %' = 100*'B'/SUM('B')
MTB > LET 'C %' = 100*'C'/SUM('C')
MTB > LET 'D %' = 100*'D'/SUM('D')
```

```
MTB > PRINT 'A %' - 'D %'

ROW       A %       B %       C %       D %

  1    46.0411   33.8542   47.2585   11.9048
  2    36.9501   42.7083   37.0757   54.7619
  3    17.0088   23.4375   15.6658   33.3333

MTB > NAME C9 'ALL %' C10 'MAKE' C11 'CAR SIZE'
MTB > STACK 'A %'-'D %', PUT IN 'ALL %';
SUBC> SUBSCRIPTS FOR MANUFACTURER IN 'MAKE'.
MTB > SET 'CAR SIZE'
DATA> 4(1:3)
DATA> END
MTB > PRINT 'ALL %' 'MAKE' 'CAR SIZE'

ROW    ALL %    MAKE    CAR SIZE

  1    46.0411    1        1
  2    36.9501    1        2
  3    17.0088    1        3
  4    33.8542    2        1
  5    42.7083    2        2
  6    23.4375    2        3
  7    47.2585    3        1
  8    37.0757    3        2
  9    15.6658    3        3
 10    11.9048    4        1
 11    54.7619    4        2
 12    33.3333    4        3

MTB > LPLOT 'ALL %' VS 'MAKE' USING LETTERS FOR 'CAR SIZE';
SUBC> TITLE 'CAR SIZE AND MANUFACTURER SALES';
SUBC> FOOTNOTE 'Letters represent size:  Small A, Intermediate B, Large C';
SUBC> XINCREMENT = 1.
```

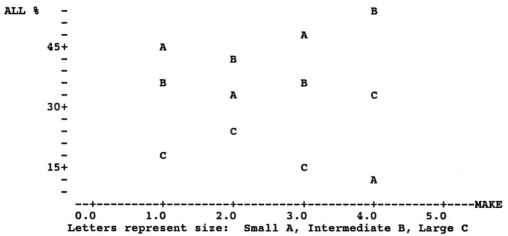

The manufacturers have substantially different percentages of total car sales for each size car. If car size and manufacturer were independent, the lines connecting the letters for each size would be approximately horizontal. The fourth manufacturer has a higher percentage of intermediate and large cars, and a smaller percentage of small cars than expected.

b. The null and alternative hypotheses are

H_0: Car size and car manufacturer are independent
H_a: Car size and car manufacturer are not independent

```
MTB > CHISQUARE FOR TABLE IN 'A' 'B' 'C' 'D'

Expected counts are printed below observed counts

              A         B         C         D     Total
    1       157        65       181        10       413
           140.83     79.30    158.18     34.69

    2       126        82       142        46       396
           135.04     76.03    151.67     33.26

    3        58        45        60        28       191
            65.13     36.67     73.15     16.04

Total       341       192       383        84      1000

ChiSq =  1.856 +  2.577 +  3.292 + 17.575 +
         0.605 +  0.468 +  0.616 +  4.876 +
         0.781 +  1.891 +  2.365 +  8.910 = 45.812
df = 6

MTB > NAME C12 'P-VALUE'
MTB > CDF 45.812, PUT IN K1;
SUBC> CHISQUARE WITH 6 DF.
MTB > LET 'P-VALUE' = 1 - K1
MTB > PRINT 'P-VALUE'

P-VALUE
   0.0000001

MTB > STOP
```

The p-value of 0 is less than $\alpha = .05$. There is sufficient evidence that the two classifications, car size and manufacturer, are dependent.

EXERCISES

1. If the sample sizes of a contingency table analysis are small, the sampling distribution of X^2 may not approximate a chi-square distribution. Consider the following contingency table which gives the number of items sold, classified by size and color.

Color	Small	Large
Red	50	1
Blue	30	3
Green	10	6

    Use Minitab to test whether size and color are not independent. Discuss the warning that is given on the computer output. Explain why this is a concern.

2. The Gallup Organization reported in 1989 that the average American spends 21 minutes traveling to work each day. Interviews conducted with part-time or full-time men and women 18 years or older showed that 14 percent traveled more than thirty minutes, 82 percent traveled less than thirty minutes, and 4 percent did not travel to work. A study of 200 part-time or full-time workers in Minneapolis 18 years or older revealed that 18 traveled more than thirty minutes, 176 traveled less than thirty minutes, and 6 did not travel.

    a. Use MPLOT to graphically compare the Minneapolis data with the Gallup poll results. Summarize.
    b. Do the data provide sufficient evidence that the percents of Minneapolis workers traveling to work differ from the percents reported in the Gallup poll? Test using $\alpha = .05$.

3. Refer to Exercise 12.14 on page 638 of the text. In the game of chess, five opening moves are highly favored by experts. A random sample of 100 grand masters was taken to determine whether one or more of these opening strategies is most preferred. The results are recorded in the following table.

Strategy	Frequency
A	17
B	27
C	22
D	15
E	19

    a. Graphically describe the differences between the observed and expected frequencies.
    b. Test whether the data indicate a preference for one or more of the strategies. Use $\alpha = .05$. Calculate and interpret the p-value.

4. The U.S. Department of Labor compiles data on the mean unemployment benefits and the percent of claims approved for each state. To study the relationship between the mean unemployment benefits and the percent of approved claims, a consultant used a random sample of data from 45 states in 1988. The following table classifies the sample data by mean unemployment benefits and percent of approved claims.

234   Chapter 12

	Mean Benefits	
Percent	Under $2,000	Over $2,000
50-65%	7	4
65-75%	9	6
75-95%	9	10

a. Graphically study the sample data.
b. Do the data provide sufficient evidence that mean unemployment benefits and percent of approved claims are dependent? Test using $\alpha = .05$.

5. The manager of Rental Property in Phoenix, Arizona wants to determine whether rental prices within the city differ from prices in suburbs. A random sample of 200 rental prices of residential three bedroom homes, classified according to price and location, is given in the following table.

	Rental Property	
Prices	City	Suburb
Under $500	48	2
$500-$599	51	11
$600-$699	30	17
$700 and over	22	19

a. Graphically analyze the sample of rental properties.
b. Do the data provide sufficient evidence that the rental prices and location are dependent? Test using $\alpha = .05$.

6. Exercise 12.28 on page 651 of the text describes a study to determine whether alcoholism is related to marital status. The following table gives a sample of 280 adults classified as a diagnosed alcoholic, undiagnosed alcoholic, or nonalcoholic, and according to marital status.

Alcoholic	Married	Not Married
Diagnosed	21	59
Undiagnosed	37	63
Nonalcoholic	58	42

a. Graphically study the sample data.
b. Do the data provide sufficient evidence of a relationship between a marital status and alcoholic classifications? Test using $\alpha = .05$.

7. Consider the class data set given in Appendix B. Test whether the gender of students is related to whether the student owns an American car, owns a foreign car, or does not own a car.

CHAPTER 13

SIMPLE LINEAR REGRESSION

Simple regression analysis is a statistical technique used to study the relationship between two variables. An objective of regression analysis is to develop a model that can be used to predict a response variable using information contained in an independent variable. In this chapter, we describe a scattergram and correlation analysis to hypothesize the form of the model, and illustrate how to use sample data to estimate the model.

NEW COMMANDS

BRIEF   REGRESS

INTRODUCTION TO REGRESSION ANALYSIS

Regression analysis is used to establish the relationship between two variables. The response variable y is the dependent variable or variable of interest, and the predictor variable x is the independent variable. For example, sales of a video cassette recorder tend to be related to advertising expenditures; interest paid on a checking account may be related to the minimum account balance; and absenteeism in a firm may be related to job satisfaction.

The objective of simple regression analysis is to develop a linear regression model, relating y to x, that can be used to explain or predict the response variable. A regression model can incorporate a quantitative or qualitative independent variable. For example, the type of advertising, such as television, radio, or newspaper, is a qualitative variable that may explain sales of a product. Qualitative independent variables are included in Chapter 15.

A scattergram of the data and a correlation analysis give information to hypothesize the linear regression model. A scattergram of the data provides information on the type and strength of the relationship between the two variables. The Minitab PLOT command, with y on the vertical axis and x on the horizontal axis, gives a scattergram of the data.

The Pearson product moment correlation coefficient r is a numerical measure of the strength of the linear relationship between y and x. The coefficient can have a value between -1 and 1. A value close to -1 or to +1 implies a strong linear relationship in the sample data. A value close to 0 implies no linear relationship. If r is positive, there is a direct relationship between y and x; that is, y tends to increase as x increases. If r is negative, there is an inverse relationship and y tends to decrease as x increases.

CORRELATION, described on page 218, calculates the Pearson product moment correlation coefficient. If the scattergram and correlation coefficient suggest a linear relationship between y and x, you can use the following simple linear regression model to relate y to x:

$$y = \beta_0 + \beta_1 x + \epsilon$$

The two parameters of the model are β_0, the y-intercept, and β_1, the slope of the line. β_0 is the mean value of y when x is 0. β_1 is the change in the mean value of y for a one unit change in x. The random error, ϵ, accounts for the variability in y that is not explained by the independent variable x. The variability could be due to other important independent variables or to some random phenomenon.

The model assumes that the errors are independent; that is, the error associated with any one observation has no effect on the error associated with any other observation. The model also assumes the probability distribution of ϵ is normal, with zero mean and a constant variance σ^2 for all values of x.

The REGRESS command fits the least squares line to the sample of observations. Use BRIEF to vary the amount of output.

REGRESS C WITH 1 PREDICTOR C (STORE STD RES IN C (FIT IN C))
  RESIDUALS IN C
  PREDICT FOR E

This command determines the simple linear regression equation using the least squares method. The output includes the regression equation; a table of coefficients, standard deviations, and t-ratios; s and r^2; and the analysis of variance table. Use the options to store the standardized residuals and fitted values, and the RESIDUALS subcommand to store the residuals in a column.

The subcommand PREDICT calculates $\hat{y}$, the standard deviation of $\hat{y}$, 95% confidence and prediction intervals for specified values of x. The values of x may be stored in a column or as constants. If one x value is specified, it may be entered directly on the subcommand line.

The BRIEF command entered before REGRESS changes the amount and type of regression output. BRIEF 2 is the default output of the REGRESS command.

> **BRIEF (OUTPUT CODE K)**
>
> This command is entered before REGRESS to control the amount and type of regression output. The larger the value of K, the more output. Use K for the following options.
>
> K = 0   No regression output. Error messages are printed.
> K = 1   The regression equation, table of coefficients, s, r^2 and r^2-adjusted, and the first part of the analysis of variance table.
> K = 2   The output of BRIEF 1, plus the second part of the analysis of variance table, and unusual observations.
> K = 3   The output of BRIEF 2, plus a table of the data, fitted values and residuals.

■ **Example 1** Refer to the experiment, described on page 669 of the text, to determine the relationship between the percentage of a certain drug x in the bloodstream and the reaction time y (in seconds) to a stimulus. The following data for five subjects are given in Table 13.1.

Subject	Amount of Drug	Reaction Time
1	1	1
2	2	1
3	3	2
4	4	2
5	5	4

a. Construct a scattergram. Find and interpret the correlation coefficient.
b. Determine the linear regression equation. Interpret the parameter estimates.
c. Find the sum of squared errors and estimates of σ^2 and σ.
d. Test the usefulness of the model. Use $\alpha = .05$.
e. Find a 95% confidence interval for the slope. Interpret.
f. What is the coefficient of determination? How is it related to the correlation coefficient?
g. Find a point estimate for the mean reaction time for all subjects that have 4% of the drug in their bloodstream.

Solution

a. PLOT gives a scattergram of reaction time versus percentage of the drug in the bloodstream.

```
MTB > NAME C1 'DRUG %' C2 'REACTION'
MTB > READ 'DRUG %' AND 'REACTION'
DATA> 1   1
DATA> 2   1
DATA> 3   2
DATA> 4   2
DATA> 5   4
DATA> END
```

238 Chapter 13

```
             5 ROWS READ
MTB > SAVE 'DRUGTIME'

Worksheet saved into file: DRUGTIME.MTW
MTB > PLOT 'REACTION' VS 'DRUG %';
SUBC> TITLE 'REACTION TIME AND DRUG PERCENTAGE';
SUBC> FOOTNOTE 'Reaction time in seconds'.
```

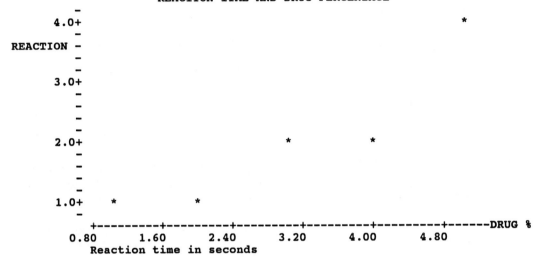

The scattergram, similar to that given in Figure 13.3 on page 669 of the text, shows a strong positive linear relationship between reaction time and drug percentage. Reaction time tends to increase as the percentage of the drug in the bloodstream increases. The slope of the least squares line is positive.

```
MTB > CORRELATION OF 'REACTION' VS 'DRUG %'

Correlation of REACTION and DRUG % = 0.904
```

The coefficient of correlation $r = .904$ confirms the positive linear relationship that we observed in the scattergram.

b. We use BRIEF 3 before REGRESS to obtain the maximum regression output.

```
MTB > BRIEF 3
MTB > REGRESS 'REACTION' ON 1 PREDICTOR 'DRUG %'

The regression equation is
REACTION = - 0.100 + 0.700 DRUG %

Predictor         Coef        Stdev      t-ratio         p
Constant       -0.1000       0.6351        -0.16     0.885
DRUG %          0.7000       0.1915         3.66     0.035

s = 0.6055      R-sq = 81.7%     R-sq(adj) = 75.6%
```

Analysis of Variance

SOURCE	DF	SS	MS	F	p
Regression	1	4.9000	4.9000	13.36	0.035
Error	3	1.1000	0.3667		
Total	4	6.0000			

Obs.	DRUG %	REACTION	Fit	Stdev.Fit	Residual	St.Resid
1	1.00	1.000	0.600	0.469	0.400	1.04
2	2.00	1.000	1.300	0.332	-0.300	-0.59
3	3.00	2.000	2.000	0.271	0.000	0.00
4	4.00	2.000	2.700	0.332	-0.700	-1.38
5	5.00	4.000	3.400	0.469	0.600	1.57

The least squares regression equation is given on the first part of the output.

```
The regression equation is
REACTION = - 0.100 + 0.700 DRUG %
```

You can plot the straight line by hand on the scattergram, as illustrated in Figure 13.5 on page 673 of the text. The y-intercept, $\hat{\beta}_0 = -0.1$, indicates the point at which the line crosses the y axis. In any application, the y-intercept has a practical interpretation only if the range of x values in the sample includes zero. Since the range of drug percentage is from 1% to 5%, the y-intercept has no practical interpretation in this example.

The slope, $\hat{\beta}_1 = .7$, is the estimated increase in the mean reaction time for every percentage increase in the drug. We estimate reaction time to increase an average of .7 seconds for every one percentage increase in the drug.

c. The sum of squared errors, SSE = 1.1000, appears in the analysis of variance table at the intersection of Error and SS. The least squares line minimizes the sum of squared errors.

SOURCE	DF	SS	MS	F	p
Regression	1	4.9000	4.9000	13.36	0.035
Error	3	1.1000	0.3667		
Total	4	6.0000			

The mean squared error, $s^2 = 0.3667$, is an estimate of the random error variance σ^2. It is located in the analysis of variance table at the intersection of Error and MS, and is calculated by dividing the SSE by $(n - 2)$ degrees of freedom.

SOURCE	DF	SS	MS	F	p
Regression	1	4.9000	4.9000	13.36	0.035
Error	3	1.1000	0.3667		
Total	4	6.0000			

An estimate of σ is s, the estimated standard error of the regression model. It is the square root of s^2, and appears on a line in the regression output before the analysis of variance table.

```
s = 0.6055
```

Since s measures the spread of the reaction times about the least squares line, most of the observations should lie within $2s = 1.21$ seconds of the line.

d. To test whether the regression model is useful for predicting reaction time, we test whether the slope is equal to 0. The null and alternative hypotheses are

H_0: $\beta_1 = 0$
H_a: $\beta_1 \neq 0$

The regression output gives the test statistic $t = 3.66$ and p-value $= 0.035$ in the columns labeled t-ratio and p in the regression coefficients table.

```
Predictor        Coef        Stdev      t-ratio          p
Constant       -0.1000      0.6351       -0.16        0.885
DRUG %          0.7000      0.1915        3.66        0.035
```

INVCDF with the T subcommand gives the critical t value for the rejection region. For $\alpha = .05$ and a two-tailed hypothesis test, the area to the left of the upper tail is .975. The degrees of freedom are $n - 2 = 3$.

```
MTB > INVCDF .975;
SUBC> T WITH 3 DF.
      0.9750    3.1824
MTB > STOP
```

The rejection region is $t < -3.18$ or $t > 3.18$. Since the t-ratio of 3.66 is greater than 3.18, we have significant evidence to conclude that the slope β_1 is not 0. The sample evidence indicates that the percentage of blood in the drugstream contributes useful information for the prediction of reaction time.

We can also use the p-value or observed significance level to interpret the hypothesis test. The p-value is the probability, assuming the slope is zero, of obtaining a t-ratio greater than 3.66 or less than -3.66. Since the p-value of .035 is less than $\alpha = .05$, we have enough evidence to reject H_0 and conclude that the model is useful.

e. A confidence interval for the slope can be calculated using the estimates of the slope and the standard error of the slope that are given on the output. The formula is

$$\hat{\beta}_1 \pm t_{\alpha/2}(\text{Stdev of } \hat{\beta}_1)$$

The estimates are given in the coefficient and standard deviation columns of the regression coefficients table.

```
Predictor        Coef        Stdev      t-ratio          p
Constant       -0.1000      0.6351       -0.16        0.885
DRUG %          0.7000      0.1915        3.66        0.035
```

We used INVCDF to find the t value for a 95% confidence interval and $n - 2 = 3$ degrees of freedom in part d. Thus a 95% confidence interval for the slope is

$0.70 \pm 3.1824(0.1915)$
0.70 ± 0.61
$0.09 \leq \beta_1 \leq 1.31$

We are 95% confident that the mean reaction time increases from .09 to 1.31 seconds for every one percentage increase in the drug in the bloodstream.

f. The coefficient of determination r^2 is the square of the correlation coefficient, $r = .904$, usually expressed as a percent, $r^2(100)\%$. The coefficient is printed on the same line as s.

   s = 0.6055   R-sq = 81.7%   R-sq(adj) = 75.6%

   About 82% of the total sum of squared errors of reaction times about their mean can be explained by using the least squares regression equation. We can reduce the sum of squares of prediction errors by about 82% by using the least squares equation $\hat{y}$, instead of $\bar{y}$, to predict reaction time. The adjusted r^2 takes into account the degrees of freedom.

g. With BRIEF 3, the regression output includes a table of the fitted values, the standard deviations of the fitted value, the residuals, and standardized residuals for each value of x in the data set.

Obs.	DRUG %	REACTION	Fit	Stdev.Fit	Residual	St.Resid
1	1.00	1.000	0.600	0.469	0.400	1.04
2	2.00	1.000	1.300	0.332	-0.300	-0.59
3	3.00	2.000	2.000	0.271	0.000	0.00
4	4.00	2.000	2.700	0.332	-0.700	-1.38
5	5.00	4.000	3.400	0.469	0.600	1.57

The point estimate for $x = 4\%$ is the fitted value, $\hat{y} = 2.7$ seconds of reaction time. ∎

ESTIMATION AND PREDICTION

If there is a significant linear relationship between two variables, the regression equation can be used for estimation and prediction. You can use it to estimate the mean value of y for a specific value of x using a $(1-\alpha)100\%$ confidence interval. Or you can use it to predict a particular value of y for a specific value of x using a $(1-\alpha)100\%$ prediction interval. The subcommand PREDICT with REGRESS gives 95% confidence and prediction intervals.

Confidence and prediction intervals for confidence coefficients other than 95% can be calculated using the PREDICT printout and the INVCDF command for t values. Using the notation on the printout, the confidence interval is

$$\text{Fit} \pm t_{\alpha/2}(\text{Stdev. Fit})$$

and the prediction interval is

$$\text{Fit} \pm t_{\alpha/2}\sqrt{s^2 + \text{Stdev.Fit}^2}$$

■ **Example 2** Refer to Example 1 of this chapter. Use the PREDICT subcommand to obtain a 95% confidence interval and prediction interval for the reaction time if the drug percentage is 4%.

Solution The data are saved in file named DRUGTIME. The predictor x_p is 4.

```
MTB > RETRIEVE 'DRUGTIME'
  WORKSHEET SAVED  1/10/1994

Worksheet retrieved from file: DRUGTIME.MTW
MTB > INFORMATION

COLUMN    NAME         COUNT
C1        DRUG %         5
C2        REACTION       5

CONSTANTS USED: NONE

MTB > REGRESS 'REACTION' ON 1 PREDICTOR 'DRUG %';
SUBC> PREDICT AT 4.

The regression equation is
REACTION = - 0.100 + 0.700 DRUG %

Predictor        Coef       Stdev     t-ratio         p
Constant       -0.1000      0.6351      -0.16      0.885
DRUG %          0.7000      0.1915       3.66      0.035

s = 0.6055      R-sq = 81.7%     R-sq(adj) = 75.6%

Analysis of Variance

SOURCE         DF         SS          MS          F         p
Regression      1       4.9000      4.9000      13.36     0.035
Error           3       1.1000      0.3667
Total           4       6.0000

     Fit   Stdev.Fit         95% C.I.            95% P.I.
   2.700      0.332     ( 1.645,  3.755)    ( 0.503,  4.897)

MTB > STOP
```

We are 95% confident that the mean reaction time for 4% of drug in the bloodstream is from 1.65 to 3.76 seconds. We are 95% confident that the reaction time for a particular person with a drug reading of 4% is from .50 to 4.87 seconds.

EXERCISES

1. The Minnesota Real Estate Research Center compiled data on the homes sold in St. Cloud, Minnesota during 1989. The selling prices y and living areas in square feet x of 20 homes are given in the following table.

Price	Area	Price	Area
$46,000	770	$ 78,900	1,652
46,600	740	85,000	1,190
52,000	832	90,600	1,200
52,500	768	99,875	1,444
53,500	700	104,400	1,516
54,000	1,030	108,000	2,024
64,000	1,120	116,000	1,840
64,500	1,068	111,900	1,684
69,900	760	119,500	1,760
71,900	1,400	123,000	1,760

   a. What is the average square footage and selling price of the homes in the sample?
   b. Construct a scattergram and find the correlation coefficient. Interpret.
   c. Determine the linear regression equation. Interpret the parameter estimates. Give a practical interpretation for s.
   d. Test whether there is a significant linear relationship between selling price and area. Use $\alpha = .05$.
   e. What is the reduction in the sum of squared errors when the fitted value $\hat{y}$, rather than $\bar{y}$, is used for estimation?

2. The Minnesota Real Estate Research Center provided the following table on the selling prices and living areas in square feet of 20 homes randomly selected from the homes sold during 1989 in Rochester, Minnesota.

Price	Area	Price	Area
$47,800	763	$ 85,900	1,060
53,000	780	89,000	1,100
54,900	1,086	92,330	1,176
58,000	960	96,000	1,044
61,500	1,040	103,925	1,146
64,000	840	135,000	1,533
66,000	1,137	144,000	2,174
69,900	872	129,900	2,008
74,500	1,153	125,500	1,736
80,750	920	129,195	2,202

   a. Construct and interpret a scattergram of selling price versus area. Calculate the correlation coefficient. Would a simple linear regression model describe the relationship between the two variables?
   b. Use the REGRESS command to find the least squares regression line. Interpret the

coefficients.
c. Test the usefulness of the model. Use $\alpha = .05$.
d. Interpret s and r^2.
e. If the model is useful, estimate the mean selling price of Rochester homes with living areas of 1,500 square feet. Use a 95% confidence interval.

3. Each month, the Federal Aviation Administration releases information on the performance of the major U.S. airlines. The following table gives the percent of flights arriving within 15 minutes of the scheduled time, the number of baggage complaints per 1,000 passengers, and the number of consumer complaints per 100,000 passengers in December, 1989.

Carrier	On-time Performance	Baggage Complaints	Consumer Complaints
Southwest	81.6	4.99	0.19
Northwest	76.7	13.59	1.53
American	74.1	11.76	0.84
Pan Am	73.5	7.96	4.09
Amer West	72.9	13.71	1.09
Continental	71.6	9.49	1.62
Delta	71.4	11.32	0.47
United	70.7	11.36	1.09
Alaska	69.7	9.03	0.89
Eastern	69.1	18.12	2.73
TWA	67.8	18.12	5.32
USAir	59.4	18.96	2.17

a. Find the mean percent of flights on-time, the mean number of baggage complaints, and the mean number of consumer complaints.
b. Construct a scattergram and calculate the coefficient of correlation for the number of consumer complaints y and on-time performance x. Interpret.
c. Find the least squares regression line relating consumer complaints to on-time performance. Do the data provide sufficient evidence to indicate that on-time performance contributes information about consumer complaints. Use $\alpha = .05$.
d. Construct a scattergram and calculate r for the number of baggage complaints y and on-time performance x. Interpret.
e. Find the least squares regression line relating baggage complaints to on-time performance. Do the data provide sufficient evidence to indicate that on-time performance contributes information about baggage complaints. Use $\alpha = .05$.
f. Estimate the change in baggage complaints when on-time performance increases by 1%. If USAir could increase its on-time performance by 10%, what is the estimated change in baggage complaints?

4. An important factor in maintaining employee satisfaction and productivity is an adequate compensation for an employee's responsibilities. The personnel manager needs to continually monitor salaries for certain job types, and should compare these salaries with the salaries of the industry. The following table gives job titles, job evaluation points, and comparative starting salaries for 11 jobs in a data processing department. Each job has been assigned points based on responsibilities and requirements. The salaries of each job are the industry benchmark weekly

starting salary from the *1988-1989 Occupational Outlook* and the actual starting salary in the department.

Job Title	Job Points	Industry Salary	Actual Salary
File Clerk	250	$240	$220
Data Entry	340	240	250
Receptionist	365	255	300
Wood Processor	390	265	245
General Office Clerk	450	250	255
Computer Operator	490	280	300
Secretary	505	290	330
Accounting Clerk	530	290	280
Statistical Clerk	540	290	300
Programmer	620	330	320
System Analyst I	770	355	400

The objectives of the study are to determine whether any actual salaries are out of line with the industry, and to analyze the relationship between job points and industry benchmark salaries.

a. Use MPLOT to plot industry salary versus job points and actual salary versus job points on the same plot. Discuss.
b. Find the least squares regression equation that relates industry salary y to job points x. Interpret the regression coefficients.
c. Do the data provide sufficient evidence of a linear relationship between industry salary and job points? Test using $\alpha = .01$.
d. Do any actual salaries in the data processing department seem out of line with industry standards? Explain.

5. A survey conducted by the UCLA Graduate School of Management on the status of computer development in business schools is reported in the *Sixth Annual UCLA Survey of Business School Computer Usage*, Jason L. Fraud and Julia A. Britt, September, 1989. The survey queries business schools on hardware, software, and resource commitments. One hundred sixty-three business schools completed the sixth annual UCLA computer survey. The following random sample of 25 schools concerns the number of microcomputers y and the number of full time equivalent faculty x.

School	PCs	FTE Faculty	School	PCs	FTE Faculty
1	109	60	14	383	106
2	30	93	15	66	44
3	103	79	16	192	209
4	60	36	17	52	119
5	241	143	18	157	82
6	119	115	19	305	84
7	282	144	20	91	50
8	534	205	21	155	60
9	87	85	22	53	24
10	354	153	23	44	56

11	92	62	24	310	105
12	92	42	25	258	94
13	173	114			

    a. Numerically and graphically summarize the sample of microcomputers (PCs) and the full time equivalent (FTE) faculty. What are the mean and standard deviation of the number of PCs and of the FTE faculty per school? Describe the distributions. Are there any outliers?

    b. Construct a scattergram of the number of PCs versus the number of FTE faculty. Calculate the correlation coefficient. Does the relationship between the two variables appear to be linear?

    c. Fit a least squares regression line. Do the data provide evidence that a linear regression model is useful? Use $\alpha = .05$.

    d. Interpret r^2.

    e. If the model is useful, predict the number of PCs at a business school with 100 FTE faculty. Use a 95% confidence level.

6. Consider the American League baseball statistics for 14 teams during the 1991 season that are given in Exercise 13.15 on page 676 of the text. The number of games won y is thought to be related to the team batting average x.

Team	Games Won	Batting Average
Cleveland	57	.254
New York	71	.256
Boston	84	.269
Toronto	91	.257
Texas	85	.270
Detroit	84	.247
Minnesota	95	.280
Baltimore	67	.254
California	81	.255
Milwaukee	83	.271
Seattle	83	.255
Kansas City	82	.264
Oakland	84	.248
Chicago	87	.262

    a. Construct and interpret a scattergram of the number of games won versus the team batting average. Calculate the correlation coefficient. Would a simple linear regression model describe the relationship between the two variables?

    b. Use the REGRESS command to find the least squares regression line. Interpret the coefficients.

    c. Test whether the simple linear regression model is useful. Use $\alpha = .05$.

    d. Interpret r^2 and s.

    e. If the model is useful, estimate the number of games won by a team with a batting average of .275. Use a 95% prediction interval.

7. Consider Exercise 13.59 on page 713 of the text. A new drug developed to reduce a smoker's reliance on nicotine may reduce the pulse rate to a dangerous level. The following table gives

the decrease in pulse rate y and the cubic centimeters of drug dosage x for six randomly selected patients.

Patient	Dosage	Pulse Decrease
1	2.0	15
2	1.5	9
3	3.0	18
4	2.5	16
5	4.0	23
6	3.0	20

a. Plot the decrease in pulse rate versus drug dosage. Calculate r. Discuss.
b. Find the least squares regression equation that relates pulse decrease to drug dosage. Interpret the regression coefficients.
c. Do the data provide sufficient evidence of a linear relationship between the variables? Test using $\alpha = .01$.
d. Find a 95% confidence interval and a 95% prediction interval for pulse decrease if the drug dosage is 3.5 cubic centimeters. Explain the difference in the two intervals.

8. Refer to the class data set given in Appendix B of this supplement.

a. Analyze the relationship between grade point average and the number of hours worked for all students. What assumptions are necessary for this statistical analysis?
b. Analyze the relationship between grade point average and the number of hours worked for female students.
c. Analyze the relationship between grade point average and the number of hours worked for male students. Compare with the analysis for female students.

CHAPTER 14

MULTIPLE REGRESSION

An objective of multiple regression analysis is to develop a model that can be used to predict a response variable using information contained in several independent variables. In this chapter, we describe a scattergram and correlation analysis to hypothesize the form of the model, and illustrate how to use sample data to estimate the model. We include first-order models, models with an interaction term, and quadratic models. The basic concepts and Minitab commands of simple linear regression analysis apply in multiple regression analysis.

NEW COMMAND

TPLOT

THE GENERAL LINEAR MODEL

Oftentimes, a simple linear regression model containing a single independent variable inadequately describes the response variable. Predictions of the response variable are too imprecise to be meaningful. In some cases, more than one independent variable may significantly affect the response variable, and a multiple linear regression model may better estimate or predict the value of a response variable. The general form is

$$y = \beta_0 + \beta_1 x_1 + \beta_2 x_2 + \ldots + \beta_k x_k + \epsilon,$$

where y is the response variable, $x_1, x_2, \ldots, x_k$ are the k independent variables, $\beta_0, \beta_1, \ldots, \beta_k$ are the parameters we wish to estimate, and ϵ is the random error term. As before, ϵ accounts for the variability in y that is not explained by the independent variables in the model. The model assumes that the errors are independent, and that the probability distribution of ϵ is normal, with zero mean and a constant variance σ^2 for all values of $x_1, x_2, \ldots, x_k$.

In multiple regression analysis, it is important to know the type and strength of the relationship between each pair of variables. A plot of the response variable versus each independent variable and a correlation analysis provide useful information about the relationships. A correlation matrix gives the correlation coefficients of the response and each independent variable, and of each pair of independent variables.

> CORRELATION COEFFICIENTS FOR C,...,C
>
> The Pearson product moment correlation coefficient is calculated between every pair of columns. The lower triangle of the correlation matrix is printed.

The REGRESS command fits the least squares line to the sample of observations. The Minitab command BRIEF described in Chapter 13 changes the amount and type of regression output. BRIEF 2 is the default output of the REGRESS command.

> REGRESS C ON K PREDICTORS C,...,C (STORE STD RESIDUALS IN C (FIT IN C))
>   RESIDUALS IN C
>   PREDICT FOR E
>
> This command determines the multiple linear regression equation using the least squares method. The number of predictors K must be followed by K columns containing the predictors.
>
> The output includes the regression equation; a table of coefficients, standard deviations, and t-ratios; s and R^2; and the analysis of variance table. Use the options to store the standardized residuals and fitted values and the RESIDUALS subcommand to store the residuals in a column.
>
> The subcommand PREDICT calculates $\hat{y}$, the standard deviation of $\hat{y}$, and 95% confidence and prediction intervals for specified values of the predictors. You need to enter a value on the subcommand line for every predictor. The values may be stored in K columns or as constants. If you want to predict a single $\hat{y}$, you can enter the values of the predictors directly on the subcommand line.

Consider the following examples of multiple regression analysis.

1. Suppose the response is in column C1 and the independent variables are in columns C2, C3, and C4. The Minitab commands for the scattergrams, correlation matrix, and a first-order linear model are

   ```
   MTB > PLOT C1 VS C2
   MTB > PLOT C1 VS C3
   MTB > PLOT C1 VS C4
   MTB > CORRELATIONS FOR C1-C4
   MTB > REGRESS C1 USING 3 PREDICTORS IN C2-C4
   ```

2. Suppose a second-order term x^2 is used to model the curvilinear relationship between y and x. If the response is in a column named 'SALES' and the independent variables are in columns

named 'X' and 'X*X', the Minitab commands for the second-order model are

```
MTB > PLOT 'SALES' VS 'X'
MTB > CORRELATION 'SALES' VS 'X'
MTB > REGRESS 'SALES' USING 2 PREDICTORS 'X' AND 'X*X'
```

We want a high correlation between y and each x. But if the correlation is high between any pair of independent variables, the regression coefficients may be statistically affected. Minitab addresses this problem, called multicollinearity, in the regression analysis.

For example, if the correlation between an independent variable PRICE and other independent variables is high, Minitab includes the variable in the equation, but prints a message:

```
* NOTE *   PRICE IS HIGHLY CORRELATED WITH OTHER PREDICTOR VARIABLES
```

If the correlation is extremely high, Minitab eliminates it from the equation and prints a message:

```
PRICE IS HIGHLY CORRELATED WITH OTHER X VARIABLES
PRICE HAS BEEN REMOVED FROM THE EQUATION
```

THE FIRST-ORDER LINEAR MODEL

The first-order model is linear in the independent variables. Use the model if the independent variables have an additive effect on the response variable; that is, if the independent variables do not interact. For example, the first-order regression model for three independent variables is

$$y = \beta_0 + \beta_1 x_1 + \beta_2 x_2 + \beta_3 x_3 + \epsilon$$

The y-intercept β_0 is the value of the mean response when each independent variable is zero. The ith parameter β_i represents the estimated change in the mean value of y for a one unit change in x_i, provided all other variables are held constant.

Analogous to simple linear regression, in which the regression equation is a line, the graph of a first-order regression equation is a hyperplane. For two independent variables, the graph is a three-dimensional plane.

■ **Example 1** Refer to Example 14.1 on page 750 of the text. An antique clock collector believes that the price y of a clock at an auction increases with the age of the clock x_1 and with the number of bidders x_2. The following table gives the data for a random sample of 32 clocks.

Clock	Age	Bidders	Price	Clock	Age	Bidders	Price
1	127	13	$1,235	17	170	14	$2,131
2	115	12	1,080	18	182	8	1,550
3	127	7	845	19	162	11	1,884
4	150	9	1,522	20	184	10	2,041
5	156	6	1,047	21	143	6	854
6	182	11	1,979	22	159	9	1,483
7	156	12	1,822	23	108	14	1,055

8	132	10	1,253	24	175	8	1,545
9	137	9	1,297	25	108	6	729
10	113	9	946	26	179	9	1,792
11	137	15	1,713	27	111	15	1,175
12	117	11	1,024	28	187	8	1,593
13	137	8	1,147	29	111	7	785
14	153	6	1,092	30	115	7	744
15	117	13	1,152	31	194	5	1,356
16	126	10	1,336	32	168	7	1,262

a. Construct scattergrams to study the relationship between the response variable and independent variables.
b. Calculate and interpret the correlation matrix.
c. Fit the first-order linear model to the data. Interpret the least squares estimates of the parameters.
d. Find the estimate of σ^2 and σ.
e. Test whether the auction price y increases as the number of bidders x_2 increases. Use $\alpha = .05$.
f. Construct and interpret a 95% confidence interval for β_2.
g. Interpret R^2.
h. Test whether the model is useful for predicting auction price.
i. If the model is useful, construct and interpret 95% confidence and prediction intervals for the price of a 150 year old clock sold at an auction with 10 bidders.

Solution

a. PLOT gives a scattergram of the response variable PRICE versus each independent variable.

```
MTB > NAME C1 'AGE' C2 'BIDDERS' C3 'PRICE'
MTB > READ 'AGE'   'BIDDERS'   'PRICE'
DATA> 127    13    1235
DATA> 115    12    1080
DATA> 127     7     845
DATA> 150     9    1522
DATA> 156     6    1047
DATA> 182    11    1979
DATA> 156    12    1822
DATA> 132    10    1253
DATA> 137     9    1297
DATA> 113     9     946
DATA> 137    15    1713
DATA> 117    11    1024
DATA> 137     8    1147
DATA> 153     6    1092
DATA> 117    13    1152
DATA> 126    10    1336
DATA> 170    14    2131
DATA> 182     8    1550
DATA> 162    11    1884
DATA> 184    10    2041
DATA> 143     6     854
DATA> 159     9    1483
DATA> 108    14    1055
DATA> 175     8    1545
DATA> 108     6     729
DATA> 179     9    1792
```

```
DATA> 111    15    1175
DATA> 187     8    1593
DATA> 111     7     785
DATA> 115     7     744
DATA> 194     5    1356
DATA> 168     7    1262
DATA> END
      32 ROWS READ
MTB > SAVE 'CLOCKS'

Worksheet saved into file: CLOCKS.MTW
MTB > PLOT 'PRICE' VS 'AGE';
SUBC> TITLE 'AUCTION PRICE VERSUS AGE OF CLOCK'.
```

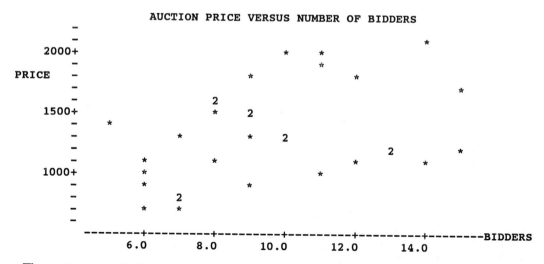

```
MTB > PLOT 'PRICE' VS 'BIDDERS';
SUBC> TITLE 'AUCTION PRICE VERSUS NUMBER OF BIDDERS'.
```

The scattergrams indicate a positive linear relationship between the auction price and each independent variable. The strongest relationship appears to be between price and clock age.

b. The CORRELATION command gives the correlation matrix.

```
MTB > CORRELATION 'PRICE' 'AGE' 'BIDDERS'

            PRICE      AGE
AGE         0.730
BIDDERS     0.395    -0.254
```

The positive correlations between auction price and the independent variables support the evidence found in the scattergrams. The highest correlation is between auction price and clock age. There is a moderate inverse relationship between age and number of bidders.

c. REGRESS uses the least squares method to fit the linear regression model to the data.

```
MTB > REGRESS 'PRICE' ON 2 PREDICTORS 'AGE' AND 'BIDDERS'

The regression equation is
PRICE = - 1337 + 12.7 AGE + 85.8 BIDDERS

Predictor       Coef       Stdev     t-ratio        p
Constant     -1336.7       173.4       -7.71    0.000
AGE          12.7362      0.9024       14.11    0.000
BIDDERS       85.815       8.706        9.86    0.000

s = 133.1      R-sq = 89.3%      R-sq(adj) = 88.5%

Analysis of Variance

SOURCE        DF          SS          MS         F          p
Regression     2     4277160     2138580    120.65      0.000
Error         29      514034       17725
Total         31     4791194

SOURCE        DF      SEQ SS
AGE            1     2554859
BIDDERS        1     1722301
```

The least squares prediction equation is

```
The regression equation is
PRICE = - 1337 + 12.7 AGE + 85.8 BIDDERS
```

The y-intercept, $\hat{\beta}_0 = -1337$, is the estimate of the auction price of a clock with $x_1 = 0$ and $x_2 = 0$. Since these values are outside the ranges of the data, the estimate of β_0 has no practical interpretation. The x_1 coefficient, $\hat{\beta}_1 = \$12.70$, is the estimated change in the mean auction price for a one year increase in age for a fixed number of bidders. The x_2 coefficient, $\hat{\beta}_2 = \$85.80$, is the estimated increase in the mean auction price per additional bidder for a fixed clock age.

d. An estimate of σ^2, the variance of the random error, is $s^2 = 17,725$, located in the analysis of variance table at the intersection of Error and MS. An estimate of σ, the standard deviation of the random error, is $s = \$133.10$. We expect the model to predict the auction price to within approximately $2s$, or $\$266$, for given values of the independent variables.

e. To test whether the auction price y increases as the number of bidders x_2 increases, the null and alternative hypotheses are

$H_0: \beta_2 = 0$
$H_a: \beta_2 > 0$

The t-ratio and p-value for the test are given in the regression coefficients table.

```
Predictor        Coef       Stdev     t-ratio       p
Constant      -1336.7       173.4       -7.71   0.000
AGE           12.7362      0.9024       14.11   0.000
BIDDERS        85.015       8.706        9.86   0.000
```

We would reject the null hypothesis that $\beta_2 = 0$ if the p-value is less than $\alpha = .05$. Since the p-value $= 0$, there is strong evidence that the auction price increases as the number of bidders increases.

f. To calculate a confidence interval for a parameter β_i, $\hat{\beta}_i$ and the standard deviation of $\hat{\beta}_i$ are located in the regression coefficients table. The estimates are labeled Coef and Stdev.

```
Predictor        Coef       Stdev     t-ratio       p
Constant      -1336.7       173.4       -7.71   0.000
AGE           12.7362      0.9024       14.11   0.000
BIDDERS        85.815       8.706        9.86   0.000
```

The confidence interval is defined

$$\text{Coef} \pm t_{\alpha/2}(\text{Stdev})$$

The INVCDF command with the T subcommand gives the $t_{\alpha/2}$ value for $[n - (k + 1)]$ degrees of freedom. Since $\alpha/2$ is the area in the upper tail of the distribution, the cumulative area to the left of the t value is $1 - \alpha/2$. For a 95% confidence interval, $1 - \alpha/2 = .975$ and degrees of freedom $= 32 - 3 = 29$.

```
MTB > INVCDF .975;
SUBC> T WITH 29 DF.
   0.9750    2.0452
```

For the number of bidders x_2, the Coef $= 85.815$, Stdev $= 8.706$, and the t value $= 2.0452$. The confidence interval is

$85.815 \pm (2.0452)8.706$
85.815 ± 17.806
$68.0 \leq \beta_2 \leq 103.6$

We are 95% confident that the increase in the mean auction price is $68 to $103.60 for each additional bidder for a fixed clock age.

g. The coefficient of determination is $R^2 = 89.3\%$. Approximately 90% of the total variation in auction prices can be explained by the regression model.

h. To test whether the model is useful for predicting clock price, the hypothesis test is

H_0: $\beta_1 = \beta_2 = 0$
H_a: At least one of β_1 and $\beta_2 \neq 0$

The test statistic, $F = 120.65$, and the p-value $= 0.000$ are given in the analysis of variance table. Since the p-value is 0, we would reject H_0 for any α level. We have strong evidence to conclude that the model is useful for predicting the auction price of clocks.

i. To construct and interpret 95% confidence and prediction intervals for the price of a 150 year old clock and 10 bidders we use PREDICT with REGRESS. We use BRIEF 1 to obtain the least amount of output.

```
MTB > BRIEF 1
MTB > REGRESSION 'PRICE' ON 2 PREDICTORS 'AGE' 'BIDDERS';
SUBC> PREDICT AT 150 YEARS AND 10 BIDDERS.

The regression equation is
PRICE = - 1337 + 12.7 AGE + 85.8 BIDDERS

Predictor        Coef       Stdev     t-ratio         p
Constant      -1336.7       173.4       -7.71     0.000
AGE           12.7362      0.9024       14.11     0.000
BIDDERS        85.815       8.706        9.86     0.000

s = 133.1      R-sq = 89.3%      R-sq(adj) = 88.5%

Analysis of Variance

SOURCE         DF          SS          MS          F          p
Regression      2     4277160     2138580     120.65      0.000
Error          29      514034       17725
Total          31     4791194

     Fit   Stdev.Fit          95% C.I.              95% P.I.
  1431.9        24.5    ( 1381.7, 1482.0)     ( 1154.9, 1708.8)

MTB > STOP
```

A 95% confidence interval for the mean auction price of 150 year old clocks with 10 bidders is from $1,382 to $1,482. A 95% prediction interval for the auction price of a particular 150 year old clock with 10 bidders is from $1,155 to $1,709.

REGRESSION MODEL WITH INTERACTION

It often happens in multiple regression analysis that the effect of one independent variable on the response variable depends on the value of another independent variable. In this case, the independent variables are said to interact, and an interaction model is fit to the data. You can use a graph to determine whether the slope of the straight line relationship between the mean response variable y and an independent variable depends on the value of the other independent variable.

An interaction model includes the independent variables and the cross product or interaction term of the independent variables. For example, the interaction model with two variables is

$$y = \beta_0 + \beta_1 x_1 + \beta_2 x_2 + \beta_3 x_1 x_2 + \epsilon$$

Use **TPLOT** to graphically determine whether an interaction term should be included in the model. TPLOT adds a quantitative third variable to a scattergram. Symbols are used to represent the distances, in terms of standard deviation, that observations are from the mean.

TPLOT Y IN C VS XONE IN C VS XTWO IN C
  YINCREMENT = K
  YSTART AT K (END AT K)
  XINCREMENT = K
  XSTART AT K (END AT K)
  TITLE = 'TEXT'
  FOOTNOTE = 'TEXT'
  XLABEL = 'TEXT'
  YLABEL = 'TEXT'

TPLOT produces a pseudo three-dimensional plot, with y on the vertical axis, x_1 on the horizontal axis, and x_2 indicated by special symbols on the plot.

The data set for x_2 is divided into four groups, and points for each group are plotted with special symbols. Let $\bar{x}_2$ and s represent the mean and standard deviation of the data. The following groups and corresponding symbols are defined:

Group	Symbol
$x_2 < \bar{x}_2 - s$	0
$\bar{x}_2 - s \leq x_2 < \bar{x}_2$	.
$\bar{x}_2 \leq x_2 \leq \bar{x}_2 + s$	/
$x_2 > \bar{x}_2 + s$	X

The values of $\bar{x}_2 - s$, $\bar{x}_2$, and $\bar{x}_2 + s$ are given on the output. If several points fall at the same location, a count is given on the plot. The subcommands are the same as those defined with MPLOT.

■ **Example 2** Refer to Example 14.3 on page 763 of the text. Suppose that a collector believes that the rate of increase in clock price with age depends on the number of bidders. For example, the price may increase at a faster rate with age for 15 bidders than for 5 bidders.

a. Graphically analyze the data to determine whether the two independent variables interact in their effect on auction price.
b. Fit a general linear model with an interaction term to the data.
c. Is there significant evidence in the sample that β_3, the interaction regression coefficient is positive? Use $\alpha = .05$.

258   *Chapter 14*

d. Use the model to predict the auction price of a 150 year old clock with 10 bidders. Use a 95% prediction interval. Compare this with the prediction interval in the previous example.

Solution

a. We use TPLOT to determine whether the mean rate of change in auction price with age depends on the number of bidders.

```
MTB > RETRIEVE 'CLOCKS'
 WORKSHEET SAVED   1/10/1994

Worksheet retrieved from file: CLOCKS.MTW
MTB > INFORMATION

COLUMN      NAME        COUNT
C1          AGE         32
C2          BIDDERS     32
C3          PRICE       32

CONSTANTS USED: NONE

MTB > TPLOT 'PRICE' VS 'AGE' VS 'BIDDERS';
SUBC> TITLE 'AUCTION PRICE VERSUS CLOCK AGE';
SUBC> TITLE 'Symbols for Number of Bidders'.
```

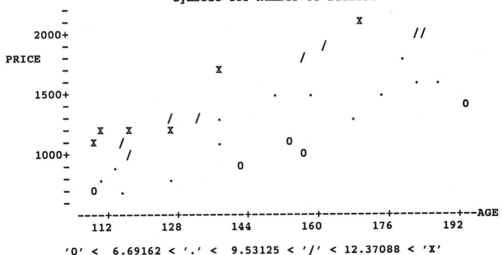

The TPLOT relates auction price y to clock age x_1 for four groups of bidders x_2. The values of $\bar{x}_2 - s = 6.69$, $\bar{x}_2 = 9.53$, and $\bar{x}_2 + s = 12.37$ years divide the number of bidders into four groups. If straight lines are drawn through the symbols for corresponding groups of bidders, the lines would not be parallel. The slope of the straight line relating auction price to age increases as the number of bidders increase. Clock age and number of bidders appear to interact in explaining clock price.

b. REGRESS fits the interaction model to the data.

```
MTB > # CALCULATE THE INTERACTION
MTB > NAME C4 'AGE*BID'
MTB > LET 'AGE*BID' = 'AGE'*'BIDDERS'
MTB > REGRESS 'PRICE' ON 3 PREDICTORS 'AGE' 'BIDDERS' 'AGE*BID'

The regression equation is
PRICE = 323 + 0.87 AGE - 93.4 BIDDERS + 1.30 AGE*BID

Predictor        Coef       Stdev     t-ratio        p
Constant        322.8       293.3        1.10    0.281
AGE             0.873       2.020        0.43    0.669
BIDDERS        -93.41       29.71       -3.14    0.004
AGE*BID        1.2979      0.2110        6.15    0.000

s = 88.37        R-sq = 95.4%     R-sq(adj) = 94.9%

Analysis of Variance

SOURCE         DF         SS           MS          F         p
Regression      3     4572548      1524183     195.19     0.000
Error          28      218646         7809
Total          31     4791194

SOURCE         DF      SEQ SS
AGE             1     2554859
BIDDERS         1     1722301
AGE*BID         1      295388

Unusual Observations
Obs.     AGE     PRICE      Fit   Stdev.Fit  Residual  St.Resid
 16      126    1336.0    1134.0      19.1     202.0    2.34R
 31      194    1356.0    1284.1      58.8      71.9    1.09 X

R denotes an obs. with a large st. resid.
X denotes an obs. whose X value gives it large influence.
```

We can use the regression equation,

```
The regression equation is
PRICE = 323 + 0.87 AGE - 93.4 BIDDERS + 1.30 AGE*BID
```

to produce a graph as shown in Figure 14.12 on page 764 of the text. We calculate auction price for $x_2 = 5$, 10, and 15 bidders using several different ages, $x_1 = 120$, 140, 160 and 180 years.

```
MTB > SET 'AGE'
DATA> 120 140 160 180
DATA> END
MTB > NAME C5 'Y:5 BIDS' C6 'Y:10BIDS' C7 'Y:15BIDS'
MTB > LET 'Y:5 BIDS'  = 323 + .87*'AGE' - 93.4*5  +1.3*'AGE'*5
MTB > LET 'Y:10BIDS'  = 323 + .87*'AGE' - 93.4*10 +1.3*'AGE'*10
MTB > LET 'Y:15BIDS'  = 323 + .87*'AGE' - 93.4*15 +1.3*'AGE'*15
```

260   Chapter 14

```
MTB > MPLOT 'Y:5 BIDS' VS 'AGE', 'Y:10BIDS' VS 'AGE', 'Y:15BIDS' VS 'AGE';
SUBC> TITLE 'AUCTION PRICE VERSUS AGE OF ANTIQUE CLOCKS';
SUBC> TITLE 'Letters for Number of Bidders'.
```

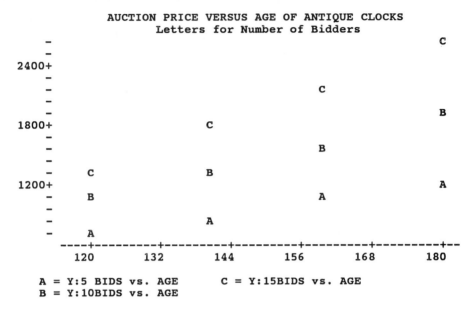

A straight line can be drawn through the letters representing 5, 10, and 15 bidders. The graph indicates the slope of the lines relating auction price to age depends on the number of bidders. The mean auction price increases at a faster rate with age as the number of bidders increases.

c. To test whether β_3 is positive, we test

   H_0: $\beta_3 = 0$
   H_a: $\beta_3 > 0$

The test statistic, $t = 6.15$, and p-value $= 0.000$ are found in the AGE*BID row of the regression coefficients table. Since the p-value of 0 is less than an α as low as .01, we have strong evidence to conclude that β_3 is positive. The rate of change of price with age increases as the number of bidders increases. The test confirms the graphical analyses given in parts a and b.

d. We use PREDICT with REGRESS to construct the 95% prediction interval. The value of the interaction term to include on the subcommand line is AGE*BID = (150)(10) = 1500.

```
MTB > BRIEF 1
MTB > REGRESS 'PRICE' ON 3 PREDICTORS 'AGE' 'BIDDERS' 'AGE*BID';
SUBC> PREDICT AT 150 10 1500.

The regression equation is
PRICE = 323 + 0.87 AGE - 93.4 BIDDERS + 1.30 AGE*BID
```

```
Predictor          Coef        Stdev      t-ratio         p
Constant          322.8        293.3         1.10     0.281
AGE               0.873        2.020         0.43     0.669
BIDDERS          -93.41        29.71        -3.14     0.004
AGE*BID          1.2979       0.2110         6.15     0.000

s = 88.37        R-sq = 95.4%      R-sq(adj) = 94.9%
```

Analysis of Variance

```
SOURCE          DF         SS         MS          F          p
Regression       3     4572548    1524183     195.19     0.000
Error           28      218646       7809
Total           31     4791194
```

```
    Fit    Stdev.Fit        95% C.I.            95% P.I.
 1466.5        17.2    ( 1431.2, 1501.8)   ( 1282.0, 1651.0)
```

MTB > STOP

We are 95% confident that the auction price of a 150 year old clock with 10 bidders is from $1,282 to $1,651. This interval is narrower ($W = \$369$) than the interval obtained in Example 1 ($W = \$553$) because another useful predictor for explaining variation in the auction prices has been added to the model.

∎

THE SECOND-ORDER LINEAR MODEL

A second-order linear model is used to model a curvilinear relationship between the response variable and an independent variable x. The model includes a second-order term, x^2. If the parameter for x^2 is positive, the graph opens upward. If the parameter for x^2 is negative, the graph opens downward. The form of the model, called a quadratic model, is

$$y = \beta_0 + \beta_1 x + \beta_2 x^2 + \epsilon$$

A scattergram gives information about a curvilinear relationship between a response and an independent variable.

∎ **Example 3** Refer to Table 14.1 on page 740 of the text. The following table gives the size x of the home in square feet and the number of kilowatt hours y of electrical usage for each of ten homes during a particular month.

Size	Usage	Size	Usage
1,290	1,182	1,840	1,711
1,350	1,172	1,980	1,804
1,470	1,264	2,230	1,840
1,600	1,493	2,400	1,956
1,710	1,571	2,930	1,954

262   *Chapter 14*

Use Minitab to model the relationship. Predict the monthly electrical usage of a 1,500 square foot home.

Solution   A scattergram and the correlation coefficient provide information on the type and strength of the relationship between electrical usage and size of a home.

```
MTB > NAME C1 'SIZE' C2 'USAGE'
MTB > SET 'SIZE'
DATA> 1290    1350    1470    1600    1710    1840    1980    2230    2400    2930
DATA> END
MTB > SET 'USAGE'
DATA> 1182    1172    1264    1493    1571    1711    1804    1840    1956    1954
DATA> END
MTB > SAVE 'E-USAGE'

Worksheet saved into file: E-USAGE.MTW
MTB > PLOT 'USAGE' VS 'SIZE';
SUBC> TITLE 'MONTHLY ELECTRICAL USAGE AND HOME SIZE';
SUBC> FOOTNOTE 'Usage in kilowatt-hours, Size in square feet'.
```

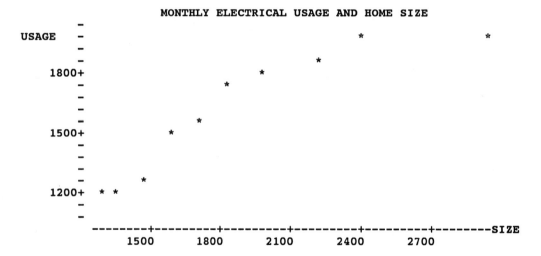

```
MTB > CORRELATION 'SIZE' AND 'USAGE'

Correlation of SIZE and USAGE = 0.912
```

The scattergram and correlation coefficient, $r = .912$, indicate a strong positive relationship between the electrical usage and size of homes. However, there appears to be some curvature in the data; electrical usage seems to increase at a faster rate for smaller size homes. A quadratic model may be a better fit than a simple linear model. Since the graph opens downward, the parameter for x^2 in the regression equation is negative.

```
MTB > NAME C3 'SQ SIZE'
MTB > LET 'SQ SIZE' = 'SIZE'**2
MTB > REGRESS 'USAGE' ON 2 PREDICTORS 'SIZE' AND 'SQ SIZE'
```

```
The regression equation is
USAGE = - 1216 + 2.40 SIZE -0.000450 SQ SIZE

Predictor        Coef        Stdev      t-ratio         p
Constant       -1216.1        242.8       -5.01      0.000
SIZE            2.3989       0.2458        9.76      0.000
SQ SIZE     -0.00045004   0.00005908      -7.62      0.000

s = 46.80       R-sq = 98.2%       R-sq(adj) = 97.7%

Analysis of Variance

SOURCE         DF         SS          MS         F         p
Regression      2      831070      415535    189.71     0.000
Error           7       15333        2190
Total           9      846402

SOURCE         DF      SEQ SS
SIZE            1      703957
SQ SIZE         1      127112

Unusual Observations
Obs.    SIZE      USAGE       Fit   Stdev.Fit   Residual   St.Resid
 10     2930     1954.0    1949.2       44.7        4.8       0.35 X

X denotes an obs. whose X value gives it large influence.
```

To determine whether the quadratic model is useful, we test

H_0: $\beta_1 = \beta_2 = 0$
H_a: At least one of β_1 and $\beta_2 \neq 0$

The test statistic is $F = 189.71$ and the p-value $= 0.000$. We have strong evidence that the model is useful for predicting the amount of electrical usage. The coefficient of determination, $R^2 = 98.2\%$, attests to the utility of the model. About 98% of the total variation in the prices is explained by the quadratic model.

To determine whether significant downward curvature exists in the model, the null and alternative hypotheses are:

H_0: $\beta_2 = 0$
H_a: $\beta_2 < 0$

The test statistic is $t = -7.62$ and the reported p-value $= 0.000$. For a one-tailed test, the correct p-value is one half of the reported p-value on the printout. There is strong statistical evidence of downward curvature in the data. Since the quadratic term in the model is significant, the test for β_1 is irrelevant. The quadratic model is useful for estimation and prediction.

To predict the monthly electrical usage of a 1,500 square foot home, we enter the size (1,500) and size squared ($1,500^2 = 2,250,000$) on the PREDICT subcommand line.

```
MTB > BRIEF 1
MTB > REGRESS 'USAGE' ON 2 PREDICTORS 'SIZE' AND 'SQ SIZE';
SUBC> PREDICT AT 1500 2250000.
```

264   Chapter 14

```
The regression equation is
USAGE = - 1216 + 2.40 SIZE -0.000450 SQ SIZE

Predictor          Coef        Stdev       t-ratio         p
Constant        -1216.1        242.8        -5.01      0.000
SIZE             2.3989       0.2458         9.76      0.000
SQ SIZE     -0.00045004   0.00005908        -7.62      0.000

s = 46.80         R-sq = 98.2%      R-sq(adj) = 97.7%

Analysis of Variance

SOURCE         DF         SS          MS          F          p
Regression      2     831070      415535     189.71      0.000
Error           7      15333        2190
Total           9     846402

     Fit   Stdev.Fit        95% C.I.             95% P.I.
  1369.7        18.9    ( 1325.0, 1414.3)    ( 1250.3, 1489.0)

MTB > STOP
```

We are 95% confident that a 1,500 square foot home will have a monthly electrical usage from 1,250 to 1,489 kilowatt-hours.

■

RESIDUAL ANALYSIS

Some uses of residual analysis are to determine whether the regression model is misspecified, whether there are are unusual observations or outliers, and whether any assumptions are violated. The residual is the difference between the observed and fitted value of the response variable. The standardized residual (St.Resid.) is the number of standard deviations that each residual is from 0. Minitab uses the following formula to calculate the standardized residual for $\hat{y}_i$.

$$\text{St.Resid.} = \text{Residual}/\sqrt{MSE - s_{\hat{y}_i}^2},$$

where $s_{\hat{y}_i}^2$ is the variance of the fitted value $\hat{y}_i$. A plot of the residuals or standardized residuals versus the fitted values or an independent variable may indicate a nonrandom pattern in the residuals. A stem and leaf display gives information on the normality of the random errors. Both graphs may indicate outliers or unusual observations.

Minitab labels unusual observations on some regression printouts. If you use BRIEF 3, Minitab prints a table of residuals for all values of the independent variables. If you use the default BRIEF 2, Minitab prints the residuals for the unusual observations only. An observation is marked X if a value of an independent variable is unusual, and marked R if the value of the response variable is unusual. Since almost all values of y should lie within 3σ of the fitted values $\hat{y}$, a standardized residual greater than 3 or less than -3 suggests an outlier.

■ **Example 4**   Refer to page 786, Example 14.4, of the text. The following table gives the size x of the home in square feet and the number of kilowatt hours y of electrical usage for each of ten homes during a particular month.

Size	Usage	Size	Usage
1,290	1,182	1,840	1,711
1,350	1,172	1,980	1,804
1,470	1,264	2,230	1,840
1,600	1,493	2,400	1,956
1,710	1,571	2,930	1,954

a. Fit a first-order linear model to the data. Plot the residuals versus the independent variable. Analyze the plot.

b. Fit a second-order linear model to the data. Plot the residuals versus the independent variable. Analyze the plot.

Solution The data were saved in a file named E-USAGE.

a. We use the RESIDUAL subcommand with REGRESS to store the residuals, and BRIEF 3 to obtain the table of residuals.

```
MTB > RETRIEVE 'E-USAGE'
 WORKSHEET SAVED   1/10/1994

Worksheet retrieved from file: E-USAGE.MTW
MTB > INFORMATION

COLUMN     NAME        COUNT
C1         SIZE         10
C2         USAGE        10

CONSTANTS USED: NONE
MTB > BRIEF 3
MTB > NAME C4 'RESIDUAL'
MTB > REGRESS 'USAGE' ON 1 PREDICTOR 'SIZE';
SUBC> RESIDUALS IN 'RESIDUAL'.

The regression equation is
USAGE = 579 + 0.540 SIZE

Predictor       Coef        Stdev      t-ratio         p
Constant       578.9        167.0        3.47       0.008
SIZE         0.54030      0.08593        6.29       0.000

s = 133.4        R-sq = 83.2%      R-sq(adj) = 81.1%

Analysis of Variance

SOURCE          DF         SS          MS          F          p
Regression       1      703957      703957      39.54      0.000
Error            8      142445       17806
Total            9      846402

Obs.    SIZE     USAGE       Fit  Stdev.Fit   Residual   St.Resid
  1     1290    1182.0    1275.9       66.0      -93.9      -0.81
  2     1350    1172.0    1308.3       62.1     -136.3      -1.15
  3     1470    1264.0    1373.2       55.0     -109.2      -0.90
  4     1600    1493.0    1443.4       48.6       49.6       0.40
  5     1710    1571.0    1502.8       44.7       68.2       0.54
  6     1840    1711.0    1573.1       42.3      137.9       1.09
  7     1980    1804.0    1648.7       43.1      155.3       1.23
```

266 Chapter 14

```
      8     2230     1840.0     1783.8     51.8      56.2      0.46
      9     2400     1956.0     1875.7     61.5      80.3      0.68
     10     2930     1954.0     2162.0     99.6    -208.0     -2.34R
```

R denotes an obs. with a large st. resid.

An analysis of the regression output suggests that the linear model is useful for the prediction of y. The test statistic is $F = 39.54$, and the p-value of 0 is less than an α as low as .01. The coefficient of determination, $R^2 = 83.2\%$, indicates that about 83% of the variation in electrical usage is explained by the first-order model.

The tenth observation is marked R in the table, signifying an unusual value of the response variable. We use a plot of the residuals versus x to check if the residuals vary randomly.

```
MTB > PLOT 'RESIDUAL' VS 'SIZE';
SUBC> TITLE 'RESIDUAL PLOT FOR ELECTRICAL USAGE';
SUBC> TITLE 'Straight-Line Model'.
```

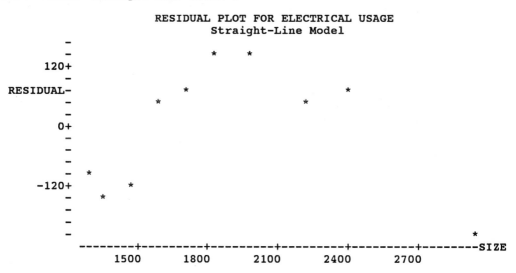

The curvilinear pattern of the residuals suggests that the model is misspecified. A quadratic model may be a better model.

b. A squared term is included in a quadratic model.

```
MTB > NAME C3 'SQ SIZE'
MTB > LET 'SQ SIZE' = 'SIZE'**2
MTB > BRIEF 3
MTB > REGRESS 'USAGE' ON 2 PREDICTORS 'SIZE' 'SQ SIZE';
SUBC> RESIDUALS IN  'RESIDUAL'.

The regression equation is
USAGE = - 1216 + 2.40 SIZE -0.000450 SQ SIZE

Predictor       Coef       Stdev      t-ratio        p
Constant      -1216.1       242.8      -5.01      0.000
SIZE           2.3989      0.2458       9.76      0.000
SQ SIZE    -0.00045004  0.00005908     -7.62      0.000
```

s = 46.80   R-sq = 98.2%   R-sq(adj) = 97.7%

Analysis of Variance

SOURCE	DF	SS	MS	F	p
Regression	2	831070	415535	189.71	0.000
Error	7	15333	2190		
Total	9	846402			

SOURCE	DF	SEQ SS
SIZE	1	703957
SQ SIZE	1	127112

Obs.	SIZE	USAGE	Fit	Stdev.Fit	Residual	St.Resid	
1	1290	1182.0	1129.6	30.1	52.4	1.46	
2	1350	1172.0	1202.2	25.9	-30.2	-0.77	
3	1470	1264.0	1337.8	19.8	-73.8	-1.74	
4	1600	1493.0	1470.0	17.4	23.0	0.53	
5	1710	1571.0	1570.1	18.0	0.9	0.02	
6	1840	1711.0	1674.2	19.9	36.8	0.87	
7	1980	1804.0	1769.4	21.9	34.6	0.84	
8	2230	1840.0	1895.5	23.3	-55.5	-1.37	
9	2400	1956.0	1949.1	23.6	6.9	0.17	
10	2930	1954.0	1949.2	44.7	4.8	0.35	X

X denotes an obs. whose X value gives it large influence.

```
MTB > PLOT 'RESIDUAL' VS 'SIZE';
SUBC> TITLE 'RESIDUAL PLOT FOR ELECTRICAL USAGE';
SUBC> TITLE 'Quadratic Model'.
```

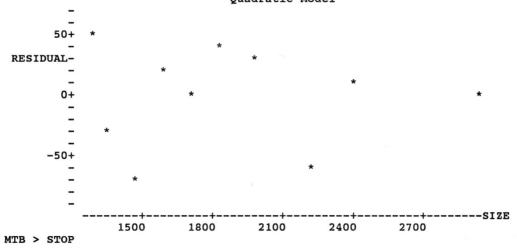

```
MTB > STOP
```

The curvilinear pattern of the straight-line residual plot has disappeared with the addition of the squared term. The quadratic model provides a better fit than the simple linear model. ∎

■ **Example 5** Consider the revised data on auction prices of clocks given in Example 14.5 on page 789 of the text. An antique clock collector believes that the price y of a clock increases with the age x_1 of a clock and with the number x_2 of bidders at an auction. The following table gives the data for a random sample of 32 clocks that we used in Examples 1 and 2 of this chapter. The price, however, of the clock numbered 17 has been changed from $2,131 to the extreme value, $1,131. The clock 17 price is called an outlier.

Clock	Age	Bidders	Price	Clock	Age	Bidders	Price
1	127	13	$1,235	17	170	14	$1,131
2	115	12	1,080	18	182	8	1,550
3	127	7	845	19	162	11	1,884
4	150	9	1,522	20	184	10	2,041
5	156	6	1,047	21	143	6	854
6	182	11	1,979	22	159	9	1,483
7	156	12	1,822	23	108	14	1,055
8	132	10	1,253	24	175	8	1,545
9	137	9	1,297	25	108	6	729
10	113	9	946	26	179	9	1,792
11	137	15	1,713	27	111	15	1,175
12	117	11	1,024	28	187	8	1,593
13	137	8	1,147	29	111	7	785
14	153	6	1,092	30	115	7	744
15	117	13	1,152	31	194	5	1,356
16	126	10	1,336	32	168	7	1,262

Fit an interaction model to the data with and without the outlier. Construct residual plots and stem and leaf displays of the residuals with and without the outlier. Discuss.

Solution The auction clock data are in CLOCKS.

```
MTB > RETRIEVE 'CLOCKS'
   WORKSHEET SAVED   1/10/1994

Worksheet retrieved from file: CLOCKS.MTW
MTB > INFORMATION

COLUMN      NAME        COUNT
C1          AGE         32
C2          BIDDERS     32
C3          PRICE       32

CONSTANTS USED: NONE

MTB > LET 'PRICE'(17) = 1131   # CHANGE THE PRICE OF THE 17TH CLOCK
MTB > NAME C4 'AGE*BID' C5 'RESIDUAL'
MTB > LET 'AGE*BID' = 'AGE'*'BIDDERS'
MTB > REGRESS 'PRICE' ON 3 PREDICTORS 'AGE' 'BIDDERS' 'AGE*BID';
SUBC> RESIDUALS IN 'RESIDUAL'.

The regression equation is
PRICE = - 511 + 8.16 AGE + 19.7 BIDDERS + 0.320 AGE*BID
```

```
Predictor        Coef      Stdev    t-ratio        p
Constant       -510.5      664.8      -0.77    0.449
AGE             8.160      4.577       1.78    0.085
BIDDERS         19.74      67.33       0.29    0.772
AGE*BID        0.3197     0.4783       0.67    0.509

s = 200.3      R-sq = 73.0%     R-sq(adj) = 70.1%

Analysis of Variance

SOURCE         DF         SS         MS         F         p
Regression      3    3029141    1009714     25.17     0.000
Error          28    1123116      40111
Total          31    4152256

SOURCE         DF     SEQ SS
AGE             1    2056592
BIDDERS         1     954626
AGE*BID         1      17922

Unusual Observations
Obs.    AGE    PRICE        Fit  Stdev.Fit   Residual   St.Resid
  17    170   1131.0     1914.0      116.5     -783.0     -4.81R
  31    194   1356.0     1481.4      133.3     -125.4     -0.84 X

R denotes an obs. with a large st. resid.
X denotes an obs. whose X value gives it large influence.

MTB > PLOT 'RESIDUAL' VS 'BIDDERS';
SUBC> TITLE 'A RESIDUAL PLOT';
SUBC> TITLE 'With Unusual Price'.
```

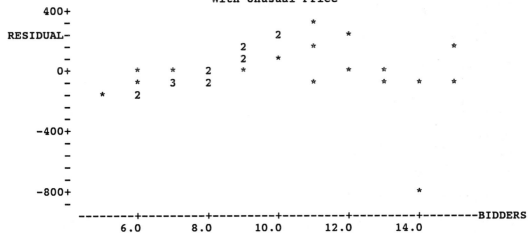

```
MTB > STEM AND LEAF 'RESIDUAL'

Stem-and-leaf of RESIDUAL   N = 32
Leaf Unit = 10

     1    -7  8
     1    -6
     1    -5
     1    -4
     1    -3
     1    -2
     6    -1  93210
    16    -0  7775544432
    16     0  0233366
     9     1  14459
     4     2  1268
```

The altered observation for 14 bidders has a residual of -.783 and a standardized residual of -4.81. It clearly is separated from the other residuals in the plot and in the stem and leaf display. To study the effect of this observation on the results of the regression analysis, we delete the outlier and redo the analysis.

```
MTB > DELETE ROW 17 FROM 'PRICE' 'AGE' 'BIDDERS' 'AGE*BID'
MTB > REGRESS 'PRICE' ON 3 PREDICTORS 'AGE' 'BIDDERS' 'AGE*BID';
SUBC> RESIDUALS IN 'RESIDUAL'.
```

The regression equation is
PRICE = 476 - 0.46 AGE - 114 BIDDERS + 1.48 AGE*BID

Predictor	Coef	Stdev	t-ratio	p
Constant	475.8	296.2	1.61	0.120
AGE	-0.465	2.093	-0.22	0.826
BIDDERS	-114.19	31.03	-3.68	0.001
AGE*BID	1.4775	0.2280	6.48	0.000

s = 85.28   R-sq = 95.2%   R-sq(adj) = 94.7%

Analysis of Variance

SOURCE	DF	SS	MS	F	p
Regression	3	3927844	1309282	180.05	0.000
Error	27	196341	7272		
Total	30	4124186			

SOURCE	DF	SEQ SS
AGE	1	2198877
BIDDERS	1	1423613
AGE*BID	1	305355

Unusual Observations

Obs.	AGE	PRICE	Fit	Stdev.Fit	Residual	St.Resid
16	126	1336.0	1137.0	18.5	199.0	2.39R
30	194	1356.0	1247.9	60.4	108.1	1.80 X

R denotes an obs. with a large st. resid.
X denotes an obs. whose X value gives it large influence.

```
MTB > PLOT 'RESIDUAL' VS 'BIDDERS';
SUBC> TITLE 'A RESIDUAL PLOT';
SUBC> TITLE 'Without Outlier'.
```

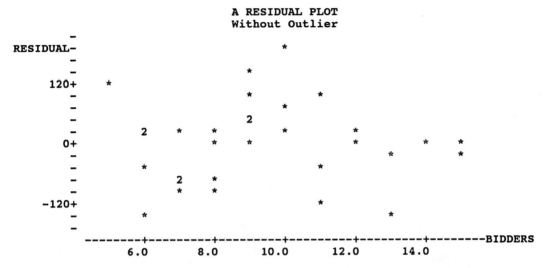

```
MTB > STEM AND LEAF 'RESIDUAL'

Stem-and-leaf of RESIDUAL   N = 31
Leaf Unit = 10

     3   -1 331
     9   -0 987765
   (7)   -0 4321000
    15    0 011223344
     6    0 79
     4    1 004
     1    1 9

MTB > STOP
```

The results of the regression analysis differ considerably when the outlier is excluded from the data. In particular, the standard deviation has decreased from 200.3 to 85.28; this will lead to more precise confidence and prediction intervals. The residuals appear random, and the stem and leaf display indicates little departure from normality. However, excluding an outlier which actually belongs to the population may lead to incorrect regression results. ∎

272   Chapter 14

EXERCISES

1. A survey conducted by the UCLA Graduate School of Management on the status of computer development in business schools is reported in the *Sixth Annual UCLA Survey of Business School Computer Usage*, Jason L. Fraud and Julia A. Britt, September, 1989. The survey queries business schools on hardware, software, and resource commitments. One hundred sixty-three business schools completed the sixth annual UCLA computer survey. The following random sample of 25 schools provides the number of microcomputers (PCs) y, and the number of the number of full time equivalent faculty (FTE) x_1, the number of full time equivalent undergraduate students x_2, and the computer operating budget per student x_3.

School	PCs	FTE Faculty	FTE Student	Budget
1	96	42	966	*
2	158	101	2150	40
3	765	215	2455	484
4	114	41	250	400
5	120	102	3171	12
6	145	95	1600	69
7	157	82	1807	17
8	14	24	525	*
9	74	43	1248	28
10	92	62	792	123
11	205	72	1423	56
12	194	105	1800	98
13	111	52	2873	22
14	164	75	700	158
15	101	79	1136	28
16	105	82	1578	63
17	254	102	2684	114
18	270	136	5753	32
19	109	52	1025	52
20	278	115	4000	*
21	421	173	3500	116
22	104	69	2947	24
23	159	64	1772	*
24	193	125	3286	165
25	370	120	4100	68

a. Construct scattergrams of the number of microcomputers versus each of the three independent variables. Calculate the correlation coefficients. Describe the relationships between the variables.

b. Fit the following multiple linear regression model to the data,

$$y = \beta_0 + \beta_1 x_1 + \beta_2 x_2 + \beta_3 x_3 + \epsilon$$

Interpret the coefficients.

c. Test whether the multiple linear regression model with the three predictors is useful in

estimating the number of microcomputers? Use $\alpha = .05$. Interpret R^2.
d. Construct a standardized residual plot and a stem and leaf display of the standardized residuals. Interpret.
e. Are all three predictors useful in the model? Test using $\alpha = .05$. If all predictors are not useful, reduce the model to a simpler form. Analyze the reduced model.
f. Use the model from part e to find a 95% confidence interval for the mean number of microcomputers for colleges with 100 full-time faculty, 3000 full-time students, and $100 per student operating budget.

2. An accountant at Fandels Department store wants to study the relationship between the amount of purchases y and customer income x. A random sample of 15 customers provided the following data on income and total purchases made at Fandels during 1992.

Customer	Purchases	Income
1	$270	$12,000
2	580	20,750
3	540	18,000
4	720	31,500
5	450	16,250
6	640	25,250
7	240	10,750
8	400	13,500
9	620	23,750
10	660	27,500
11	680	31,000
12	340	14,000
13	460	18,500
14	600	20,500
15	670	9,000

a. Analyze the relationship between the amount of purchases and income. What model is appropriate?
b. Fit the model to the data. Test whether the overall model is useful. Test whether each predictor is useful. Use $\alpha = .05$.
c. Find and interpret a 95% prediction interval for a customer with income $20,000.

3. The following data, compiled by the Minnesota Real Estate Research Center, gives the area in square feet x_1, the age x_2, and price y of a random sample of 20 homes that were sold during 1988 in the St. Cloud area.

Home	Area	Age	Price
1	770	1	$ 46,000
2	740	40	46,600
3	832	40	52,000
4	768	16	52,500
5	700	*	53,500
6	1,030	27	54,000
7	1,120	24	64,000

8	1,068	11	64,500
9	760	0	69,900
10	1,400	26	71,900
11	1,652	13	78,900
12	1,190	3	85,000
13	1,200	11	90,600
14	1,444	0	99,875
15	1,516	0	104,400
16	2,024	16	108,000
17	1,840	0	116,000
18	1,684	0	111,900
19	1,760	0	119,500
20	1,870	0	123,000

    a. Graphically analyze the data to determine whether the two independent variables interact in their effect on the price of homes.

    b. Fit the interaction model to the data. Test whether the model is useful for predicting the price of a home. Use $\alpha = .05$. Interpret R^2.

    c. Is there significant evidence that the interaction term contributes to the prediction of the price of a home?

    d. If the model is useful, find the 95% prediction interval for a 1,500 square foot home that is 10 years old.

4. The manufacturing operation data given in Exercise 14.7 on page 755 of the text was from a time study of a sample of 15 employees on an automobile assembly line. Management wants to study the relationship between the number of minutes y to complete a task and the months x of experience.

Minutes	Experience	Minutes	Experience
10	24	17	3
20	1	18	1
15	10	16	7
11	15	16	9
11	17	17	7
19	3	18	5
11	20	10	20
13	9		

    a. Construct a scattergram of the time to complete the task and the months of experience. Calculate the correlation coefficient. Discuss.

    b. Fit a straight-line linear model to the data. Interpret the least squares estimates of the parameters. Test whether the model is useful. Use $\alpha = .10$.

    c. Interpret the estimate of σ. Interpret R^2.

    d. Use a 95% prediction interval for the number of minutes that it would take an employee with 12 months of experience to complete the assembly line task.

    e. Add a quadratic term to the model. Is there significant evidence in the sample that the quadratic term is useful? Use $\alpha = .05$.

    f. Use the quadratic model to predict the task completion time for a person with 12 months of experience. Use a 95% prediction interval. Compare this result to the prediction interval in part d.

g. Construct residual plots and stem and leaf displays of the straight-line model and the quadratic model to check the regression model assumptions.

5. A real estate appraiser provided the following data given in Exercise 14.11 on page 758 of the text on apartment buildings sold during 1990 in Minneapolis. The random sample of 25 apartment buildings gives information on sales prices y, number of apartments x_1, age of structure x_2, lot size x_3, number of on-site parking spaces x_4, and gross area x_5. Use regression analysis to explore the relationship between sales price of apartment buildings in Minneapolis during 1990 and the given characteristics of the property. Include appropriate plots. Summarize the analysis.

Code No.	Sale Price	No. of Apartments	Age of Structure	Lot Size	Parking Spaces	Building Area
0229	$ 90,300	4	82	4,635	0	4,266
0094	384,000	20	13	17,798	0	14,391
0043	157,500	5	66	5,913	0	6,615
0079	676,200	26	64	7,750	6	34,144
0134	165,000	5	55	5,150	0	6,120
0179	300,000	10	65	12,506	0	14,552
0087	108,750	4	82	7,160	0	3,040
0120	276,538	11	23	5,120	0	7,881
0246	420,000	20	18	11,745	20	12,600
0025	950,000	62	71	21,000	3	39,448
0015	560,000	26	74	11,221	0	30,000
0131	268,000	13	56	7,818	13	8,088
0172	290,000	9	76	4,900	0	11,315
0095	173,200	6	21	5,424	6	4,461
0121	323,650	11	24	11,834	8	9,000
0077	162,500	5	19	5,246	5	3,828
0060	353,500	20	62	11,223	2	13,680
0174	134,400	4	70	5,834	0	4,680
0084	187,000	8	19	9,075	0	7,392
0031	155,700	4	57	5,280	0	6,030
0019	93,600	4	82	6,864	0	3,840
0074	110,000	4	50	4,510	0	3,092
0057	573,200	14	10	11,192	0	23,704
0104	79,300	4	82	7,425	0	3,876
0024	272,000	5	82	7,500	0	9,542

6. Consider Exercise 14.23 on page 771 of the text. A concessionaire wants to determine the relationship between the number of hot dogs y that are purchased at a baseball stadium during a week and x_1, the advanced sale tickets, and x_2, the visiting team's standing. The following number of hot dogs (in thousands), advanced sales (in $1,000), and standings were obtained for eight games.

Hot Dogs	Advanced Sales	Standing
54.2	55.6	1
48.9	63.5	3
50.3	60.1	2
56.1	58.9	1
20.0	45.4	6
53.8	66.6	2
46.4	59.3	4
52.7	64.8	2

    a. Construct scattergrams to study the relationship between the response variable and each independent variable. Calculate the correlations between all pairs of variables. Discuss.
    b. Fit the first-order linear model to the data. Interpret the least squares estimates of the parameters. Test whether the model is useful. Use $\alpha = .10$.
    c. Interpret the estimate of σ. Interpret R^2.
    d. Use the first-order linear model to predict the number of hot dogs for a week with $50,000 advanced ticket sales with a team in first place. Use a 95% prediction interval.
    e. Use TPLOT to graphically determine whether the two independent variables interact in their effect on the number of hot dogs.
    f. Fit a general linear model with an interaction term to the data. Is there sufficient evidence in the sample that the interaction term is useful? Use $\alpha = .05$.
    g. Use the interaction model to predict the number of hot dogs for a week with $50,000 advanced ticket sales with a team in first place. Use a 95% prediction interval. Compare this result to the prediction interval in part d.
    h. Construct residual plots and stem and leaf displays for both models to check the regression model assumptions. Discuss.

7. Consider the class data set given in Appendix B of this supplement. Model the relationship between grade point average y and the following independent variables: Age x_1, distance in miles from school x_2, and numbers of hours worked x_3.

    a. Construct scattergrams to study the relationship between grade point average and each independent variable. Calculate the correlations between all pairs of variables. Discuss.
    b. Fit the first-order linear model to the data. Interpret the least squares estimates of the parameters.
    c. Interpret the estimate of σ. Interpret R^2.
    d. Use $\alpha = .05$ to test whether the model is useful for predicting grade point average.
    e. Perform a residual analysis. Discuss.
    f. If the model is useful and the model assumptions are satisfied, construct and interpret a 95% prediction interval for a 25 year old student who works 20 hours a week and lives 10 miles from class.

CHAPTER 15
MODEL BUILDING

Simple and multiple linear regression models were presented in Chapters 13 and 14. This chapter describes steps in determining an appropriate regression model. We examine different regression models, suggest a model for a given set of data, and then use the data to confirm the appropriateness of the model.

NEW COMMANDS

INDICATOR   STEPWISE

MODELS WITH A SINGLE QUANTITATIVE VARIABLE

A polynomial model is commonly used to relate the response variable y to one quantitative independent variable x. The model of degree p is of the form,

$$y = ß_0 + ß_1 x + ß_2 x^2 + \ldots + ß_p x^p + \epsilon$$

Generally, models of degree 1 or 2 are useful in practice. If $p = 1$, the model is the simple linear regression model. The quadratic model of degree 2 is used to model a curvilinear relationship between y and x. The graph of a quadratic model is a parabola, where $ß_0$ is the y-intercept, $ß_1$ is the linear effect, and $ß_2$ describes the direction and rate of curvature.

■ **Example 1** Refer to Example 15.2 on page 822 of the text. A power company wants to determine the relationship between the daily high temperature x and the peak power load y on that day. The peak power load is the maximum power that must be generated to meet demand. The following table gives the results for a random sample of 25 summer days.

278 Chapter 15

Day	Temperature	Peak Load	Day	Temperature	Peak Load
1	94	136.0	14	79	106.2
2	96	131.7	15	97	153.2
3	95	140.7	16	98	150.1
4	108	189.3	17	87	114.7
5	67	96.5	18	76	100.9
6	88	116.4	19	68	96.3
7	89	118.5	20	92	135.1
8	84	113.4	21	100	143.6
9	90	132.0	22	85	111.4
10	106	178.2	23	89	116.5
11	67	101.6	24	74	103.9
12	71	92.5	25	86	105.1
13	100	151.9			

a. Analyze the relationship between peak power load and daily high temperature.
b. Fit a first-order model to the data. Perform a residual analysis. Discuss.
c. Fit a second-order model to the data. Test whether the rate of increase in peak power load is higher at higher temperatures. Use $\alpha = .05$. Perform a residual analysis. Discuss.

Solution

a. PLOT gives a scattergram of peak power load versus daily high temperature.

```
MTB > NAME C1 'TEMP' C2 'PEAKLOAD'
MTB > SET 'TEMP'
DATA>  94  96  95 108  67  88  89  84  90 106  67  71 100
DATA>  79  97  98  87  76  68  92 100  85  89  74  86
DATA> END
MTB > SET 'PEAKLOAD'
DATA> 136.0 131.7 140.7 189.3  96.5 116.4 118.5 113.4 132.0 178.2 101.6
DATA>  92.5 151.9 106.2 153.2 150.1 114.7 100.9  96.3 135.1 143.6 111.4
DATA> 116.5 103.9 105.1
DATA> END
MTB > SAVE 'POWER'

Worksheet saved into file:  POWER.MTW
```

```
MTB > PLOT 'PEAKLOAD' VS 'TEMP';
SUBC> TITLE 'PEAK POWER LOAD'.
```

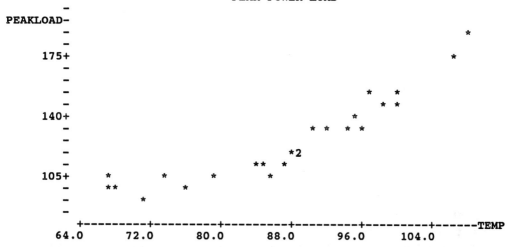

```
MTB > CORRELATION 'PEAKLOAD' 'TEMP'

Correlation of PEAKLOAD and TEMP = 0.918
```

The scattergram and correlation coefficient, $r = .918$, indicate a strong positive relationship between peak power load and daily high temperature. There appears to be some curvature; peak power load increases at a faster rate for higher temperatures.

b. The first-order model is

$$y = \beta_0 + \beta_1 x + \epsilon$$

We store the standardized residuals, fitted values of y, and residuals.

```
MTB > NAME C3 'STD RES' C4 'FIT' C5 'RESIDUAL'
MTB > REGRESS 'PEAKLOAD' ON 1 PREDICTOR 'TEMP', STORE 'STD RES' 'FIT';
SUBC> RESIDUALS IN 'RESIDUAL'.

The regression equation is
PEAKLOAD = - 47.4 + 1.98 TEMP

Predictor       Coef       Stdev     t-ratio         p
Constant      -47.39       15.67       -3.02     0.006
TEMP          1.9765      0.1776       11.13     0.000

s = 10.32     R-sq = 84.3%     R-sq(adj) = 83.7%

Analysis of Variance

SOURCE         DF          SS          MS         F         p
Regression      1       13196       13196    123.82     0.000
Error          23        2451         107
Total          24       15648
```

```
Unusual Observations
Obs.    TEMP   PEAKLOAD       Fit  Stdev.Fit   Residual   St.Resid
  4      108    189.30      166.06      4.20      23.24      2.46R
```
R denotes an obs. with a large st. resid.

The F and t test statistics suggest that this linear model is useful for predicting peak power load. The coefficient of determination indicates that 84% of the variation in peak power load can be explained using this linear model. A residual plot gives more information on the appropriateness of the model.

```
MTB > PLOT 'RESIDUAL' VS 'FIT';
SUBC> TITLE 'A RESIDUAL PLOT';
SUBC> TITLE 'First-Order Model'.
```

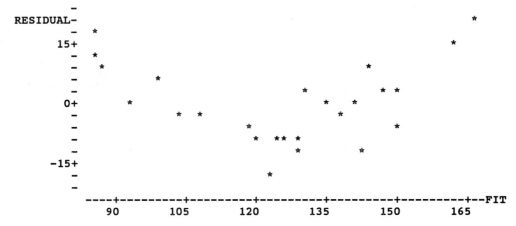

The plot exhibits a strong curvilinear relationship in the residuals. The second-order linear model should provide a better fit to the data.

c. The second-order model includes a squared term,

$$y = \beta_0 + \beta_1 x + \beta_2 x^2 + \epsilon$$

```
MTB > NAME C6 'TEMP SQ'
MTB > LET 'TEMP SQ' = 'TEMP'**2
MTB > REGRESS 'PEAKLOAD' ON 2 PREDICTORS 'TEMP' 'TEMP SQ'&
MTB > STORE 'STD RES' 'FIT';
SUBC> RESIDUALS IN 'RESIDUAL'.
* NOTE *    TEMP is highly correlated with other  predictor variables
* NOTE *    TEMP SQ is highly correlated with other  predictor variables

The regression equation is
PEAKLOAD = 385 - 8.29 TEMP + 0.0598 TEMP SQ

Predictor         Coef        Stdev      t-ratio         p
Constant        385.05        55.17         6.98     0.000
TEMP            -8.293         1.299       -6.38     0.000
TEMP SQ         0.059823      0.007549      7.93     0.000
```

```
s = 5.376        R-sq = 95.9%     R-sq(adj) = 95.6%

Analysis of Variance

SOURCE         DF         SS           MS        F         p
Regression      2      15011.8      7505.9    259.69    0.000
Error          22        635.9        28.9
Total          24      15647.7

SOURCE         DF       SEQ SS
TEMP            1      13196.4
TEMP SQ         1       1815.4

Unusual Observations
Obs.    TEMP   PEAKLOAD        Fit  Stdev.Fit   Residual   St.Resid
  4      108    189.30      187.23      3.45       2.07      0.50 X
 21      100    143.60      154.03      1.65     -10.43     -2.04R

R denotes an obs. with a large st. resid.
X denotes an obs. whose X value gives it large influence.
```

As in the first-order model, the F statistic indicates that this model is useful. The square of the temperature t-ratio = 7.93 (p-value = 0.000) is an important predictor in the model. The coefficient of determination indicates that about 96% of the variation in peak power loads is explained by the second-order model. We use a residual plot to check for patterns in the residuals.

```
MTB > PLOT 'RESIDUAL' VS 'FIT';
SUBC> TITLE 'A RESIDUAL PLOT';
SUBC> TITLE 'Second-Order Model'.
```

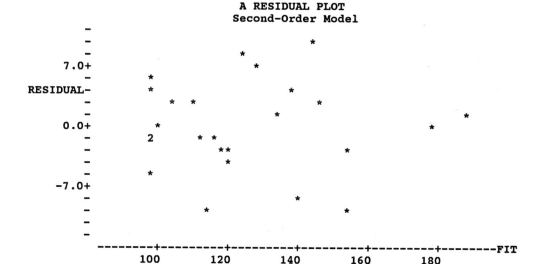

There does not appear to be any discernable pattern in the residuals. The assumption of constant variance appears to be satisfied.

282   *Chapter 15*

```
MTB > STEM AND LEAF OF 'RESIDUAL'

Stem-and-leaf of RESIDUAL   N = 25
Leaf Unit = 1.0

    1     -1  0
    3     -0  98
    3     -0
    5     -0  54
    8     -0  222
   (5)    -0  11110
   12      0  01
   10      0  222333
    4      0  4
    3      0  6
    2      0  89

MTB > STOP
```

There does not seem to be a serious departure from normality evident in the stem and leaf display. Since we have found no serious violations of the regression assumptions, we conclude that the second-order model can be used for estimation and prediction. ∎

MODELS WITH TWO QUANTITATIVE INDEPENDENT VARIABLES

The complete second-order model is

$$y = \beta_0 + \beta_1 x_1 + \beta_2 x_2 + \beta_3 x_1 x_2 + \beta_4 x_1^2 + \beta_5 x_2^2 + \epsilon$$

where β_0 is the y-intercept, $\beta_1 x_1$ and $\beta_2 x_2$ are the first-order terms, $\beta_3 x_1 x_2$ is the interaction or cross product term, and $\beta_4 x_1^2$ and $\beta_5 x_2^2$ are the quadratic terms. Two variables interact if the effect of changes in one variable on the mean response variable depends on the value of the other independent variable. As before, ϵ accounts for the variability in y that is not explained by the independent variables in the model. The multiple regression model assumes that the errors are independent, and that the probability distribution of ϵ is normal, with zero mean and a constant variance σ^2 for any set of values of the independent variables.

The graph of a complete second-order regression equation is generally a complex three-dimensional response surface, such as a paraboloid or a saddle surface. There are numerous variations to the complete model; a comprehensive graphical and numerical data analysis will help you choose an appropriate model.

■ **Example 2** In Example 15.4 on page 843 of the text, the quality y of a particular finished product is thought to be dependent on the Fahrenheit temperature x_1, and pounds per square inch of pressure x_2, at which the product is produced. The results of an experiment conducted at various temperatures and pressures are given in the following table.

Temp	Pressure	Quality	Temp	Pressure	Quality	Temp	Pressure	Quality
80	50	50.8	90	50	63.4	100	50	46.6
80	50	50.7	90	50	61.6	100	50	49.1
80	50	49.4	90	50	63.4	100	50	46.4
80	55	93.7	90	55	93.8	100	55	69.8
80	55	90.9	90	55	92.1	100	55	72.5
80	55	90.9	90	55	97.4	100	55	73.2
80	60	74.5	90	60	70.9	100	60	38.7
80	60	73.0	90	60	68.8	100	60	42.5
80	60	71.2	90	60	71.3	100	60	41.4

a. Graphically study the relationships between y and each independent variable. Calculate the correlation coefficients. Discuss.

b. Fit a first-order model and a second-order model to the data. Which model provides a better fit? Explain.

Solution

a. We do a scattergram of y versus each x.

```
MTB > NAME C1 'TEMP' C2 'PRESSURE' C3 'QUALITY'
MTB > SET 'TEMP'
DATA> (80:100/10)9
DATA> END
MTB > SET 'PRESSURE'
DATA> 3(50:60/5)3
DATA> END
MTB > SET 'QUALITY'
DATA> 50.8 50.7 49.4 93.7 90.9 90.9 74.5 73.0 71.2
DATA> 63.4 61.6 63.4 93.8 92.1 97.4 70.9 68.8 71.3
DATA> 46.6 49.1 46.4 69.8 72.5 73.2 38.7 42.5 41.4
DATA> END
MTB > PLOT 'QUALITY' VS 'TEMP';
SUBC> TITLE 'PRODUCT QUALITY VERSUS PRODUCTION TEMPERATURE'.
```

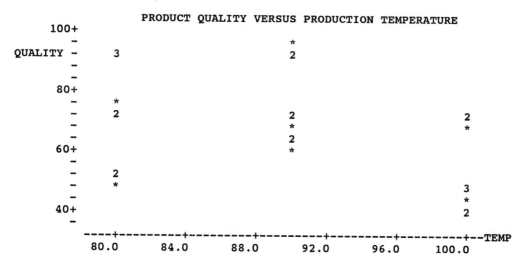

```
MTB > PLOT 'QUALITY' VS 'PRESSURE';
SUBC> TITLE 'PRODUCT QUALITY VERSUS PRODUCTION PRESSURE'.
```

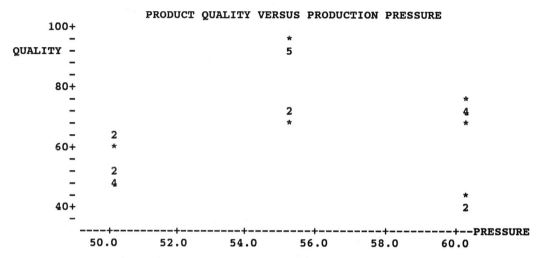

```
MTB > CORRELATION 'QUALITY' 'TEMP' 'PRESSURE'

           QUALITY       TEMP
TEMP       -0.423
PRESSURE    0.182       0.000
```

Both scattergrams indicate some curvature. The correlations do not show a strong linear relationship between quality and temperature or between quality and pressure. It does not appear that a first-order model will provide a good fit.

b. The first-order model for this example is

$$y = \beta_0 + \beta_1 x_1 + \beta_2 x_2 + \epsilon$$

```
MTB > # FIRST-ORDER MODEL
MTB > REGRESS 'QUALITY' ON 2 PREDICTORS 'TEMP' 'PRESSURE'

The regression equation is
QUALITY = 106 - 0.916 TEMP + 0.788 PRESSURE

Predictor        Coef        Stdev       t-ratio         p
Constant        106.09       55.95         1.90       0.070
TEMP           -0.9161        0.3930      -2.33       0.028
PRESSURE        0.7878        0.7860       1.00       0.326

s = 16.67      R-sq = 21.2%     R-sq(adj) = 14.6%

Analysis of Variance

SOURCE         DF          SS          MS         F         p
Regression      2       1789.9       895.0      3.22     0.058
Error          24       6671.5       278.0
Total          26       8461.4

SOURCE         DF        SEQ SS
TEMP            1        1510.7
PRESSURE        1         279.3
```

The null and alternative hypotheses to test the usefulness of the first-order model are

H_0: $\beta_1 = \beta_2 = 0$
H_a: At least one of β_1 and $\beta_2 \neq 0$

The test statistic, $F = 3.22$, has a p-value of 0.058, which is greater than $\alpha = .05$. We do not have sufficient evidence to conclude that at least one parameter differs from zero. We cannot conclude that this model is useful for estimating the product quality.

The second-order model is

$$y = \beta_0 + \beta_1 x_1 + \beta_2 x_2 + \beta_3 x_1 x_2 + \beta_4 x_1^2 + \beta_5 x_2^2 + \epsilon$$

```
MTB > # COMPLETE SECOND-ORDER MODEL
MTB > NAME C4 'TEMP*PR' C5 'TEMP SQ' C6 'PRES SQ'
MTB > LET 'TEMP*PR' = 'TEMP'*'PRESSURE'
MTB > LET 'TEMP SQ' = 'TEMP'**2
MTB > LET 'PRES SQ' = 'PRESSURE'**2
MTB > REGRESS 'QUALITY' ON 5 PREDICTORS 'TEMP' 'PRESSURE'&
CONT> 'TEMP*PR' 'TEMP SQ' 'PRES SQ'
* NOTE *     TEMP is highly correlated with other  predictor variables
* NOTE * PRESSURE is highly correlated with other  predictor variables
* NOTE *  TEMP*PR is highly correlated with other  predictor variables
* NOTE *  TEMP SQ is highly correlated with other  predictor variables
* NOTE *  PRES SQ is highly correlated with other  predictor variables

The regression equation is
QUALITY = - 5128 + 31.1 TEMP + 140 PRESSURE - 0.145 TEMP*PR - 0.133 TEMP SQ
          - 1.14 PRES SQ

Predictor        Coef       Stdev      t-ratio         p
Constant       -5127.9      110.3      -46.49       0.000
TEMP            31.096      1.344       23.13       0.000
PRESSURE       139.747      3.140       44.50       0.000
TEMP*PR        -0.145500    0.009692   -15.01       0.000
TEMP SQ        -0.133389    0.006853   -19.46       0.000
PRES SQ        -1.14422     0.02741    -41.74       0.000

s = 1.679         R-sq = 99.3%      R-sq(adj) = 99.1%

Analysis of Variance

SOURCE         DF        SS         MS         F          p
Regression      5      8402.3     1680.5     596.32     0.000
Error          21        59.2        2.8
Total          26      8461.4

SOURCE         DF      SEQ SS
TEMP            1      1510.7
PRESSURE        1       279.3
TEMP*PR         1       635.1
TEMP SQ         1      1067.6
PRES SQ         1      4909.7

MTB > STOP
```

The null and alternative hypotheses to test the usefulness of the second-order model are

H_0: $\beta_1 = \beta_2 = \beta_3 = \beta_4 = \beta_5 = 0$
H_a: At least one $\beta_i \neq 0$, for $i = 1, 2, \ldots, 5$

The p-value $= 0$ for the overall F test. We have sufficient evidence to conclude that at least one parameter differs from zero. The model is useful for estimating the product quality. The parameters for all five predictors have p-values equal to 0; each predictor provides useful information. The multiple coefficient, $R^2 = 99.3\%$, means that over 99% of the variation in the quality values is explained by the second-order model.

∎

TESTING PORTIONS OF A MODEL

Often times, we want to test whether several regression coefficients simultaneously equal zero. That is, we want to test whether a reduced model is a better model than a complete model. The F test statistic is used to test the null hypothesis that a particular set of $(k - g)$ parameters simultaneously equals 0. To do the test, the complete model with k predictors and the reduced model with g predictors are fit to the data. The difference in the sums of squares for error is used to test if the particular set of parameters contributes significant information to predict y. The form of the hypothesis test is

H_0: The set of $(k - g)$ parameters simultaneously equal 0
H_a: At least one of the parameters in the set is not 0

The test statistic is

$$F = \frac{(SSE_1 - SSE_2)/(k - g)}{SSE_2/[n - (k+1)]}$$

where SSE_1 and SSE_2 are the sums of squares for error for the reduced and complete models, respectively. When H_0 is true, and the assumptions of the regression model are satisfied, this test statistic has an F distribution with $(k - g)$ and $[n - (k + 1)]$ degrees of freedom.

■ **Example 3** Refer to Example 2 of this chapter. Use the F test to determine whether the second-order terms contribute information for the prediction of y. Test using $\alpha = .05$.

Solution The null and alternative hypotheses for testing the usefulness of the second-order terms are

H_0: $\beta_3 = \beta_4 = \beta_5 = 0$
H_a: At least one of $\beta_3, \beta_4, \beta_5 \neq 0$

The output from Example 2 contains the necessary values to calculate the F test statistic. The complete model is the second-order model and the reduced model is the first-order model.

Sum of squares for error for reduced model, $SSE_1 = 6671.5$
Sum of squares for error for complete model, $SSE_2 = 59.2$
Number of predictors in the complete model, $k = 5$
Number of predictors in the reduced model, $g = 2$

Sample size, $n = 27$

We use Minitab to calculate the F test statistic and p-value.

```
MTB > LET K1 = ((6671.5 - 59.2)/(5-2))/(59.2/(27 - (5+1))) # THE F STATISTIC
MTB > PRINT K1
K1        781.860
MTB > CDF K1 PUT IN K2;
SUBC> F WITH 3 AND 21 DF.
MTB > LET K3 = 1 - K2    # THE P-VALUE
MTB > PRINT K1 K2 K3
K1        781.860
K2        1.00000
K3        0
MTB > STOP
```

The test statistic, $F = 781.86$, has a p-value equal to 0. At least one of the second-order terms contributes information for predicting quality of a finished product. This F test statistic differs from the one that is calculated on page 846 of the text because of rounding differences.

■

MODELS WITH ONE QUALITATIVE INDEPENDENT VARIABLE

A linear regression model can contain a qualitative independent variable, which is one that is not measured on a numerical scale. Examples include gender, education level, sales regions, and type of product. Indicator or dummy variables with values (0,1) can be used to represent levels of qualitative variables, where each indicator variable x_i is assigned the value 1 for level i, and 0 otherwise. The INDICATOR command creates the indicator variables.

INDICATOR VARIABLES FOR C, PUT VALUES IN C,...,C

This command is used to create (0,1) indicator or dummy variables. Enter consecutive integers, usually 1 for the first level, 2 for the second, and so on, in the first column specified on the command line. Use the same number of columns to store the indicator variables as the number of levels. The command puts a 1 or a 0 in the columns.

Comment *The INDICATOR command is not available in the student version of Minitab. Use READ or SET command to directly enter the indicator variables.*

Suppose a qualitative variable has k levels. Assign one level as the base level, and use the indicator variables corresponding to the other $(k - 1)$ levels in the model. Then $\hat{\beta}_0$ is the mean value of y at the base level, and $\hat{\beta}_i$ is the difference in the mean value of y at level i and at the base level.

For example, consider the class data set given in Appendix B of this supplement. Suppose ten randomly selected students have the following types of cars: No car, foreign car, or US car.

Student	Car
1	No Car
2	Foreign
3	Foreign
4	US Car
5	US Car
6	US Car
7	No Car
8	US Car
9	No Car
10	Foreign

We need three dummy variables for the INDICATOR command. We use the code, 1 for No car, 2 for foreign, and 3 for US car.

```
MTB > SET C1
DATA> 1 2 2 3 3 3 1 3 1 2
DATA> END
MTB > INDICATOR C1 PUT IN C2 C3 C4
MTB > PRINT C1-C4

  ROW    C1    C2    C3    C4

    1     1     1     0     0
    2     2     0     1     0
    3     2     0     1     0
    4     3     0     0     1
    5     3     0     0     1
    6     3     0     0     1
    7     1     1     0     0
    8     3     0     0     1
    9     1     1     0     0
   10     2     0     1     0
```

■ **Example 4** Refer to Example 15.5 on page 857 of the text. A sociologist wants to study the relationship between the dollar amounts owed by delinquent credit card customers and the customers income. The following samples of amounts owed by 10 delinquent customers were selected from each of three income categories.

	Income Category	
Under $12,000	$12,000-$25,000	Over $25,000
$148	$513	$335
76	264	643
393	433	216
520	94	536
236	535	128
134	327	723
55	214	258
166	135	380
415	280	594
153	304	465

a. Graphically analyze the data. Calculate descriptive measures. Does the dollar amount owed by delinquent customers depend on income?
b. Create indicator variables for the income categories. Find the regression equation for estimating the mean dollar amounts owed.
c. Test whether the mean dollar amounts owed by customers differ for the three income groups. Use $\alpha = .05$.
d. If the model is useful, estimate the mean dollar amounts owed by each customers in the three income categories.

Solution

a. We stack the data in a column, and enter codes for the categories in another column. DOTPLOT and DESCRIBE summarize the categories.

```
MTB > NAME C1 '$ OWED'   C2 'CATEGORY'
MTB > SET '$ OWED'
DATA> 148    76   393   520   236   134    55   166   415   153
DATA> 513   264   433    94   535   327   214   135   280   304
DATA> 335   643   216   536   128   723   258   380   594   465
DATA> END
MTB > SET 'CATEGORY'
DATA> (1:3)10
DATA> END
MTB > SAVE 'CREDIT'

Worksheet saved into file:   CREDIT.MTW
MTB > DOTPLOT '$ OWED';
SUBC> BY 'CATEGORY'.

CATEGORY
1
          ..  .:.     .         .  .       .
        +---------+---------+---------+---------+---------+-------$ OWED
CATEGORY
2
             .   .   .   ...  .       .       . .
        +---------+---------+---------+---------+---------+-------$ OWED
CATEGORY
3
                .   .    .     .   .     .    .   .    .     .
        +---------+---------+---------+---------+---------+-------$ OWED
        0        150       300       450       600       750

MTB > DESCRIBE '$ OWED';
SUBC> BY 'CATEGORY'.

            CATEGORY      N      MEAN    MEDIAN    TRMEAN    STDEV    SEMEAN
$ OWED             1     10     229.6     159.5     215.1    158.2      50.0
                   2     10     309.9     292.0     308.8    147.9      46.8
                   3     10     427.8     422.5     428.4    196.8      62.2


            CATEGORY       MIN       MAX        Q1        Q3
$ OWED             1      55.0     520.0     119.5     398.5
                   2      94.0     535.0     194.2     453.0
                   3     128.0     723.0     247.5     606.2
```

The dot plots and descriptive measures show that the mean amount owed increases as the customer income increases. The standard deviations indicate considerable variation within income categories.

b. The INDICATOR command creates dummy variables for the three income categories. The regression model is

$$y = \beta_0 + \beta_1 x_1 + \beta_2 x_2 + \epsilon$$

where $x_1 = 1$ if income is from \$12,000 to \$25,000 and 0 otherwise, and $x_2 = 1$ of income is over \$25,000 and 0 otherwise. The base income category is Under \$12,000.

```
MTB > NAME C3 'UNDER$12' C4 '$12-$25' C5 'OVER $25'
MTB > INDICATOR 'CATEGORY' PUT IN C3-C5
MTB > PRINT '$ OWED' 'CATEGORY' INDICATORS IN C3-C5
```

ROW	$ OWED	CATEGORY	UNDER$12	$12-$25	OVER $25
1	148	1	1	0	0
2	76	1	1	0	0
3	393	1	1	0	0
4	520	1	1	0	0
5	236	1	1	0	0
6	134	1	1	0	0
7	55	1	1	0	0
8	166	1	1	0	0
9	415	1	1	0	0
10	153	1	1	0	0
11	513	2	0	1	0
12	264	2	0	1	0
13	433	2	0	1	0
14	94	2	0	1	0
15	535	2	0	1	0
16	327	2	0	1	0
17	214	2	0	1	0
18	135	2	0	1	0
19	280	2	0	1	0
20	304	2	0	1	0
21	335	3	0	0	1
22	643	3	0	0	1
23	216	3	0	0	1
24	536	3	0	0	1
25	128	3	0	0	1
26	723	3	0	0	1
27	258	3	0	0	1
28	380	3	0	0	1
29	594	3	0	0	1
30	465	3	0	0	1

```
MTB > REGRESS '$ OWED' ON 2 PREDICTORS '$12-$25' AND 'OVER $25'

The regression equation is
$ OWED = 230 + 80.3 $12-$25 + 198 OVER $25

Predictor       Coef        Stdev      t-ratio        p
Constant      229.60        53.43         4.30    0.000
$12-$25        80.30        75.56         1.06    0.297
OVER $25      198.20        75.56         2.62    0.014

s = 168.9        R-sq = 20.5%        R-sq(adj) = 14.6%
```

Analysis of Variance

SOURCE	DF	SS	MS	F	p
Regression	2	198772	99386	3.48	0.045
Error	27	770671	28543		
Total	29	969443			

SOURCE	DF	SEQ SS
$12-$25	1	2356
OVER $25	1	196416

c. Testing whether the mean dollar amounts owed by customers differ for the three income groups is equivalent to testing whether at least one of the parameters, $ß_1$ and $ß_2$, differs from 0. The null and alternative hypotheses are

H_0: $ß_1 = ß_2 = 0$
H_a: At least one of $ß_1$ and $ß_2 \neq 0$

The test statistic, $F = 3.48$, has a p-value of 0.045, which is less than $\alpha = .05$. We have sufficient evidence to conclude that at least one of the parameters differs from zero. The mean dollar amounts owed by customers differ for the three income categories.

d. To estimate the mean dollar amounts, we use 0 or 1 to represent the income categories for the PREDICT subcommand. Recall that $x_1 = 1$ for $12,000 to $25,000 and 0 otherwise, and $x_2 = 1$ for Over $25,000 and 0 otherwise, and the base income category is Under $12,000. For a customer in the Under $12,000 category, we enter $x_1 = 0$, $x_2 = 0$; for a customer in the $12,000 to $25,000 category, $x_1 = 1$, $x_2 = 0$; and for a customer in the Over $25,000, $x_1 = 0$, $x_2 = 1$.

```
MTB > BRIEF 1
MTB > REGRESS '$ OWED' ON 2 PREDICTORS '$12-$25' AND 'OVER $25';
SUBC> PREDICT AT 0 0;
SUBC> PREDICT AT 1 0;
SUBC> PREDICT AT 0 1.
```

The regression equation is
$ OWED = 230 + 80.3 $12-$25 + 198 OVER $25

Predictor	Coef	Stdev	t-ratio	p
Constant	229.60	53.43	4.30	0.000
$12-$25	80.30	75.56	1.06	0.297
OVER $25	198.20	75.56	2.62	0.014

s = 168.9   R-sq = 20.5%   R-sq(adj) = 14.6%

Analysis of Variance

SOURCE	DF	SS	MS	F	p
Regression	2	198772	99386	3.48	0.045
Error	27	770671	28543		
Total	29	969443			

Fit	Stdev.Fit	95% C.I.	95% P.I.
229.6	53.4	(120.0, 339.2)	(-134.1, 593.3)
309.9	53.4	(200.3, 419.5)	(-53.8, 673.6)
427.8	53.4	(318.2, 537.4)	(64.1, 791.5)

MTB > STOP

For customers with income under $12,000, $\hat{y} = \$229.60$. We can say that we are 95% confident that the mean dollar amount owed by customers with income under $12,000 is from $120 to $339. The other confidence intervals have similar interpretations. Because there is only one qualitative predictor in this model, each point estimate $\hat{y}$ of the mean dollar amount owed by customers is the same as the sample mean obtained with DESCRIBE in part a.

■

COMPARING THE SLOPES OF TWO OR MORE LINES

Suppose you want to study the relationship between a response variable y and two independent variables, one quantitative variable and one qualitative. There are several possible straight-line models. Consider the following cases. For each, x_1 is a quantitative variable and x_2 and x_3 are indicator variables for a qualitative variable with three levels.

1. A single straight line describes the relationship. The relationship is the same for all levels of the qualitative variable. The regression model is

    $$y = \beta_0 + \beta_1 x_1 + \epsilon$$

2. The straight lines differ from one level to another. The y-intercepts differ, but the slopes of the lines are the same; that is, the lines are parallel. The indicator terms are added to the above regression model,

    $$y = \beta_0 + \beta_1 x_1 + \beta_2 x_2 + \beta_3 x_3 + \epsilon$$

3. The straight lines differ in terms of the y-intercept and slope. Interaction terms are added to the second model,

    $$y = \beta_0 + \beta_1 x_1 + \beta_2 x_2 + \beta_3 x_3 + \beta_4 x_1 x_2 + \beta_5 x_1 x_3 + \epsilon$$

The LPLOT command adds information about a third variable to a scattergram. Letters are used to plot different observations of the third variable. The observations may be levels of a quantitative or a qualitative variable.

■ **Example 5** Refer to Example 15.7 on page 868 of the text. An industrial psychologist studied the relationship between worker productivity y, salary incentive x_1, and the type of production plant x_2. The incentive is a bonus paid for excess castings produced by the worker, and the plant types are union and nonunion representation. The table gives productivity results for 18 workers.

	$0.20 Bonus			$0.30 Bonus			$0.40 Bonus		
Union	1,435	1,512	1,491	1,583	1,529	1,610	1,601	1,574	1,636
Nonunion	1,575	1,512	1,488	1,635	1,589	1,661	1,645	1,616	1,689

a. Plot the productivity data. Use TABLE to compare the productivity means of workers for each salary incentive level at each production plant.
b. Find the regression equation that best describes the relationship.

Solution

a. We use a labeled plot to graph the data.

```
MTB > NAME C1 'CASTINGS' C2 'INCENT' C3 'UNION'
MTB > SET 'CASTINGS'
DATA> 1435 1512 1491 1583 1529 1610 1601 1574 1636
DATA> 1575 1512 1488 1635 1589 1661 1645 1616 1689
DATA> END
MTB > SET 'INCENT'
DATA> 2(20 30 40)3
DATA> END
MTB > SET 'UNION'
DATA> 9(0) 9(1)
DATA> END
MTB > SAVE 'INDUSTRY'

Worksheet saved into file: INDUSTRY.MTW
MTB > LPLOT 'CASTINGS' VS 'INCENT' USING LABELS FOR 'UNION';
SUBC> TITLE 'INDUSTRIAL PRODUCTIVITY';
SUBC> XINCREMENT 10.
```

```
                    INDUSTRIAL PRODUCTIVITY
            -
      1700+
            -                                            A
  CASTINGS -                              A
            -                              A            2
            -                              Z            A
      1600+                                              Z
            -                   A          2            Z
            -
            -                   2          Z
      1500+                     Z
            -                   A
            -
            -                   Z
      1400+
            ---------+---------+---------+---------+---------+--------INCENT
                    10        20        30        40        50
```

```
MTB > TABLE 'UNION' BY 'INCENT';
SUBC> MEANS 'CASTINGS'.

   ROWS: UNION    COLUMNS: INCENT

             20        30        40       ALL

    0     1479.3    1574.0    1603.7    1552.3
    1     1525.0    1628.3    1650.0    1601.1
  ALL     1502.2    1601.2    1626.8    1576.7

  CELL CONTENTS --
            CASTINGS:MEAN
```

Both the graph and table of means indicate that productivity generally increases as salary incentive increases. At fixed incentive levels, nonunion worker productivity (A on the LPLOT) tends to be higher than union worker productivity. It appears that we need separate lines to characterize this relationship for the two plants.

b. If we assign union as the base level, and define $x_2 = 1$ for nonunion and 0 otherwise, UNION contains the correct coding for the (0,1) indicator variable. To model separate lines for the two plants, we use

$$y = \beta_0 + \beta_1 x_1 + \beta_2 x_2 + \beta_3 x_1 x_2 + \epsilon$$

```
MTB > NAME C4 'INTERACT'
MTB > LET 'INTERACT' = 'INCENT'*'UNION'
MTB > # FULL MODEL
MTB > REGRESS 'CASTINGS' ON 3 PREDICTORS 'INCENT' 'UNION' 'INTERACT'

The regression equation is
CASTINGS = 1366 + 6.22 INCENT + 47.8 UNION + 0.03 INTERACT

Predictor         Coef       Stdev     t-ratio         p
Constant       1365.83       51.84       26.35     0.000
INCENT           6.217       1.667        3.73     0.002
UNION           47.78       73.31        0.65     0.525
INTERACT         0.033       2.358        0.01     0.989

s = 40.84        R-sq = 71.1%      R-sq(adj) = 64.9%

Analysis of Variance

SOURCE        DF          SS          MS         F          p
Regression     3        57332       19111     11.46     0.000
Error         14        23349        1668
Total         17        80682

SOURCE        DF       SEQ SS
INCENT         1        46625
UNION          1        10707
INTERACT       1            0
```

To determine whether the relationship between mean productivity and salary incentives differs at the two types of plants, the hypothesis test is

H_0: $\beta_2 = \beta_3 = 0$
H_a: At least one of β_2 and $\beta_3 \neq 0$

If there is no difference, a single regression line characterizes the relationship at both plants. We need the reduced model to conduct the test:

$$y = \beta_0 + \beta_1 x_1 + \epsilon$$

```
MTB > # REDUCED MODEL
MTB > REGRESS 'CASTINGS' ON 1 PREDICTOR 'INCENT'

The regression equation is
CASTINGS = 1390 + 6.23 INCENT

Predictor         Coef       Stdev     t-ratio         p
Constant       1389.72       41.41       33.56     0.000
INCENT           6.233       1.332        4.68     0.000

s = 46.14        R-sq = 57.8%      R-sq(adj) = 55.2%
```

Analysis of Variance

SOURCE	DF	SS	MS	F	p
Regression	1	46625	46625	21.91	0.000
Error	16	34056	2129		
Total	17	80682			

```
MTB > LET K1 = ((34056 - 23349)/2)/(23349/14)   # F TEST STATISTIC
MTB > CDF K1, STORE IN K2;   # CUMULATIVE PROBABILITY
SUBC> F WITH 2 AND 14 DF.
MTB > LET K3 = 1 - K2   # THE P-VALUE
MTB > PRINT K1-K3
K1        3.20994
K2        0.928793
K3        0.0712070
MTB > STOP
```

The test statistic, $F = 3.21$, has a p-value of 0.07, which is greater than $\alpha = .05$. We do not have sufficient evidence that at least one of β_2 and β_3 differs from 0. The single straight-line model is the most appropriate. The relationship between productivity and incentive is not significantly different for union and nonunion plants. ∎

MODEL BUILDING: STEPWISE REGRESSION

Stepwise regression is a screening process that selects a useful set of predictors for a response variable y. To use STEPWISE, identify the set of all possible predictors that you want to include in the regression analysis. This set may include interaction and higher order terms. STEPWISE uses an F test to determine which predictors in this set are useful.

STEPWISE REGRESSION ON C, PREDICTORS IN C,...,C
   FENTER = K
   FREMOVE = K
   FORCE C...C
   ENTER C...C
   REMOVE C...C
   BEST K
   STEPS = K
   NOCONSTANT

Stepwise regression identifies a useful subset of predictors from up to 100 predictors listed on the command line. An F test determines if a predictor is to be entered or removed from the equation. Use the FENTER and FREMOVE subcommands to set the critical values for the F test. The default value is 4. For each step, Minitab prints the constant term (unless you have used the NOCONSTANT subcommand), the coefficient and t-ratio for each predictor in the model, s and R^2. You can intervene this procedure with one or more subcommands, or you can terminate at any step.

Chapter 15

■ **Example 6** Refer to Example 14.1 on page 750 of the text. An antique clock collector believes that the auction price y of a clock increases with the age x_1 of a clock and the number x_2 of bidders. The following table gives the data for a random sample of 32 clocks.

Clock	Age	Bidders	Price	Clock	Age	Bidders	Price
1	127	13	$1,235	17	170	14	$2,131
2	115	12	1,080	18	182	8	1,550
3	127	7	845	19	162	11	1,884
4	150	9	1,522	20	184	10	2,041
5	156	6	1,047	21	143	6	854
6	182	11	1,979	22	159	9	1,483
7	156	12	1,822	23	108	14	1,055
8	132	10	1,253	24	175	8	1,545
9	137	9	1,297	25	108	6	729
10	113	9	946	26	179	9	1,792
11	137	15	1,713	27	111	15	1,175
12	117	11	1,024	28	187	8	1,593
13	137	8	1,147	29	111	7	785
14	153	6	1,092	30	115	7	744
15	117	13	1,152	31	194	5	1,356
16	126	10	1,336	32	168	7	1,262

Consider a complete second-order model for predicting the auction price of a clock. State the model. Use stepwise regression to determine which predictors might be useful in the model. Summarize the steps.

Solution The data are saved in a file named CLOCKS. The complete second-order model is

$$y = \beta_0 + \beta_1 x_1 + \beta_2 x_2 + \beta_3 x_1 x_2 + \beta_4 x_1^2 + \beta_5 x_2^2 + \epsilon$$

```
MTB > RETRIEVE 'CLOCKS'
WORKSHEET SAVED   1/10/1994

Worksheet retrieved from file:    CLOCKS.MTW
MTB > INFORMATION

COLUMN     NAME         COUNT
C1         AGE            32
C2         BIDDERS        32
C3         PRICE          32

CONSTANTS USED: NONE

MTB > NAME C4 'INTERACT' C5 'AGE SQ' C6 'BIDDERSQ'
MTB > LET 'INTERACT' = 'AGE'*'BIDDERS'
MTB > LET 'AGE SQ' = 'AGE'**2
MTB > LET 'BIDDERSQ' = 'BIDDERS'**2
MTB > # FORWARD STEPWISE REGRESSION
MTB > STEPWISE 'PRICE' ON 'AGE' 'BIDDERS' 'INTERACT' 'AGE SQ' 'BIDDERSQ'

   STEPWISE REGRESSION OF   PRICE    ON   5 PREDICTORS, WITH N =    32
```

```
    STEP            1        2
CONSTANT       219.64    -11.52

INTERACT        0.813     1.342
T-RATIO          9.16     23.80

BIDDERSQ                  -4.96
T-RATIO                  -12.26

S                 205      83.9
R-SQ            73.67     95.74
 MORE? (YES, NO, SUBCOMMAND, OR HELP)
SUBC> NO

MTB > STOP
```

At step 1, the stepwise procedure added the interaction term. The regression equation is

$$\hat{y} = 219.64 + 0.813 x_1 x_2$$

The t-ratio for testing the usefulness of the interaction term is 9.16. The standard deviation, $s = \$205$, and the regression equation accounts for about 74% of the variation in auction prices. About 26% of the variation in auction prices still remains unexplained.

Step 2 added the square of the number of bidders. The resulting equation is

$$\hat{y} = -11.52 + 1.342 x_1 x_2 - 4.96 x_2^2$$

The t-ratio for testing the useful of the interaction term is now 23.80 and of x_2^2 is -12.26. The standard deviation decreased to $s = \$83.90$, and the regression equation accounts for about 96% of the variation in auction prices. About 4% of the variation in auction price is not explained by the two predictors. The stepwise procedure terminated after step 2. We could intervene with a subcommand to force Minitab to add more predictors.

■

298 Chapter 15

EXERCISES

1. Exercise 15.14 on page 828 of the text considers the relationship between highway deaths and the number of licensed vehicles. The following table gives the number of highway deaths (to the nearest hundred) and the number of licensed vehicles (in hundreds of thousands) for the years 1950-1985 (coded 1-36). The 55-mile-per-hour speed limit was in effect during the years 1974-1985 (coded 25-36).

Year	Deaths	Vehicles	Year	Deaths	Vehicles
1	34.8	49.2	19	54.9	103.1
2	37.0	51.9	20	55.8	107.4
3	37.8	53.3	21	54.6	111.2
4	38.0	56.3	22	54.4	116.3
5	35.6	58.6	23	56.3	122.3
6	38.4	62.8	24	55.5	129.8
7	39.6	65.2	25	46.4	134.9
8	38.7	67.6	26	45.9	137.9
9	37.0	68.8	27	47.0	143.5
10	37.9	72.1	28	49.5	148.8
11	38.1	74.5	29	52.4	153.6
12	38.1	76.4	30	53.5	159.6
13	40.8	79.7	31	53.1	161.6
14	43.6	83.5	32	51.4	164.1
15	47.7	87.3	33	45.8	165.2
16	49.2	91.8	34	44.5	169.4
17	53.0	95.9	35	46.2	172.0
18	52.9	98.9	36	45.6	175.7

   a. Obtain a scattergram of highway deaths versus licensed vehicles. Calculate the correlation coefficient. Discuss.
   b. Fit a first-order model to the data. Is the model useful for explaining the variation in highway deaths? Test using $\alpha = .05$.
   c. Fit a second-order model to the data. Use $\alpha = .05$ to test whether this model is useful.
   d. Test whether the second-order term can be removed from the model. Use $\alpha = .05$.
   e. Which model do you recommend using for predicting highway deaths for a given number of licensed vehicles? Discuss.

2. A survey conducted by the UCLA Graduate School of Management on the status of computer development in business schools is reported in the *Sixth Annual UCLA Survey of Business School Computer Usage*, Jason L. Fraud and Julia A. Britt, September, 1989. The survey queries business schools on hardware, software, and resource commitments. One hundred sixty-three business schools completed the sixth annual UCLA computer survey. The following random sample of 25 schools provides the number of microcomputers y, the number of full time equivalent faculty x_1, the number of full time equivalent undergraduate students x_2, and the computer operating budget per student x_3.

School	PCs	FTE Faculty	FTE Student	Budget
1	96	42	966	*
2	158	101	2150	40
3	765	215	2455	484
4	114	41	250	400
5	120	102	3171	12
6	145	95	1600	69
7	157	82	1807	17
8	14	24	525	*
9	74	43	1248	28
10	92	62	792	123
11	205	72	1423	56
12	194	105	1800	98
13	111	52	2873	22
14	164	75	700	158
15	101	79	1136	28
16	105	82	1578	63
17	254	102	2684	114
18	270	136	5753	32
19	109	52	1025	52
20	278	115	4000	*
21	421	173	3500	116
22	104	69	2947	24
23	159	64	1772	*
24	193	125	3286	165
25	370	120	4100	68

a. Fit a complete second-order model to the data.
b. Test if the second-order terms are useful for predicting the number of microcomputers.
c. Use stepwise regression to find a useful subset of predictors from all the predictors in a complete second-order model. Explain each step in the procedure.

3. A real estate appraiser in Minneapolis provided the following data on apartment buildings sold during 1990 in Exercise 15.36 on page 853 of the text. The random sample of 25 apartment buildings includes information on sales prices y, number of apartments x_1, age of structure x_2, lot size x_3, number of on-site parking spaces x_4, and gross area x_5. Use appropriate plots and stepwise regression to find a useful model to predict sale price. Analyze the model. Include a residual analysis.

Code No.	Sale Price	No. of Apartments	Age of Structure	Lot Size	Parking Spaces	Building Area
0229	$ 90,300	4	82	4,635	0	4,266
0094	384,000	20	13	17,798	0	14,391
0043	157,500	5	66	5,913	0	6,615
0079	676,200	26	64	7,750	6	34,144
0134	165,000	5	55	5,150	0	6,120
0179	300,000	10	65	12,506	0	14,552
0087	108,750	4	82	7,160	0	3,040

0120	276,538	11	23	5,120	0	7,881
0246	420,000	20	18	11,745	20	12,600
0025	950,000	62	71	21,000	3	39,448
0015	560,000	26	74	11,221	0	30,000
0131	268,000	13	56	7,818	13	8,088
0172	290,000	9	76	4,900	0	11,315
0095	173,200	6	21	5,424	6	4,461
0121	323,650	11	24	11,834	8	9,000
0077	162,500	5	19	5,246	5	3,828
0060	353,500	20	62	11,223	2	13,680
0174	134,400	4	70	5,834	0	4,680
0084	187,000	8	19	9,075	0	7,392
0031	155,700	4	57	5,280	0	6,030
0019	93,600	4	82	6,864	0	3,840
0074	110,000	4	50	4,510	0	3,092
0057	573,200	14	10	11,192	0	23,704
0104	79,300	4	82	7,425	0	3,876
0024	272,000	5	82	7,500	0	9,542

4. To determine the best suited peas for production, a large agribusiness cooperative tested five varieties of peas. A field was divided into 20 sections, and each of the five varieties were planted in four plots. The following bushels of peas produced from each plot are given in Exercise 15.43 on page 861 of the text.

Variety of Peas

A	B	C	D	E
26.2	29.2	29.1	21.3	20.1
24.3	28.1	30.8	22.4	19.3
21.8	27.3	33.9	24.3	19.9
28.1	31.2	32.8	21.8	22.1

a. Plot the yields versus the five varieties of peas. Obtain numerical descriptive measures of the yields. Summarize the differences in the varieties.
b. Fit the regression model

$$y = \beta_0 + \beta_1 x_1 + \beta_2 x_2 + \beta_3 x_3 + \beta_4 x_4 + \epsilon$$

where y = yield, and coding $x_1 = 1$ for variety A, $x_2 = 1$ for variety B, $x_3 = 1$ for variety C, and $x_4 = 1$ for variety D. Interpret the parameters.
c. Interpret s and R^2.
d. Test whether the model is useful for predicting yield. Use $\alpha = .05$.
e. Construct and interpret a 95% confidence interval for the difference between the mean yields of varieties D and E.
f. Use the model to predict the yield for each variety of peas. Compare the fitted value with the mean yield for each variety.

5. The accountant at Fandels Department Store wants to determine the relationship between customer's purchases y, and the two independent variables, yearly income and type of residence.

A sample of 15 customers provided the following data.

Customer	Purchases	Income	Residence
1	$270	$12,000	Rent home
2	580	20,750	Own Home
3	540	18,000	Own Home
4	720	31,500	Own Home
5	450	16,250	Apartment
6	640	25,250	Apartment
7	240	10,750	Apartment
8	400	13,500	Own Home
9	620	23,750	Apartment
10	660	27,500	Apartment
11	680	31,000	Own Home
12	340	14,000	Rent home
13	460	18,500	Rent home
14	600	20,500	Own Home
15	670	9,000	Apartment

a. Create indicator variables for the type of residence. Fit a first-order model to the data.
b. Test whether the overall model is useful. Use $\alpha = .05$.
c. Test whether the type of residence is useful in the model. Use $\alpha = .05$.
d. Find and interpret a 95% prediction interval for a homeowner with $20,000 annual income.

6. An investor is interested in the relationship between the average annual investment yield for three types of mutual funds, and the size of the funds, the fees charged, and the annual expenses. The accompanying table provides the data for a random sample of 43 mutual funds, as reported in Forbes, September 4, 1989. The three types of funds are stock (1), stock and bond (2), and foreign (3). Yield is the average annual percent return on investment over the past ten years. Assets gives the size of the funds in millions of dollars as of June 30, 1989. The sales fee is the maximum percentage sales charge on money invested in a mutual fund. Annual expense is the annual percentage charge for operating the fund.

Fund	Type	Yield	Assets	Fee	Expenses
1	1	14.8	$ 738	0.00	0.80
2	1	16.2	149	4.50	1.00
3	1	14.5	546	0.00	1.31
4	1	14.7	72	4.00	1.10
5	1	8.5	161	0.00	1.02
6	1	14.8	991	0.00	0.67
7	1	12.0	833	0.00	0.47
8	1	5.0	19	8.50	1.30
9	1	15.3	464	5.75	0.71
10	1	12.5	32	15.00	0.68
11	1	6.8	184	0.00	1.35
12	1	20.1	534	0.00	0.92
13	1	12.2	233	7.25	0.89
14	1	7.0	1019	8.50	0.85

15	1	16.2	529	1.00	1.01
16	1	11.0	644	8.50	0.68
17	1	8.4	275	0.00	1.31
18	1	11.3	79	5.75	0.82
19	1	16.3	186	0.00	1.14
20	1	1.9	161	0.00	0.87
21	2	16.2	143	5.50	1.42
22	2	12.0	6	3.00	1.23
23	2	4.4	46	0.00	0.73
24	2	12.6	164	8.00	0.80
25	2	14.3	266	7.25	0.91
26	2	5.7	1826	0.00	0.45
27	2	9.7	139	0.00	0.89
28	2	13.1	873	8.50	0.92
29	2	3.7	1052	4.00	0.59
30	2	12.9	72	4.00	1.12
31	3	11.5	31	5.50	1.93
32	3	0.7	528	0.00	1.40
33	3	4.7	280	4.00	0.82
34	3	17.0	89	4.75	2.20
35	3	0.5	303	0.00	1.01
36	3	12.4	114	4.00	2.19
37	3	12.5	62	0.00	2.31
38	3	3.8	288	6.50	1.02
39	3	16.0	100	5.75	0.69
40	3	14.7	455	8.50	1.89
41	3	7.3	466	8.50	1.61
42	3	16.0	564	0.00	1.22
43	3	3.5	727	8.50	0.83

a. Construct the scattergrams of the response variable versus each independent variable. Find the correlation matrix. Summarize the relationships between the variables.

b. Fit the regression model

$$y = \beta_0 + \beta_1 x_1 + \beta_2 x_2 + \beta_3 x_3 + \beta_4 x_4 + \beta_5 x_5 + \epsilon$$

where y = Average annual yield, x_1 = Assets, x_2 = Sales fee, x_3 = Annual expenses, x_4 = 1 if stock fund, 0 otherwise, and x_5 = 1 if stock and bond fund, 0 otherwise.

c. Interpret s and R^2.

d. Test whether the model is useful for predicting fund yield. Use $\alpha = .05$.

e. Use STEPWISE to select a subset of useful predictors.

f. Write a short summary of the implications of this data analysis.

7. Consider Exercise 15.53 on page 874 of the text. A real estate appraiser in Minneapolis provided the following data on apartment buildings sold during 1990. The random sample of 25 apartment buildings includes information on sale price, number of apartment units, and the physical condition of the apartments. The condition is listed as excellent (E), good (G), and fair (F).

Code	Sale Price	Number	Condition
0229	$ 90,300	4	F
0094	384,000	20	G
0043	157,500	5	G
0079	676,200	26	E
0134	165,000	5	G
0179	300,000	10	G
0087	108,750	4	G
0120	276,538	11	G
0246	420,000	20	G
0025	950,000	62	G
0015	560,000	26	G
0131	268,000	13	F
0172	290,000	9	E
0095	173,200	6	G
0121	323,650	11	G
0077	162,500	5	G
0060	353,500	20	F
0174	134,400	4	E
0084	187,000	8	G
0031	155,700	4	E
0019	93,600	4	F
0074	110,000	4	G
0057	573,200	14	E
0104	79,300	4	F
0024	272,000	5	E

a. Construct a labeled plot of sale price y versus number of apartment units x_1, using letters for the condition of building. Discuss the plot. Does it appear that a model which describes the relationship between sale price and number of apartment units as three parallel lines is appropriate?

b. Create indicator variables for the building condition categories. Find the regression equation for estimating the mean apartment sales price using number of units and building condition. Interpret the least squares estimates. Evaluate the overall regression model.

c. Test whether the relationship between the mean apartment sale price and number of units depends on the building condition. Use $\alpha = .05$.

d. Obtain the three prediction equations for the three building condition categories.

8. In Exercise 15.84 on page 900 of the text, a consumer advocacy organization conducted an experiment to compare the effectiveness of three weight-reducing diets. Ten people were assigned to each diet. The following before-diet weights and weight losses after a month were recorded.

Diet A		Diet B		Diet C	
Weight	Loss	Weight	Loss	Weight	Loss
227	14	255	19	206	7
286	16	193	8	222	9
180	-2	186	4	168	2
176	8	145	15	132	0
204	15	219	16	173	-3
155	5	273	19	210	8
303	17	289	25	269	10
146	7	168	6	275	15
215	15	194	12	241	8
187	6	248	21	219	5

a. Use a labeled plot to graphically study the relationship between weight loss y and weight before the diet program and the type of diet. Describe the plot.
b. Fit a first-order model to the data. Summarize the results.
c. Add interaction terms to the model of part b. Fit the model to the data. Test whether there is significant interaction between initial weight and type of diet. Use $\alpha = .05$.
d. Use the best model from parts b and c to determine whether there is a difference in mean weight loss for the three diets. Use $\alpha = .05$.

CHAPTER 16

SURVEY ANALYSIS

In this chapter we describe a survey project that we generally assign near the end of a statistics course. The objectives of the project are to use Minitab to analyze a questionnaire and to provide a review of statistical methods covered in the course. The project illustrates a realistic approach to analyze a survey containing many variables. It is a survey for Fandels Department Store. The following questionnaire was distributed to 120 customers at the store. All or part of the data may be used for the project. The data file is on the computer diskette available with this supplement.

QUESTIONNAIRE

We are interested in your opinion about Fandels Department Store. By completing this questionnaire, you will be providing information so that we can better serve your needs. Thank you for your cooperation.

1. Which of the following services, not now available at Fandels, would you most like to see offered? Check only one.

    1. __ Luncheon counter/restaurant
    2. __ Check cashing
    3. __ Child care
    4. __ Free gift wrapping

2. Overall, how would you rate the personnel at Fandels as compared to other department stores in the area?

    1. __ Definitely better
    2. __ Somewhat better
    3. __ No different
    4. __ Somewhat worse
    5. __ Definitely worse
    6. __ No opinion

305

306 Chapter 16

3. For each of the following, please check the blank which best describes your feelings about Fandels store.

                                                1 2 3 4 5 6

   a. Poor Value   _ _ _ _ _ _   Good Value
   b. Low Prices   _ _ _ _ _ _   High Prices
   c. Unfriendly Atmosphere   _ _ _ _ _ _   Friendly Atmosphere
   d. Old-Fashioned   _ _ _ _ _ _   Modern
   e. Poor Selection   _ _ _ _ _ _   Good Selection
   f. Poor Location   _ _ _ _ _ _   Good Location
   g. Lower Class Clientele   _ _ _ _ _ _   Higher Class Clientele

4. What is your age?

   1. __ 18-34
   2. __ 35-49
   3. __ 50-64
   4. __ 65 plus

5. Gender

   1. __ Male
   2. __ Female

6. Type of residence

   1. __ Own a home
   2. __ Rent an apartment
   3. __ Rent a home
   4. __ Other
   If Other, describe the type_____

7. Annual income to the nearest one hundred dollars (If married, your household income)
_____

8. Approximately, how much did you spend in Fandels Department Store in 1990?
_____

Data from 120 Questionnaires

Row	Q1	Q2	Q3 a	b	c	d	e	f	g	Q4	Q5	Q6	Q7	Q8
1	2	1	3	2	2	6	1	2	5	4	1	2	22700	180
2	3	2	2	2	3	2	6	6	4	3	1	1	20200	320
3	4	3	3	3	1	6	2	5	4	1	2	2	21900	780
4	4	1	2	2	1	5	6	4	6	3	1	1	38400	650
5	2	3	2	2	3	2	2	3	3	3	2	3	33200	560
6	2	4	2	3	2	3	1	5	3	4	2	1	23600	330
7	4	5	5	4	3	6	1	3	3	3	1	1	25000	410
8	1	2	3	2	3	6	5	4	4	3	2	3	16500	260

						Q3								
Row	Q1	Q2	a	b	c	d	e	f	g	Q4	Q5	Q6	Q7	Q8
9	2	1	5	5	4	6	6	5	3	4	2	3	21300	270
10	3	1	4	3	5	3	4	5	3	3	2	2	7800	140
11	4	4	2	2	5	5	6	3	3	4	1	3	27400	240
12	2	3	4	2	4	5	6	6	3	4	1	1	28800	370
13	3	4	4	2	3	4	3	6	5	1	2	2	28500	650
14	3	6	2	3	3	5	3	1	4	2	1	2	34000	620
15	3	4	1	2	3	6	5	2	4	4	2	2	28200	250
16	2	4	3	3	3	4	1	1	1	3	2	2	19100	320
17	3	5	2	2	5	1	5	4	3	1	2	1	37300	890
18	4	2	3	2	2	6	1	4	2	3	1	1	21400	410
19	1	3	2	3	1	3	6	3	4	2	2	1	36100	640
20	3	1	2	4	3	1	1	5	3	4	2	3	30300	260
21	3	2	2	1	2	2	6	6	3	3	1	1	3400	70
22	2	2	3	1	3	4	5	5	4	4	1	4	45700	680
23	4	2	2	3	3	1	6	2	3	4	2	4	31200	380
24	4	2	2	3	5	6	4	6	3	1	2	1	23800	610
25	3	4	5	2	3	5	6	6	4	3	1	2	28300	460
26	3	3	3	3	3	5	2	2	5	3	2	4	27500	480
27	3	6	5	4	4	1	4	2	3	2	2	3	11600	340
28	2	1	1	1	3	6	2	1	3	4	1	1	22500	190
29	4	3	1	2	3	3	6	5	4	3	2	2	27900	490
30	3	2	3	1	3	5	5	3	2	1	1	1	19100	670
31	1	3	3	2	2	5	3	2	2	3	1	1	25400	460
32	3	3	3	2	4	6	1	6	2	3	1	2	31300	500
33	2	2	1	4	3	1	5	1	4	3	1	1	11200	180
34	1	3	4	4	2	1	1	5	5	4	2	1	29700	390
35	2	3	4	2	4	1	3	2	3	3	2	1	9100	80
36	2	2	3	4	3	3	1	2	4	3	2	3	36200	610
37	3	5	3	2	3	3	6	1	4	1	1	1	38100	920
38	2	3	2	2	2	3	6	5	5	2	1	1	16000	380
39	3	5	2	2	3	6	6	3	5	1	1	2	34500	740
40	2	1	2	1	4	5	3	6	1	3	1	2	25300	420
41	2	3	2	1	1	3	3	4	4	1	1	1	29800	630
42	2	4	1	1	4	1	1	3	3	3	1	4	16700	300
43	3	3	6	2	2	1	6	6	5	4	2	4	32700	390
44	1	3	3	2	2	1	1	3	3	1	2	1	23400	670
45	3	6	2	2	4	1	1	3	1	1	2	1	25200	730
46	4	6	2	1	1	6	5	5	5	2	1	1	33900	610
47	4	1	3	3	3	3	1	1	2	3	2	1	26100	440
48	3	3	4	1	2	1	1	2	5	4	1	2	17100	190
49	1	3	3	1	2	4	6	1	5	1	1	1	39800	780
50	3	2	2	4	2	5	6	1	4	3	2	2	11800	280
51	4	1	1	2	2	5	6	4	5	4	1	2	22600	300
52	3	1	2	3	2	1	6	1	2	1	1	1	35800	780
53	3	6	3	4	5	4	6	3	3	3	1	1	16900	340
54	3	2	3	3	1	1	3	1	5	1	1	1	22900	530
55	3	2	2	2	4	5	1	1	3	2	1	1	21600	490
56	3	3	2	3	1	5	6	5	4	4	2	1	41300	620
57	2	2	3	1	5	5	6	1	3	4	1	2	31000	210
58	4	1	1	3	4	1	2	4	6	4	2	1	30600	390
59	3	3	2	2	3	4	4	3	2	3	1	3	31000	540
60	1	5	2	3	4	1	4	5	4	2	1	1	22600	300
61	3	4	4	5	3	6	4	2	6	1	1	1	33500	690
62	1	1	2	2	4	6	6	6	3	1	1	2	30100	790
63	3	2	2	4	3	6	1	1	5	2	1	2	25000	530
64	1	3	2	3	4	6	3	4	3	4	1	1	34300	170
65	4	4	1	3	2	1	5	1	4	1	1	4	16200	490
66	2	3	1	3	3	3	5	1	5	4	2	3	29900	250
67	2	2	3	1	4	2	5	4	5	1	2	2	29900	690
68	4	1	1	3	3	5	5	3	4	1	2	2	36100	860

308   Chapter 16

Row	Q1	Q2	Q3 a	b	c	d	e	f	g	Q4	Q5	Q6	Q7	Q8
69	1	1	2	3	3	6	3	1	2	1	2	2	21600	760
70	2	2	3	4	4	6	3	3	4	1	2	1	15500	670
71	3	4	2	2	2	2	5	3	3	3	2	4	9400	190
72	3	3	5	3	1	6	6	3	3	4	2	1	20000	60
73	4	3	3	1	2	2	2	3	5	4	1	1	32900	260
74	4	3	2	2	5	1	1	5	4	1	2	1	30100	730
75	2	1	4	4	3	5	5	3	2	1	1	1	27100	62
76	3	2	4	5	3	2	6	2	3	2	2	2	35500	640
77	3	1	3	2	3	6	5	4	4	1	1	3	30200	790
78	1	6	3	2	3	1	4	3	4	3	2	1	37600	650
79	6	4	3	2	3	4	2	5	4	4	3	2	29900	500
80	1	1	1	2	3	4	1	3	2	4	1	1	38200	440
81	2	4	2	2	2	6	4	1	6	4	1	1	50000	770
82	3	2	3	1	3	6	5	4	3	3	1	3	15100	290
83	4	2	4	2	3	1	1	6	5	3	1	1	24800	480
84	4	1	2	5	2	4	3	4	3	4	1	3	20500	230
85	4	2	2	2	2	5	4	5	4	4	2	1	18300	240
86	1	1	2	3	5	1	6	2	3	3	2	1	30200	510
87	2	3	5	4	4	3	4	3	3	1	2	2	22100	620
88	2	3	5	4	3	6	5	4	4	4	2	1	22500	320
89	2	2	1	4	2	1	6	2	3	1	1	2	27500	620
90	3	2	3	3	2	1	6	2	2	4	1	2	29300	370
91	2	2	1	5	2	5	4	1	5	4	2	1	40400	560
92	3	3	4	2	3	1	5	6	6	4	1	4	15400	150
93	2	5	2	3	3	6	5	6	4	1	2	2	32900	680
94	2	3	3	3	3	5	5	2	3	3	2	2	21100	340
95	4	1	1	3	3	5	6	1	3	1	1	2	32500	730
96	1	3	3	1	2	6	1	2	4	4	2	2	20400	100
97	3	1	2	1	2	5	2	3	3	3	1	2	25400	470
98	4	2	3	1	2	3	1	4	2	2	1	2	24900	510
99	3	4	1	2	3	1	6	6	3	3	1	3	16700	300
100	2	3	2	2	5	5	5	4	4	3	2	3	13800	220
101	1	5	3	2	2	6	5	6	3	4	1	1	27500	370
102	4	4	1	2	1	6	6	6	5	3	1	2	25400	470
103	2	3	2	2	4	6	3	5	4	3	2	1	30000	480
104	1	3	2	2	3	6	6	5	3	4	2	1	33300	470
105	3	3	2	4	3	1	6	6	4	2	2	1	33500	610
106	2	2	1	3	2	5	3	6	4	3	1	2	29200	500
107	3	2	2	3	4	2	6	2	4	4	1	4	35600	470
108	3	6	3	3	6	6	6	6	4	4	2	2	33400	430
109	2	2	1	2	3	4	6	5	2	4	1	2	34600	450
110	1	1	4	3	3	1	6	3	3	4	2	1	40200	450
111	4	3	4	4	3	2	3	5	2	2	1	1	29500	560
112	2	2	1	3	3	5	5	6	4	4	2	1	26700	390
113	2	3	3	3	2	6	2	4	3	3	2	2	33400	530
114	2	2	4	4	2	5	3	6	3	4	1	1	31100	410
115	2	1	3	3	3	2	6	4	3	4	1	3	42700	570
116	1	1	2	1	2	1	5	1	4	4	1	2	20600	180
117	3	1	4	2	4	6	2	3	2	1	1	2	39600	950
118	2	1	2	2	4	1	1	2	3	3	2	1	22000	360
119	2	4	1	2	5	5	5	5	3	3	2	4	7900	110
120	2	5	4	2	3	1	2	6	3	6	2	2	25500	290

EXERCISES

1. Create a file containing the data set. Check the file for errors and make corrections if necessary. It is important that the data file be free of errors before beginning the analysis. Name the columns and save an error-free copy of the data set.

2. The questionnaire has been designed to provide answers to the following questions requested by Fandels Department Store.

   a. What percent of the customers would most like to have a child care service? A check cashing service? Which service not now available is most desired by Fandels customers?
   b. How does Fandels personnel rate when compared with other department stores?
   c. What is the overall feeling toward Fandels Department Store. If Fandels wanted to improve their image, where might they direct their efforts?
   d. Is there a relationship between the age of customers and the services desired but not now available?
   e. Is there a difference in the average amount spent for the four groups formed by the most desired service? Why is it important to know this?
   f. What are the average amounts spent by customers when they are cross-classified by sex and service desired? Discuss how this information can be used.
   g. What is the relationship, if any, between the amount spent and the two variables, annual income and age? How can this information be used?
   h. Estimate the average amount spent by all Fandels customers.
   i. Use regression to estimate the average amount spent by customers with income of $25,000.

   Outline the statistical procedure or procedures that you would use to answer each question. Hint: It is important to distinguish between qualitative or quantitative variables.

3. Write a Minitab program that will provide the output for the statistical procedures identified in Exercise 2.

4. Run the Minitab program. Write a summary report of your survey analysis for Fandels Department Store.

APPENDIX

A   MICROCOMPUTER INFORMATION

This appendix describes some features of Minitab on a DOS microcomputer system. Throughout, we assume that Minitab is installed onto drive C of your system according to the instructions in the *Minitab User Guide*.

Minitab Access

The usual way to access Minitab from drive C is to enter the following commands which we have printed in italics. The CD is a DOS command to change directory. You can use either uppercase or lowercase letters.

   C:\> *CD\Minitab*
   C:\MINITAB> *Minitab*

If you have a MENU program on your system, choose the Minitab option. You could also create a AUTOEXEC.BAT file which would load Minitab with one command at a DOS prompt.

Using Menus

In this supplement, Minitab commands are entered on a command line following the MTB > prompt. Commands may also be entered using menus and dialog boxes. To open a menu, press the ALT and highlighted letter keys simultaneously. Within menus, select a menu command by pressing the appropriate highlighted letter to obtain the command's dialog entry screen. Press the TAB key to move around the entry boxes of a dialog screen. Information may be entered into the dialog boxes from the keyboard or, if columns are to be specified, entered from the list of columns in the variable list box. To select a variable from the list box, press F2, use the arrow keys to select and the space bar to highlight the variable, and press F2 again to move the variable to the dialog box.

If using a mouse, click on the appropriate menu and command names. Within a dialog screen, move to a box by clicking it.

Files

You can specify any defined directory or available drive to store and retrieve files. Some file handling examples follow.

1. MTB > *SAVE 'DATA10-3'*
   Saves the worksheet file DATA10-3.MTW in the current directory.

2. MTB > *JOURNAL 'A:PRG10-3'*
   Saves the program as PRG10-3.MTJ on the disk in drive A.

3. MTB > *RETRIEVE 'A:DATA10-3'*
   Retrieves the worksheet file DATA10-3.MTW from the disk in drive A.

4. MTB > *OUTFILE 'D:\MINFILES\EX10-3'*
   Saves the output file EX10-3.LIS in the MINFILES directory on drive D.

Output

You can print commands and output during or after a session. To obtain a printout during a session, enter the Minitab PAPER command. All subsequent Minitab lines and output are sent to the printer; printing ends when you enter NOPAPER or STOP to end the Minitab session.

For example, the following program prints the MEAN and STDEV command lines and results.

```
MTB > NAME C1 'SCORES'
MTB > SET 'SCORES'
DATA> 79 95 76 91 84 79 51 72 89 81
DATA> 93 97 68 72 85 99
DATA> END
MTB > PAPER
MTB > MEAN 'SCORES'
   MEAN     =        81.937
MTB > STDEV 'SCORES'
   ST.DEV.  =        12.540
MTB > NOPAPER
MTB > STOP
```

Use OUTFILE to save the program if you want to print it after a Minitab session. All subsequent Minitab lines and results are sent to the output file; saving ends when you enter NOOUTFILE or STOP to end the Minitab session. For example, the following program saves the Minitab lines and output between the OUTFILE and STOP commands in a file named TEST1.LIS.

```
MTB > OUTFILE 'TEST1'
MTB > NAME C1 'SCORES'
MTB > SET 'SCORES'
DATA> 79 95 76 91 84 79 51 72 89 81
DATA> 93 97 68 72 85 99
DATA> END
MTB > MEAN 'SCORES'
   MEAN     =        81.937
```

```
MTB > STDEV 'SCORES'
   ST.DEV. =        12.540
MTB > STOP
```

You can print the output file using the command

    C> COPY \MINITAB\TEST1.LIS PRN

Or you can use a word processing package to edit and print the output. This allows you to add, delete, or modify the file before printing.

DOS Commands Accessible from Minitab

Use the DIR (Directory) command to locate a file, or to obtain a listing of all or only certain files on a disk or directory. Consider the following examples:

1. MTB > *DIR*
   Lists all the files in the current directory.

2. MTB > *DIR *.MTW*
   Lists all Minitab worksheet files in the current directory.

3. MTB > *DIR A:*
   Lists all files on the disk in drive A.

4. MTB > *DIR D:\MINFILES\EX*.LIS*
   Lists all files starting with EX having extension LIS in directory MINFILES on drive D.

The CD (Change Directory) command displays or changes the current directory. Some examples include:

1. MTB > *CD*
   Displays the current directory.

2. MTB > *CD C:\MINITAB\DATA*
   Changes the current directory to the DATA subdirectory in the MINITAB directory on drive C.

The TYPE command displays ASCII files on the computer screen. Use TYPE to view a file before entering it into a Minitab program, or to check the output from a previous Minitab program. For example, the following command displays the SURVEY.DAT file that is on the disk in drive A:

    MTB > *TYPE A:SURVEY.DAT*

If the file is not an ASCII file, it is not readable. For example, worksheet files with the SAVE command can only be read by a Minitab computer package.

Data Editor

The Data Editor provides a full screen view of your Minitab worksheet which allows you to quickly enter and edit data. Some special features include automatic column width adjustments, automatic data formatting, and the ability to insert or delete items and rows. To access the Data Editor, use Alt and D simultaneously; to exit, use Alt and M simultaneously. The Data Editor is very similar to a spreadsheet program, such as Lotus 1-2-3. Documentation on all the features is available through the on-line Help (F1 key).

```
                  SAMPLE DATA EDITOR SCREEN

              SCORES
    ->         C1     C2    C3    C4    C5    C6  ...
    1          79
    2          95
    3          76
    4          91
    5          84
    .           .
    .           .
    .           .
```

Cells are located at the intersection of rows and columns. The active cell containing the cursor is shown in reverse video. Use the four arrow keys to move to the cell at which you want to enter data. Type in the data, and use the ENTER key or arrow keys to move to the next data entry cell. Enter column names without quotes in the same way that you enter data. The column name area is above the first row of the worksheet. Press the up arrow key from row one to enter the column name area.

If you make a mistake while typing an entry, use the backspace key to delete the error. If you notice a mistake after leaving a cell, use the arrow keys to return to the cell and type in the correct entry. Press the ENTER key after making corrections. The following include additional notes on the Data Editor.

1. Your Minitab worksheet is not saved by the Data Editor. You need to use the SAVE command at the MTB > prompt.

2. A Data Editor menu is provided by pressing the F10 function key.

3. The default column width is four characters. The column width automatically increases as needed.

Lotus and Minitab Interface

Data can easily be transferred between a Minitab worksheet and a Lotus or Symphony spreadsheet. The Lotus program is specified if you used the MSETUP program to install Minitab on your system.

1. Lotus to Minitab

Use the LOTUS subcommand with RETRIEVE to enter a Lotus spreadsheet in a Minitab worksheet. For example, if the Lotus file EX10-3 is in the current directory, the following command lines will enter it in a Minitab worksheet:

    MTB > *RETRIEVE 'EX10-3';*
    SUBC> *LOTUS.*

2. Minitab to Lotus

Use the LOTUS subcommand with SAVE to save a Minitab worksheet as a Lotus spreadsheet. For example, to save a Minitab worksheet that can be retrieved by Lotus as EX10-3, enter:

    MTB > *SAVE 'EX10-3';*
    SUBC> *LOTUS.*

The default conversion from Lotus to Minitab changes alpha data to missing values. Use named ranges to convert alpha data.

B DATA SETS

CLASS PROJECT
File Name: CLASSDAT

The class data set was created as a class project in a business statistics course. To collect the data, a questionnaire containing the following variables was given to 200 students enrolled in the course. The codes used to create the Minitab worksheet are given with qualitative variables.

1. Gender _____ Male (1) Female (0)
2. Number of credits earned prior to this quarter _____
3. Number of credits this quarter _____
4. Marital status
   Single _____ (0)
   Married _____ (1)
   Other _____ (2)
5. Age _____
6. Distance you live from class (in miles) _____
7. Number of hours per week you work (on a job) _____
8. Grade point average _____
9. Type of car
   US _____ (0)
   Foreign _____ (1)
   No car _____ (2)
10. Major program:
    Business _____ (0)
    Economics _____ (1)
    Other _____ (2)

ROW	C1	C2	C3	C4	C5	C6	C7	C8	C9	C10
1	1	81	16	0	21	1.00	25.0	2.800	0	0
2	0	92	16	0	21	4.00	20.0	2.630	0	0
3	1	66	16	0	19	5.00	10.0	2.850	1	0
4	1	64	16	0	19	1.00	0.0	3.500	0	0
5	1	72	*	0	19	1.20	20.0	4.000	0	0
6	1	66	16	0	20	1.00	0.0	3.450	0	0
7	1	108	17	0	20	0.20	5.0	3.000	2	0
8	1	64	16	0	20	0.50	0.0	3.000	0	0
9	1	64	16	0	19	0.10	12.0	3.800	0	0
10	0	101	16	0	23	3.50	*	3.360	0	0
11	0	64	16	0	20	0.50	10.0	4.000	2	0
12	0	72	16	1	30	8.00	0.0	3.294	0	0
13	0	68	16	0	20	0.50	20.0	3.441	0	0
14	0	103	16	1	23	0.50	20.0	3.200	0	0
15	0	64	16	0	19	1.00	8.0	3.500	2	0
16	0	67	16	0	19	0.20	10.0	3.680	0	0
17	1	64	16	0	19	0.20	10.0	3.330	0	0
18	1	88	16	0	21	0.30	0.0	3.900	1	0
19	1	101	12	0	21	0.30	15.0	2.950	1	0
20	0	64	17	0	20	0.10	0.0	3.125	0	0
21	1	69	12	0	20	10.00	25.0	3.941	0	0
22	0	117	17	0	21	0.00	0.0	3.500	2	0

318 *Appendix B*

ROW	C1	C2	C3	C4	C5	C6	C7	C8	C9	C10
23	1	64	16	0	19	0.25	0.0	3.400	0	0
24	1	84	12	0	20	0.50	12.0	3.400	1	0
25	0	62	16	0	20	1.00	28.0	3.533	0	0
26	1	121	8	0	26	50.00	40.0	3.022	0	0
27	0	67	16	0	19	0.50	23.0	3.312	1	0
28	0	129	12	0	21	2.00	12.0	2.860	0	0
29	1	108	12	1	34	4.00	18.0	3.200	0	0
30	1	101	13	0	20	1.00	13.0	2.550	2	0
31	1	126	12	0	23	3.00	25.0	3.100	1	0
32	0	84	15	0	21	20.00	20.0	3.300	0	0
33	0	68	16	0	19	1.00	0.0	3.940	2	0
34	1	66	16	0	20	0.20	0.0	3.600	2	0
35	1	106	16	0	20	1.50	17.5	*	*	*
36	1	150	12	1	37	0.50	32.0	3.140	0	0
37	1	132	12	0	23	30.00	20.0	2.800	0	1
38	1	99	9	0	20	0.25	20.0	2.650	0	0
39	0	92	8	0	21	6.00	25.0	2.650	0	0
40	1	103	8	0	23	2.00	40.0	2.470	0	0
41	1	106	12	0	21	0.50	15.0	3.428	0	0
42	1	65	17	0	19	1.00	25.0	3.672	0	0
43	1	110	16	0	20	1.00	30.0	3.000	0	0
44	1	103	12	0	21	1.00	0.0	3.000	0	0
45	0	86	12	0	20	0.10	8.0	2.630	2	0
46	0	113	13	0	21	2.00	20.0	3.650	1	2
47	1	103	16	0	20	0.50	5.0	2.650	0	0
48	1	90	12	0	20	1.00	0.0	3.500	1	0
49	1	96	16	0	20	1.00	18.0	2.670	0	0
50	0	144	9	0	21	3.00	0.0	2.890	2	0
51	1	85	12	0	22	0.10	20.0	2.588	1	0
52	1	100	12	0	22	1.00	0.0	2.850	0	0
53	1	102	16	0	20	0.40	0.0	3.010	0	2
54	0	64	12	0	19	0.20	25.0	3.250	0	0
55	1	110	14	0	22	0.50	12.0	2.900	0	0
56	1	136	12	0	24	0.50	16.0	3.670	1	0
57	0	87	12	0	20	0.50	0.0	2.800	2	0
58	0	86	17	0	22	1.50	20.0	2.850	0	0
59	0	98	8	0	25	5.00	60.0	2.790	0	0
60	0	91	12	0	19	8.00	20.0	3.000	0	0
61	1	104	12	0	21	0.30	21.0	3.440	0	0
62	1	140	16	0	23	1.00	8.0	2.750	1	2
63	1	90	16	0	21	1.00	*	3.024	2	0
64	0	89	12	0	20	0.00	20.0	2.750	1	0
65	0	108	12	0	20	4.00	0.0	2.730	0	0
66	1	110	12	0	20	4.00	40.0	3.100	0	0
67	1	90	12	0	21	0.40	0.0	3.200	0	0
68	1	100	12	0	21	4.00	22.0	2.850	1	0
69	1	97	12	0	21	1.00	35.0	3.290	0	0
70	1	121	14	0	21	1.00	0.0	2.500	0	0
71	1	96	17	0	21	0.50	20.0	3.350	0	0
72	0	136	16	0	22	3.00	30.0	2.660	0	0
73	0	125	12	0	21	1.00	13.0	*	1	3
74	1	109	19	0	20	0.30	22.0	3.380	0	0
75	1	139	8	1	22	60.00	12.0	2.500	0	0
76	0	161	16	1	23	30.00	20.0	2.560	0	0
77	1	115	16	0	22	5.00	10.0	3.690	2	0
78	0	102	12	0	20	1.00	0.0	2.700	0	0
79	1	129	12	0	21	0.10	0.0	2.600	0	0
80	1	106	12	0	22	0.10	12.0	2.000	0	0
81	0	145	16	1	28	45.00	15.0	3.000	0	0
82	0	103	12	0	21	0.50	12.5	3.460	0	0
83	0	110	12	0	20	30.00	0.0	3.890	1	0

ROW	C1	C2	C3	C4	C5	C6	C7	C8	C9	C10
84	0	114	16	0	21	0.10	20.0	3.000	0	0
85	1	110	12	0	19	8.00	0.0	3.400	0	0
86	1	96	18	0	21	0.50	10.0	3.000	1	0
87	1	109	12	0	21	0.50	7.0	3.666	2	0
88	1	106	16	0	21	0.50	20.0	2.950	0	0
89	1	126	16	0	21	0.12	7.0	3.000	0	0
90	0	136	12	0	21	0.50	10.0	3.800	2	2
91	0	124	16	0	22	2.00	0.0	3.100	0	0
92	0	116	16	0	21	0.50	22.0	3.700	1	0
93	0	106	12	0	20	0.12	14.0	*	0	0
94	0	126	12	1	30	75.00	0.0	3.900	0	0
95	1	114	12	0	20	0.50	0.0	3.500	2	0
96	1	94	16	0	21	0.50	0.0	3.000	0	0
97	1	131	16	0	22	0.75	15.0	3.000	1	0
98	1	120	12	0	22	0.50	0.0	2.550	0	0
99	0	110	15	0	20	1.00	10.0	3.400	0	0
100	1	96	12	0	20	1.00	0.0	3.000	0	0
101	1	117	12	0	21	0.25	0.0	3.000	0	0
102	0	152	16	0	21	0.00	5.0	3.100	0	2
103	0	107	12	0	20	2.00	30.0	3.000	0	0
104	1	8	8	0	26	0.10	12.0	3.030	0	0
105	0	102	12	0	20	1.00	30.0	2.690	0	0
106	0	151	17	0	21	0.50	8.0	3.200	0	0
107	0	100	15	0	20	1.00	27.0	2.800	0	0
108	1	104	12	0	21	0.25	12.0	2.900	0	0
109	1	105	12	0	20	1.00	0.0	2.800	0	0
110	1	118	16	0	21	1.50	28.0	2.900	0	0
111	1	93	14	0	20	0.50	25.0	2.666	0	0
112	1	92	13	0	21	2.50	35.0	3.390	0	0
113	1	98	12	0	22	0.25	0.0	2.530	0	0
114	0	115	16	0	20	0.25	0.0	3.000	2	0
115	1	99	14	0	21	0.30	8.0	2.734	0	0
116	1	118	16	0	21	0.50	12.0	3.625	2	0
117	0	105	12	0	20	25.00	27.0	2.560	0	0
118	0	96	12	0	20	0.50	15.0	2.660	0	0
119	0	100	12	0	21	0.50	30.0	2.780	1	0
120	1	133	16	0	22	2.00	25.0	2.600	0	0
121	1	102	12	0	20	1.00	25.0	2.650	1	0
122	0	100	12	0	21	4.00	25.0	2.750	0	0
123	0	103	12	0	21	4.00	17.0	3.100	0	0
124	0	146	12	2	36	15.00	25.0	3.200	1	0
125	0	98	12	0	21	0.00	19.0	3.000	2	0
126	1	113	17	0	24	0.30	25.0	3.522	1	0
127	1	102	12	0	20	18.90	23.0	2.768	2	0
128	1	110	16	0	20	0.25	0.0	3.163	2	0
129	0	103	15	0	21	0.25	16.0	3.200	1	0
130	0	110	12	0	20	0.50	10.0	2.780	0	0
131	0	112	16	0	21	0.10	18.0	3.353	2	0
132	1	156	16	0	24	0.10	0.0	2.680	0	0
133	1	194	4	0	25	2.00	45.0	2.900	0	0
134	0	102	16	0	21	1.00	20.0	3.040	0	0
135	0	102	16	1	21	18.00	0.0	4.000	1	0
136	1	120	12	0	21	1.00	5.0	3.000	1	1
137	0	109	17	0	21	0.12	0.0	2.642	2	0
138	1	109	14	0	20	0.50	20.0	3.265	0	0
139	1	125	14	2	28	5.00	15.0	3.010	1	0
140	1	109	12	0	21	0.25	0.0	3.384	2	0
141	1	130	13	0	22	0.25	14.0	4.000	0	0
142	1	130	12	0	23	0.25	25.0	2.520	0	0
143	1	132	16	0	21	0.25	0.0	3.100	2	0
144	0	140	14	0	22	1.00	0.0	3.000	0	0

Appendix B

ROW	C1	C2	C3	C4	C5	C6	C7	C8	C9	C10
145	1	109	12	0	21	0.25	0.0	3.330	0	0
146	1	106	16	0	21	0.25	10.0	3.000	0	0
147	0	136	12	0	20	4.00	30.0	2.970	0	0
148	1	99	12	0	26	57.00	22.0	3.100	0	0
149	0	121	16	0	25	5.00	0.0	3.000	2	0
150	0	12	12	0	23	1.00	20.0	3.000	0	0
151	0	106	12	0	21	0.25	25.0	3.320	0	0
152	1	108	12	0	20	0.75	20.0	3.500	0	2
153	1	96	16	2	20	2.00	13.0	2.650	0	0
154	0	101	12	0	20	0.10	10.0	2.650	2	0
155	0	152	14	0	21	2.50	16.0	3.788	0	0
156	1	108	15	0	20	0.25	8.0	2.330	0	0
157	1	141	16	0	23	0.50	0.0	3.166	0	0
158	1	115	13	0	21	0.25	50.0	3.500	0	0
159	1	111	12	0	21	0.10	10.0	3.330	2	0
160	1	115	16	0	20	1.10	0.0	3.200	0	0
161	1	111	12	0	21	1.00	23.0	3.790	0	0
162	1	127	16	0	22	0.40	30.0	2.724	0	0
163	0	109	16	0	21	2.00	12.0	3.750	0	0
164	0	100	17	0	20	0.40	11.0	2.540	2	0
165	0	138	16	0	21	1.00	13.0	3.800	0	0
166	1	112	12	1	30	*	*	2.700	1	0
167	0	120	12	0	21	0.50	30.0	3.300	2	2
168	0	157	16	1	31	40.00	20.0	3.900	0	2
169	0	112	12	0	22	36.00	20.0	2.600	0	0
170	0	112	12	0	22	4.00	29.0	3.300	0	0
171	0	200	16	0	25	15.00	27.0	2.760	0	0
172	0	165	12	0	23	1.00	25.0	3.000	0	0
173	0	114	12	0	21	0.20	10.0	3.500	2	0
174	0	112	16	0	20	1.00	0.0	3.200	0	0
175	0	190	12	1	26	25.00	25.0	3.000	0	0
176	0	121	16	1	25	3.00	1.0	3.400	0	0
177	0	144	16	0	22	0.10	21.0	3.400	0	0
178	0	96	12	1	20	0.25	12.0	2.900	0	0
179	1	105	16	2	29	16.00	12.0	3.450	0	0
180	1	98	16	0	21	2.00	20.0	2.790	0	0
181	1	111	16	0	21	.50	10.0	*	2	0
182	0	116	12	0	22	.10	15.0	3.250	2	0
183	0	109	14	0	21	5.00	20.0	3.500	1	0
184	1	98	16	0	20	1.50	0.0	2.900	0	0
185	0	100	18	0	21	4.00	0.0	2.540	0	1
185	0	132	16	1	23	10.00	0.0	2.750	0	2
187	0	108	16	0	20	.80	10.0	3.000	0	0
188	1	122	12	0	21	4.00	20.0	3.110	0	0
189	0	116	8	1	20	45.00	0.0	3.900	1	0
190	1	103	12	1	22	2.50	15.0	2.960	0	0
191	1	112	16	0	21	35.00	0.0	2.870	0	0
192	1	176	16	1	32	3.50	5.0	2.650	0	2
193	1	108	16	0	20	4.00	10.0	3.100	0	0
194	1	111	16	0	21	2.50	0.0	3.000	0	0
195	0	68	14	0	20	16.00	7.5	*	0	0
196	1	96	16	0	19	.10	0.0	3.750	2	0
197	1	92	12	0	19	36.00	15.0	3.800	1	0
198	1	110	12	1	21	1.00	30.0	3.400	2	0
199	1	131	16	0	24	18.00	0.0	2.980	0	1
200	1	104	16	0	19	.50	0.0	3.200	2	0

SELLING PRICES OF HOMES
File Name: HOMES
Source: Minnesota Real Estate Research Center, College of Business, St. Cloud State University, St. Cloud, MN. Dr. George Karvel, Director

The table gives information on residential homes sold during 1988 for one Minnesota community.

Column	Description
C1(AREA)	Location within the community. Area 1 is within the city, area 2 is suburbs, and area 3 is in the country surrounding the city.
C2(BEDROOMS)	The number of bedrooms.
C3(LIST PR)	List price.
C4(SOLD PR)	Price at which the home sold.
C5(FINANCE)	Type of financing: 1 = Assumed Seller's Financing 2 = Cash 3 = Contract for Deed 4 = Conventional Loan 5 = FHA Loan 6 = VA Loan 7 = Other type of Financing
C6(DAYS)	Days on the marketing.
C7(MTH SOLD)	Month sold: 1 = January, 2 = February, Etc.
C8(DAY SOLD)	Day of the month sold

```
MTB > INFORMATION

COLUMN    NAME        COUNT    MISSING
C1        AREA         197
C2        BEDROOMS     197
C3        LIST PR      197
C4        SOLD PR      197
C5        FINANCE      197
C6        DAYS         197        2
C7        MTH SOLD     197
C8        DAY SOLD     197

CONSTANTS USED: NONE
```

HOME	AREA	BEDROOMS	LIST PR	SOLD PR	FINANCE	DAYS	MTH SOLD	DAY SOLD
1	1	1	17900	17900	3	606	6	1
2	1	2	32000	29000	5	93	11	9
3	1	2	35900	32000	2	22	3	14
4	1	2	34875	34400	3	24	9	1
5	1	2	41900	37000	5	150	12	30
6	1	2	39700	39200	6	239	12	1
7	1	2	41900	40000	4	42	7	29
8	1	2	44900	41900	5	281	9	9
9	1	2	46500	45000	5	28	1	28
10	1	2	49900	48500	4	161	10	31
11	1	2	54900	53000	1	17	4	11
12	1	2	58900	56500	5	2	4	11
13	1	2	67875	61000	5	407	5	16
14	1	2	69900	69000	5	7	9	10

HOME	AREA	BEDROOMS	LIST PR	SOLD PR	FINANCE	DAYS	MTH SOLD	DAY SOLD
15	1	2	74900	72500	4	26	9	2
16	1	2	76900	76000	4	62	3	8
17	1	2	85000	83000	4	14	4	20
18	1	3	26000	22000	3	363	7	8
19	1	3	36900	35000	4	76	8	29
20	1	3	40875	40875	5	153	8	29
21	1	3	45900	42500	5	95	9	14
22	1	3	45800	45800	5	*	10	27
23	1	3	49500	49500	5	114	3	8
24	1	3	54900	52000	5	85	6	20
25	1	3	53000	53000	5	67	10	28
26	1	3	54900	53900	6	16	7	29
27	1	3	56900	54900	4	13	10	6
28	1	3	58875	55600	4	136	10	7
29	1	3	57900	57900	6	16	5	18
30	1	3	61500	59500	5	5	7	28
31	1	3	61875	61000	2	66	6	1
32	1	3	62875	62000	5	30	4	8
33	1	3	63875	63000	5	140	4	22
34	1	3	63900	63900	6	14	8	31
35	1	3	69000	65000	4	44	8	10
36	1	3	67200	67200	5	13	9	1
37	1	3	69900	69000	4	17	6	23
38	1	3	84900	80000	4	10	6	1
39	1	3	90900	88900	4	147	9	30
40	1	4	45000	41000	4	166	9	30
41	1	4	53900	52250	4	90	12	16
42	1	4	61500	58000	1	100	12	1
43	1	4	81750	81750	2	32	10	17
44	1	4	116900	116900	4	164	8	1
45	1	2	45900	45900	5	16	3	31
46	1	3	42900	41900	3	419	8	26
47	1	3	56900	56000	3	178	9	30
48	1	3	46900	44800	5	33	1	31
49	1	1	30000	27000	4	92	6	21
50	1	2	52900	47000	5	72	2	1
51	1	3	39900	38400	3	79	6	28
52	1	3	47950	44000	5	343	5	27
53	1	3	57500	54000	1	92	10	31
54	1	3	63900	61000	5	45	5	23
55	1	3	71900	70950	4	91	1	29
56	1	4	39450	34000	4	18	3	31
57	1	4	115000	110000	4	2	3	31
58	1	5	189000	187000	4	256	5	22
59	3	2	34900	32000	1	127	10	27
60	1	2	39900	39000	5	15	4	29
61	1	2	44900	43000	5	74	9	28
62	1	2	53900	52000	5	4	5	27
63	1	2	129900	117000	4	143	3	2
64	1	3	43900	42000	3	1	1	1
65	1	3	49900	49900	4	40	5	19
66	1	3	59900	56500	6	73	5	16
67	1	3	61400	61400	5	41	4	27
68	1	3	69900	63900	4	26	7	29
69	1	3	82900	79000	4	21	3	14
70	1	3	97400	92000	6	48	5	25
71	1	3	127900	116000	2	32	6	30
72	1	4	45900	45900	5	45	6	1
73	1	4	55900	54500	5	48	12	23
74	1	4	71900	69000	4	140	8	15
75	1	4	89875	87000	4	23	4	28

HOME	AREA	BEDROOMS	LIST PR	SOLD PR	FINANCE	DAYS	MTH SOLD	DAY SOLD
76	1	4	114900	107000	6	107	7	15
77	1	4	167875	160000	4	119	8	22
78	1	2	41700	40000	5	10	6	16
79	1	3	42500	40000	5	89	12	12
80	1	5	61875	61500	2	10	10	14
81	1	3	97500	88000	4	47	11	9
82	1	2	16000	16000	2	354	3	31
83	1	2	32000	28800	4	127	1	29
84	1	1	36900	35900	6	43	12	21
85	1	2	39200	39200	6	9	9	28
86	1	2	46875	46000	5	11	2	16
87	1	2	49900	48300	5	143	2	4
88	1	2	61900	57500	3	55	3	30
89	1	2	68900	68900	5	38	4	28
90	1	2	70800	70800	5	4	8	1
91	1	2	88900	88900	4	454	6	22
92	1	3	38500	38500	5	19	5	31
93	1	3	43200	43200	2	*	11	23
94	1	3	49900	48000	5	115	5	26
95	1	3	55900	54425	6	47	8	22
96	1	3	57900	56000	6	4	4	24
97	1	3	60000	57000	4	130	11	30
98	1	3	59900	59900	4	1	6	21
99	1	3	66900	65500	5	46	5	31
100	1	3	69900	69900	5	82	3	18
101	1	3	89900	89900	3	74	12	22
102	1	4	37900	31200	1	169	4	29
103	1	4	55800	54000	5	14	4	20
104	1	4	74900	70000	5	85	8	25
105	1	4	84900	84784	5	29	10	17
106	2	2	70500	70000	5	180	12	30
107	2	3	89900	85500	4	6	12	6
108	2	2	65900	62500	4	240	3	1
109	2	2	64475	64475	5	141	9	30
110	2	2	71875	68500	5	189	5	20
111	2	2	75900	73000	5	133	10	3
112	2	3	43900	42250	6	39	9	30
113	2	3	65900	56850	5	76	7	20
114	2	3	63750	61000	4	32	4	15
115	2	3	63900	63000	5	11	1	21
116	2	3	74750	66000	5	93	11	15
117	2	3	75900	72000	6	43	11	15
118	2	3	79900	77000	4	138	5	23
119	2	3	79900	79000	5	124	6	23
120	2	3	92900	91000	4	50	2	25
121	2	3	108900	104500	4	294	6	1
122	2	4	61875	61875	5	234	11	1
123	2	4	76900	71000	4	24	7	22
124	2	4	79900	76550	5	80	12	22
125	2	4	87500	84500	6	74	8	26
126	2	4	129900	116000	2	125	8	12
127	2	4	199500	191000	4	135	8	23
128	2	3	55900	55600	5	142	5	31
129	2	3	69900	69000	1	50	6	6
130	2	2	39900	39000	5	88	8	23
131	2	2	69900	69900	4	11	8	4
132	2	3	47500	42000	3	41	5	31
133	2	3	54250	53000	5	45	6	15
134	2	3	58500	56000	4	35	5	10
135	2	3	65500	64000	5	282	5	11
136	2	3	71700	69900	5	98	6	4

HOME	AREA	BEDROOMS	LIST PR	SOLD PR	FINANCE	DAYS	MTH SOLD	DAY SOLD
137	2	4	65900	64900	5	17	5	27
138	2	4	89800	88000	4	44	4	20
139	2	2	59900	58900	5	39	5	27
140	2	2	74900	74900	4	232	9	16
141	2	3	56900	54300	5	201	12	30
142	2	3	69200	69200	5	1	3	10
143	2	3	89500	84000	5	111	5	6
144	2	3	105900	105900	4	2	3	5
145	2	3	179000	159000	3	79	12	9
146	2	4	59900	57000	3	148	9	19
147	2	4	81900	79900	4	57	6	29
148	2	4	98500	96900	4	50	11	2
149	2	2	73875	70000	5	63	8	5
150	2	3	75900	73650	1	39	3	25
151	3	2	24900	19500	2	51	4	20
152	3	3	27900	20500	4	408	4	14
153	3	3	49900	48900	4	61	12	1
154	3	3	62900	60000	4	37	3	27
155	3	4	29900	27500	3	77	9	1
156	3	3	47000	42000	5	84	7	29
157	3	4	28500	26000	3	135	3	8
158	3	2	63900	61900	2	139	5	12
159	3	3	59900	55500	5	57	1	7
160	3	3	69900	69000	5	134	4	22
161	3	4	84900	82500	5	343	9	22
162	3	2	49000	47500	4	59	12	15
163	3	3	47500	46900	5	86	7	27
164	3	3	79900	75900	5	73	7	28
165	3	4	47900	47900	2	44	11	28
166	3	4	84900	83000	6	82	6	16
167	3	2	41100	38000	2	118	10	10
168	3	2	51900	50900	5	441	5	27
169	3	3	24900	22000	3	454	5	20
170	3	3	47900	47900	5	12	7	26
171	3	3	57300	55000	4	501	11	18
172	3	3	64900	60000	5	101	6	24
173	3	3	86900	86900	4	108	2	26
174	3	4	54500	50000	3	29	5	27
175	3	4	59900	57500	4	18	10	31
176	3	4	84900	84900	5	22	4	28
177	3	1	17000	17000	4	94	7	29
178	3	2	49900	47900	6	219	12	29
179	3	3	34900	34900	7	105	6	24
180	3	3	47900	47500	5	323	8	2
181	3	3	55900	54187	1	14	12	29
182	3	3	59900	58500	2	233	8	2
183	3	3	74900	71500	5	135	12	29
184	3	4	69900	68000	4	25	4	29
185	3	5	34900	34900	3	142	11	2
186	3	2	41900	37800	5	157	2	5
187	3	2	48500	47500	5	160	9	12
188	3	2	69900	69000	5	20	8	29
189	3	3	35400	35000	2	69	8	12
190	3	3	48900	48000	1	24	9	2
191	3	3	54200	54000	1	84	5	21
192	3	3	67500	63500	4	218	3	1
193	3	3	79900	68500	5	261	2	2
194	3	4	35500	32500	3	175	1	25
195	3	4	55000	55000	5	25	9	30
196	3	5	59900	57500	4	102	12	4
197	3	5	159000	150000	3	220	5	26

C LIST OF MINITAB COMMANDS

This appendix summarizes some Minitab commands. Minitab only reads the first four letters of a command and then looks for an argument, which may be a column number or constant. Most commands have extra text for explanatory purposes. Commands that have the same subcommands are grouped together in this summary list. For additional information on the commands, please refer to this supplement or Minitab's HELP facility. All of these commands are available with Minitab Release 8, and most are available with earlier releases.

Notation

E	Either a column (C) or a constant (K)
C	A column; can be a number (C1) or a name ('SALES')
K	A constant; can be a number (6.2) or a stored constant (K1)
M	Matrix (M1)
( )	Options

General Information

```
INFORMATION (C,...,C)              Gives status of the current worksheet
HELP (COMMAND (SUBCOMMAND))        Explains commands
NAME C 'NAME1' C 'NAME2'...        Names columns
NOTE                               Annotates a program; begins a command line
#                                  Similar to NOTE; use anywhere on a line
STOP                               Ends a Minitab session
RESTART                            Begins a new Minitab session
```

Data Input and Output

```
READ THE FOLLOWING DATA ('FILENAME') IN C,...,C
SET THE FOLLOWING DATA ('FILENAME') IN C
INSERT DATA ('FILENAME') BETWEEN ROWS K AND K+1 OF COLUMNS C,...,C
   NOBS = K
   FORMAT (SPECIFICATION)

DELETE ROWS K,...,K FROM C,...,C

WRITE THE DATA IN 'FILENAME' FROM C,...,C
PRINT E,...,E
   FORMAT (SPECIFICATION)
OH = K LINES    (PC ONLY)
OW = K SPACES
NEWPAGE
PAPER
NOPAPER
```

```
SAVE THE WORKSHEET (IN 'FILENAME')
RETRIEVE THE WORKSHEET STORED IN 'FILENAME'
   PORTABLE
   LOTUS [PC ONLY]

JOURNAL ('FILENAME')
NOJOURNAL

EXECUTE 'FILENAME'

OUTFILE 'FILENAME'
NOOUTFILE
```

Editing and Manipulating Data

```
CODE (K,...,K) TO K,...,(K,...K,) TO K FOR C,...,C PUT IN C,...,C

COPY E,...,E IN E,...,E
   USE ROWS K,...,K
   USE ROWS WHERE C = K,...,K
   OMIT ROWS K,...,K
   OMIT ROWS WHERE C = K,...,K

ERASE E,...,E

INDICATOR VARIABLES FOR C, PUT VALUES IN C,...,C

RANK C, PUT IN C

SIGN OF E (PUT IN E)

SORT THE DATA IN C,...,C PUT IN C,...,C
   BY C,...,C
   DESCENDING C,...,C

STACK (E,...,E),...,(E,...,E) IN (C,...,C)
UNSTACK (C,...,C) IN (E,...,E),...,(E,...,E)
   SUBSCRIPTS IN C
```

Arithmetic

```
LET E = ARITHMETIC EXPRESSION

ABSOLUTE OF E, PUT IN E
SQRT OF E, PUT IN E
LOGE OF E, PUT IN E
LOGTEN OF E, PUT IN E
EXPONENTIATE E, PUT IN E
ROUND E, PUT IN E
SIN OF E, PUT IN E
COS OF E, PUT IN E
TAN OF E, PUT IN E
ASIN OF E, PUT IN E
ACOS OF E, PUT IN E
ATAN OF E, PUT IN E
PARPRODUCTS OF C, PUT IN C
PARSUMS OF C, PUT IN C
```

Statistics

```
DESCRIBE THE DATA IN C,...,C
   BY C
N OF C (PUT IN K)              Number of nonmissing values
NMISS OF C (PUT IN K)          Number of missing values
MEAN OF C (PUT IN K)           Sample average
MEDIAN OF C (PUT IN K)         Center value in ordered array
STDEV OF C (PUT IN K)          Sample standard deviation
MAXIMUM OF C (PUT IN K)        Largest value
MINIMUM OF C (PUT IN K)        Smallest value
SUM OF C (PUT IN K)            Sum of all the values
SSQ OF C (PUT IN K)            Sum of squares of the values
COUNT OF C (PUT IN K)          Total number of observations
RANGE OF C (PUT IN K)          Largest value minus smallest value
NSCORES OF C, PUT IN C         Normal scores

RN E,...,E, PUT THE NUMBER OF NONMISSING VALUES IN C
RNMISS E,...,E, PUT THE NUMBER OF MISSING VALUES IN C
RMEAN E,...,E, PUT THE MEAN IN C
RMEDIAN E,...,E, PUT THE MEDIAN IN C
RSTDEV E,...,E, PUT THE STANDARD DEVIATION IN C
RMAX E,...,E, PUT THE MAXIMUM IN C
RMIN E,...,E, PUT THE MINIMUM IN C
RSUM E,...,E, PUT THE SUM IN C
RSSQ E,...,E, PUT THE SUM OF SQUARES IN C
RCOUNT E,...,E, PUT THE NUMBER OF COLUMNS IN C
RRANGE E,...,E, PUT THE RANGE IN C

TINTERVAL (K PERCENT CONFIDENCE) FOR C,...,C
ZINTERVAL (K PERCENT CONFIDENCE) SIGMA = K, FOR C,...,C

TTEST (OF MU = K) ON C,...,C
ZTEST (OF MU = K) SIGMA = K ON C,...,C
   ALTERNATIVE = K

TWOSAMPLE-T (K PERCENT CONFIDENCE) DATA IN C, C
TWOT (K PERCENT CONFIDENCE) DATA IN C, CODE IN C
   POOLED
   ALTERNATIVE
```

Graphs

```
BOXPLOT OF DATA IN C
   INCREMENT = K
   START AT K (END AT K)
   BY C

DOTPLOT OF DATA IN C,...,C
HISTOGRAM OF C,...,C
   INCREMENT = K
   START AT K (END AT K)
   SAME
   BY C
```

```
LPLOT C VS C USING LETTERS IN C
MPLOT OF C VS C, C VS C,..., C VS C
PLOT C VERSUS C
TPLOT Y IN C VS XONE IN C VS XTWO IN C
   XINCREMENT = K
   XSTART AT K (END AT K)
   YINCREMENT = K
   YSTART AT K (END AT K)
   SYMBOL = 'SYMBOL'   (PLOT only)
   TITLE = 'TEXT'
   FOOTNOTE = 'TEXT'
   XLABEL = 'TEXT'
   YLABEL = 'TEXT'

STEM AND LEAF OF C,...,C
   INCREMENT = K
   TRIM OUTLIERS
   BY C
```

Tables

```
CHISQUARE ANALYSIS OF THE TABLE IN C,...,C

TABLE THE DATA CLASSIFIED BY C,...,C
   COUNTS
   ROWPERCENTS
   COLPERCENTS
   TOTPERCENTS
   MEANS FOR C,...,C
   MEDIANS FOR C,...,C
   SUMS FOR C,...,C
   MINIMUMS FOR C,...,C
   MAXIMUMS FOR C,...,C
   STDEV FOR C,...,C
   STATS FOR C,...,C
   DATA FOR C,...,C
   N FOR C,...,C
   NMISS FOR C,...,C
   CHISQUARE
   PROPORTION OF K FOR C,...,C
   NOALL
   ALL FOR C,...,C

TALLY THE DATA IN C,...,C
   COUNTS
   CUMCOUNTS
   PERCENTS
   CUMPERCENTS
   ALL
   STORE C,...,C
```

Probability Distributions and Random Data

```
CDF (FOR VALUES IN E) (STORE PROBABILITIES IN E)
INVCDF FOR PROBABILITIES IN E (STORE RESULTS IN E)
PDF (FOR VALUES IN E) (STORE RESULTS IN E)
```

```
RANDOM K OBSERVATIONS IN EACH OF C,...,C
   BERNOULLI P = K     (RANDOM only)
   BINOMIAL FOR N = K P = K
   CHISQUARE V = K
   DISCRETE X IN C P(X) IN C
   EXPONENTIAL MU = K
   F WITH VONE = K AND VTWO = K
   INTEGER FOR A = K B = K
   LOGNORMAL MU = K SIGMA = K
   NORMAL WITH MU = K AND SIGMA = K
   T WITH V = K
   POISSON MU = K
   UNIFORM A = K B = K

BASE = K

SAMPLE K ROWS FROM C,...,C, PUT IN C,...,C
   REPLACE
```

Storing Commands

```
STORE THE FOLLOWING COMMANDS ('FILENAME')
END OF STORED COMMANDS
ECHO THE COMMANDS THAT FOLLOW
NOECHO THE COMMANDS THAT FOLLOW
EXECUTE THE STORED COMMANDS ('FILENAME') (K TIMES)
```

Analysis of Variance and Regression

```
AOVONEWAY ON THE DATA IN C,...,C
ONEWAY ANOVA FOR DATA IN C, LEVELS IN C
   TUKEY
   FISHER K

ANOVA C = C,...,C
   MEANS C,...,C
   RANDOM C,...,C
   RESIDUALS C,...,C
   FITS C,...,C

TWOWAY ANOVA FOR DATA IN C, LEVELS IN C AND C
   ADDITIVE
   MEANS

CORRELATION COEFFICIENTS FOR C,...,C

BRIEF K

REGRESS C ON K PREDICTORS C,...,C (STORE STD RESIDS IN C (FIT IN C))
   RESIDUALS IN C
   PREDICT FOR E
   DW
```

```
STEPWISE REGRESSION ON C, PREDICTORS IN C,...,C
   FENTER = K
   FREMOVE = K
   FORCE C...C
   ENTER C...C
   REMOVE C...C
   BEST K
   STEPS = K
   NOCONSTANT
```

Control Charts

```
XBARCHART FOR DATA IN C,...,C, SAMPLES IN E
   MU = K
   SIGMA = K
   SUBGROUP SIZE, K
   RSPAN = K
   TEST K,...,K

RCHART FOR DATA IN C...C, SAMPLES IN E
   SIGMA = K
   SUBGROUP SIZE IS K

PCHART NONCONFORMITIES IN C,...,C, SAMPLE SIZE = E
   P = K
   SUBGROUP SIZE IS K
   TEST K...K
```

Common Subcommands

```
      SLIMITS ARE K...K
      HLINES AT E...E
      ESTIMATE USING SAMPLES K...K
      XSTART AT K (END AT K)
      YSTART AT K (END AT K)
      YINCREMENT = K
      XLABEL 'TEXT'
      YLABEL 'TEXT'
      TITLE 'TEXT'
      FOOTNOTE 'TEXT'
```

Time Series

```
TSPLOT (PERIOD = K) TIME SERIES DATA IN C
   INCREMENT K
   START AT K (END AT K)
   TSTART AT K (END AT K)
   ORIGIN K

MTSPLOT (PERIOD=K) TIME SERIES DATA IN C,...,C
   INCREMENT = K
   START AT K (END AT K)
   TSTART AT K (END AT K)
   ORIGIN K (FOR C,...,C) (K FOR C,...,C)...
```

Nonparametric Statistics

```
MANN-WHITNEY (PERCENT CONFIDENCE K) C,C
STEST OF MEDIAN = K ON DATA IN C
WTEST (CENTER = K) DIFFERENCES IN C,...,C
   ALTERNATIVE = K

FRIEDMAN DATA IN C, TREATMENT IN C,BLOCK IN C

KRUSKAL-WALLIS FOR DATA IN C, LEVELS IN C
```

INDEX

The Minitab commands and subcommands are printed in capital letters in this index. Subcommands are listed with their respective commands.

#, 27
ABSOLUTE, 326
ADDITIVE (TWOWAY), 188
ALL (TABLE), 75
ALL (TALLY), 36
Alpha α, 156
Alpha data, 13
ALTERNATIVE (STEST), 208
ALTERNATIVE (ZTEST), 152
ALTERNATIVE (TTEST), 152
ALTERNATIVE (TWOSAMPLE), 166
ALTERNATIVE (MANN-WHITNEY), 209
ALTERNATIVE (TWOT), 166
ALTERNATIVE (WTEST), 211
Ampersand &, 30
ANOVA, 188
AOVONEWAY, 182
Arithmetic, 25
ASCII File, 313
BASE, 70
BERNOULLI (RANDOM), 70
BEST (STEPWISE), 295
Beta β, 157
BINOMIAL (RANDOM), 70
BINOMIAL (PDF), 84
BINOMIAL (CDF), 84
Bivariate Table, 74
Bonferroni Comparison, 192
BOXPLOT, 57
BRIEF, 237
BY (SORT), 22
BY (DESCRIBE), 52
BY (BOXPLOT), 57
BY (HISTOGRAM), 41
BY (STEM AND LEAF), 39
BY (DOTPLOT), 45
CDF, 84, 91, 98, 100, 109
Central Limit Theorem, 125
Change Directory, 313
Chebyshev's Rule, 55
CHISQUARE (RANDOM), 70
CHISQUARE (TABLE), 75
CHISQUARE, 230
CODE, 21
COLPERCENTS (TABLE), 75
Columns, 3

Command Structure, 3
Completely Randomized Design, 181
Confidence Interval, 137
Constants, 3
Contingency Table, 229
Continuation ++, 30
COPY, 19
CORRELATION, 218, 250
COUNT, 52
COUNTS (TALLY), 36
COUNTS (TABLE), 75
CUMCOUNTS (TALLY), 36
CUMPERCENTS (TALLY), 36
Data Editor, 314
Data Type, 13
DATA (TABLE), 75
DELETE, 30
DESCENDING (SORT), 22
DESCRIBE, 52
DISCRETE (RANDOM), 70
DOS Commands, 313
DOTPLOT, 45
Dummy Variable, 287
DW (REGRESS), 329
ECHO, 117
Empirical Rule, 55
END, 13
ENTER (STEPWISE), 295
Error Correction, 29
ESTIMATE (Control Charts), 330
EXECUTE, 10, 117
Expected Value, 81
Experiment, 69
EXPONENTIAL (CDF), 109
EXPONENTIAL (PDF), 109
EXPONENTIAL (RANDOM), 70
F (RANDOM), 70
Factorial Design, 193
FENTER (STEPWISE), 295
Finite Population, 115
FISHER (ONEWAY), 182
FITS (ANOVA), 188
FOOTNOTE (MPLOT), 106
FOOTNOTE (TPLOT), 257
FOOTNOTE (LPLOT), 189
FOOTNOTE (PLOT), 43

Index

FOOTNOTE (Control Charts), 330
FORCE (STEPWISE), 295
FORMAT (SET), 13
FORMAT (INSERT), 18
FORMAT (PRINT), 24
FORMAT (WRITE), 12
FORMAT (READ), 16
FREMOVE (STEPWISE), 295
Frequency Distributions, 35
FRIEDMAN, 216
HELP, 4
HISTOGRAM, 41
HLINES (Control Charts), 330
Hypothesis Test, 151
INCREMENT (TSPLOT), 330
INCREMENT (DOTPLOT), 45
INCREMENT (BOXPLOT), 57
INCREMENT (STEM AND LEAF), 39
INCREMENT (MTSPLOT), 330
INCREMENT (HISTOGRAM), 41
Independent Variable, 235
INDICATOR, 287
Infinite Population, 119
INFORMATION, 6
INSERT, 12, 18
INTEGER (RANDOM), 70
Interaction, 256
Interval Data, 38
INVCDF, 101
JOURNAL, 9
KRUSKAL-WALLIS, 214
LET, 26
LOGE, 326
LOGNORMAL (RANDOM), 70
LOTUS (INSERT), 11
LOTUS (RETRIEVE), 11
Lotus Interface, 315
LPLOT, 189
MANN-WHITNEY, 209
MAXIMUM, 52
MAXIMUMS (TABLE), 75
MEAN, 52
MEANS (TABLE), 75
MEANS (TWOWAY), 188
MEANS (ANOVA), 188
MEDIAN, 52
MEDIANS (TABLE), 75
Menus, 311
MINIMUM, 52
MINIMUMS (TABLE), 75
Missing Data, 16
MPLOT, 106
MTSPLOT, 330
MU (XBARCHART), 330
Multinomial Distribution, 227
Multiple Comparison, 184
N, 52
N (TABLE), 75
NAME, 28
NEWPAGE, 25
NMISS, 52
NMISS (TABLE), 75

NOALL (TABLE), 75
NOBS (READ), 16
NOBS (SET), 13
NOBS (INSERT), 18
NOCONSTANT (STEPWISE), 295
NOECHO, 117
NOJOURNAL, 9
Nominal Data, 35
NOOUTFILE, 10
NOPAPER, 312
NORMAL (PDF), 100
NORMAL (CDF), 100
NORMAL (RANDOM), 70
NORMAL (INVCDF), 101
NOTE, 27
OH, 25
OMIT (COPY), 19
ONEWAY, 182
Ordinal Data, 35
ORIGIN (MTSPLOT), 330
ORIGIN (TSPLOT), 330
OUTFILE, 10
Outliers, 56
Output File, 10
OW, 25
P (PCHART), 330
p-value, 153
Paired Difference, 169, 211
PAPER, 312
PARSUMS, 24
PCHART, 330
PDF, 84, 91, 98, 100, 109
PERCENTS (TALLY), 36
PLOT, 43
POISSON (PDF), 91
POISSON (CDF), 91
POISSON (RANDOM), 70
Polynomial Model, 277
POOLED (TWOSAMPLE), 166
POOLED (TWOT), 166
PORTABLE (SAVE), 11
PORTABLE (RETRIEVE), 11
PREDICT (REGRESS), 236, 250
PRINT, 24
Probability, 69
Program File, 9
PROPORTIONS (TABLE), 75
Qualitative Data, 35
Quantitative Data, 38
RANDOM (ANOVA), 188
RANDOM, 70
Randomized Block Design, 187
RANGE, 52
RANK, 23
Ratio Data, 38
RCHART, 330
RCOUNT, 59
READ, 12, 16
REGRESS, 236, 250
REMOVE (STEPWISE), 295
REPLACE (SAMPLE), 116
RESIDUALS (REGRESS), 236, 250

Index

RESIDUALS (ANOVA), 188
Response Variable, 235
RETRIEVE, 11
RMAX, 59
RMEAN, 59
RMEDIAN, 59
RMIN, 59
RN, 59
RNMISS, 59
ROWPERCENTS (TABLE), 75
RRANGE, 59
RSPAN (XBARCHART), 330
RSSQ, 59
RSTDEV, 59
RSUM, 59
SAME (HISTOGRAM), 41
SAME (DOTPLOT), 45
SAMPLE, 116
Sampling Distribution, 121
SAVE, 11
SET, 12, 13
SIGMA (RCHART), 330
SIGMA (XBARCHART), 330
Sign Test, 207
SIGN, 21
Simple Random Sampling, 69, 115
Simulation, 143, 156
SLIMITS (Control Charts), 330
SORT, 22
Spearman's Rank Correlation, 218
SSQ, 52
STACK, 20
START (MTSPLOT), 330
START (HISTOGRAM), 41
START (TSPLOT), 330
START (BOXPLOT), 57
START (DOTPLOT), 45
STATS (TABLE), 75
STDEV, 52
STDEV (TABLE), 75
STEM AND LEAF, 39
STEPS (STEPWISE), 295
STEPWISE, 295
STEST, 208
STOP, 29
STORE, 117
Subcommands, 4
SUBGROUP (PCHART), 330
SUBGROUP (XBARCHART), 330
SUBGROUP (RCHART), 330
SUBSCRIPTS (STACK), 20
SUBSCRIPTS (UNSTACK), 20
SUM, 52
SUMS (TABLE), 75
SYMBOL (PLOT), 43
Symphony Interface, 315
T (RANDOM), 70
TABLE, 75
TALLY, 36
TEST (PCHART), 330
TEST (XBARCHART), 330
TINTERVAL, 138
TITLE (TPLOT), 257
TITLE (Control Charts), 330
TITLE (LPLOT), 189
TITLE (MPLOT), 106
TITLE (PLOT), 43
TOTPERCENTS (TABLE), 75
TPLOT, 257
TRIM (STEM AND LEAF), 39
TSPLOT, 330
TSTART (MTSPLOT), 330
TSTART (TSPLOT), 330
TTEST, 152
TUKEY (ONEWAY), 182
TWOSAMPLE, 166
TWOT, 166
TWOWAY, 188
Type II error, 157
Type I error, 156
UNIFORM (PDF), 98
UNIFORM (CDF), 98
UNIFORM (RANDOM), 70
UNSTACK, 20
USE (COPY), 19
Wilcoxon Signed Ranks Test, 211
Wilcoxon Rank Sum Test, 209
Worksheet, 2
Worksheet File, 10
WRITE, 12
WTEST, 211
XBARCHART, 330
XINCREMENT (TPLOT), 257
XINCREMENT (MPLOT), 106
XINCREMENT (LPLOT), 189
XINCREMENT (PLOT), 43
XLABEL (PLOT), 43
XLABEL (TPLOT), 257
XLABEL (MPLOT), 106
XLABEL (Control Charts), 330
XLABEL (LPLOT), 189
XSTART (LPLOT), 189
XSTART (PLOT), 43
XSTART (TPLOT), 257
XSTART (Control Charts), 330
XSTART (MPLOT), 106
YINCREMENT (PLOT), 43
YINCREMENT (Control Charts), 330
YINCREMENT (LPLOT), 189
YINCREMENT (TPLOT), 257
YINCREMENT (MPLOT), 106
YLABEL (MPLOT), 106
YLABEL (TPLOT), 257
YLABEL (LPLOT), 189
YLABEL (PLOT), 43
YLABEL (Control Charts), 330
YSTART (MPLOT), 106
YSTART (Control Charts), 330
YSTART (TPLOT), 257
YSTART (LPLOT), 189
YSTART (PLOT), 43
z-score, 56
ZINTERVAL, 138
ZTEST, 152